Two Manuals Combined into One Book

Operator's Manual for Machine Gun 7.62-MM, M60

And

Technical Manual for 5.56MM, M16A2, M4 and M4A1 Carbine

Department of Defense

Pentagon
Publishing

TECHNICAL MANUAL
NO. 9-1005-224-10

*TM 9-1005-224-10
T.O. 11W2-6-4-11
HEADQUARTERS
DEPARTMENTS OF THE ARMY
AND AIR FORCE
Washington D.C., 2 April 1998

Operator's Manual
for
MACHINE GUN, 7.62-MM, M60 W/E

REPORTING ERRORS AND RECOMMENDING IMPROVEMENTS
You can help improve this manual. If you find any mistakes or if you know of a way to Improve the procedures, please let us know. Mail your letter or DA Form 2026 (Recommended Changes to Equipment Publications and Blank Forms) direct to: Director, Armament and Chemical Acquisition and Logistics Activity, ATTN: AMSTA-AC-NML, Rock Island, IL 61299-7630. A reply will be furnished to you.

*This manual supersedes TM 9-1005-224-10, dated 30 July 1985; TM 9-1005-224-10HR, dated 18 May 1979; TB 9-1005-224-50-1, including all changes.

WARNING SUMMARY

Be sure to clear weapon before starting inspection.

If cover is opened on a hot cartridge (hot barrel), an open cover cookoff could occur and result in serious injury or death. Evacuate area for 15 minutes and then do REMEDIAL ACTION. Refer to WP 00011 00.

Never reload a runaway machine gun until it has been repaired. Be sure machine gun is cleared before moving it.

When removing a stuck unfired cartridge, stay clear of the muzzle. Do not allow cartridge to contact any hard surface. Cartridge may fire on contact. Remove a stuck unfired cartridge using the same procedures for removing a stuck or ruptured cartridge case.

Before field stripping check that the bolt is in forward position.

To prevent possible bodily injury, or aircraft/weapon damage, personnel should not stand the weapon on its barrel assembly when disassembling or assembling the weapon.

Bolt assembly in under spring tension; it can twist and injure your hand.

Barrels issued for a specific gun will not be changed from gun to gun. Each barrel and bolt assembly should be tagged during cleaning with the serial number of the receiver (gun) and should remain together as initially assigned. Interchanging barrels/bolts may result in injury or death.

The climate temperature in different regions will make a difference as to what constitutes a hot gun. A hot, sunny day can cause a cookoff within 50 rounds, weapon and ammunition in the sun.

Insert the hinge pin from one side and the hinge pin latch from the opposite side.

After unloading dummy round, be sure the barrel is clear.

Ammunition to be used in machine gun is shown in WP 0031 00.

Keep weapon pointed at target - down range/impact area.

Keep safety on 'S' (safe) until your ready to fire.

Always check the chamber/bore after clearing the weapon

Check barrel bore and chamber before firing.

Do not interchange bolt/barrel assemblies from one weapon to another. This may cause injury or death to personnel.

Change hot barrels with your heat protective mitten only or by holding bipod legs. (Refer to WP 0017 00 for additional hot barrel procedures.)

Do not interchange flat leaf springs from one model weapon to another.

Check that assigned and spare barrels have been headspaced and tagged to your receiver. Rotate usage of the barrel.

WARNING SUMMARY (cont)

Do not fire blank ammunition toward personnel within 20 feet of the muzzle. Fragments of a closure wad or particles of unburned propellant might inflict injury within that range.

Dry cleaning solvent is FLAMMABLE and TOXIC and must be kept away from open flames and used in a well ventilated area. Use of rubber gloves is necessary to protect the skin when washing machine gun parts.

Appropriate eye protection is recommended when cleaning your weapon and/or its parts.

For first aid information see FM 21-11.

TABLE OF CONTENTS

WP Sequence No.

WP Sequence No.

CHAPTER 1

INTRODUCTION

GENERAL INFORMATION 0001 00

SCOPE

Type of Manual: Operator's Manual.

Model Numbers and Equipment Names: M60, 7.62mm, Machine Gun, and M60D, 7.62mm, Machine Gun.

Purpose of Equipment:

The M60, 7.62mm, Machine Gun is a general purpose weapon capable of being fired from several mounts or handheld. The weapons primary use is for ground operations.

The M60D, 7.62mm, Machine Gun is a general purpose weapon capable of being fired from several mounts. The weapons primary use is for support of ground operations. The M60D is an aircraft door-mounted or vehicle-mounted machine gun.

MAINTENANCE FORMS AND RECORDS

Department of the Army forms and procedures used for equipment maintenance will be those prescribed by DA PAM 738-750, the Army Maintenance Management System (TAMMS).

REPORTING EQUIPMENT IMPROVEMENT RECOMMENDATIONS (EIR's).

If your machine needs improvement, let us know. Send us an EIR. You, the user, are the only one who can tell us what you don't like about your equipment. Let us know why you don't like the design or performance. Put it on an SF 368 (Quality Deficiency Report). Mail it to us at Commander, US Army Armament Research, Development and Engineering Center, ATTN: AMSTA-AR-QAW-A, Rock Island, IL 61299-7630. We'll send you a reply.

EQUIPMENT DESCRIPTION AND DATA 0002 00

EQUIPMENT CHARACTERISTICS, CAPABILITIES AND FEATURES.

The M60 and M60D are air-cooled and have fixed headspace allowing quick barrel changes for cooling and maintenance when required. In order to extend the life of the barrels, retain accuracy, and allow for continuous firing over long periods of time, two barrel assemblies are issued with each gun. The machine guns are gas-operated and fire from open bolt position.

LOCATION AND DESCRIPTION OF MAJOR COMPONENTS.

M60 and M60D Machine Gun.

A **COVER ASSEMBLY** - Positions and holds cartridge in place for stripping, feeding link belt, and chambering rounds.

B **COCKING HANDLE ASSEMBLY** - Provides a means to manually move the bolt assembly to the rear.

C **TRIGGER MECHANISM AND GRIP ASSEMBLY (M60 ONLY)** - Controls the firing of the machine gun.

D **SEAR AND SAFETY HOUSING (M60D ONLY)** - Controls the firing of the machine gun.

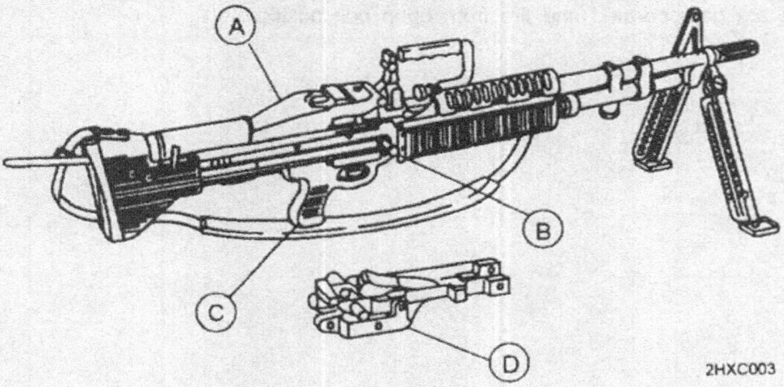

LOCATION AND DESCRIPTION OF MAJOR COMPONENTS (cont)

E **BARREL AND BIPOD ASSEMBLY -** The barrel houses cartridges for firing. The bipod assembly provides a semistable platform when the machine gun is fired from the prone position.

F **CARRYING HANDLE ASSEMBLY** - Provides a means to carry the machine gun with one hand. The carrying handle assembly folds down when the rear sight is used and the machine gun is fired.

G **RECEIVER ASSEMBLY** - Supports all major components. Houses internal parts and, through a series of cam ways, controls operation of weapon.

H **SHOULDER GUN STOCK (M60 ONLY)** - Provides a suitable surface to stabilize the weapon against the shoulder while firing the machine gun from any position except from the hip.

I **GRIP AND TRIGGER ASSEMBLY (M60D ONLY)** - Provides handles to move machine gun and houses the machine gun trigger.

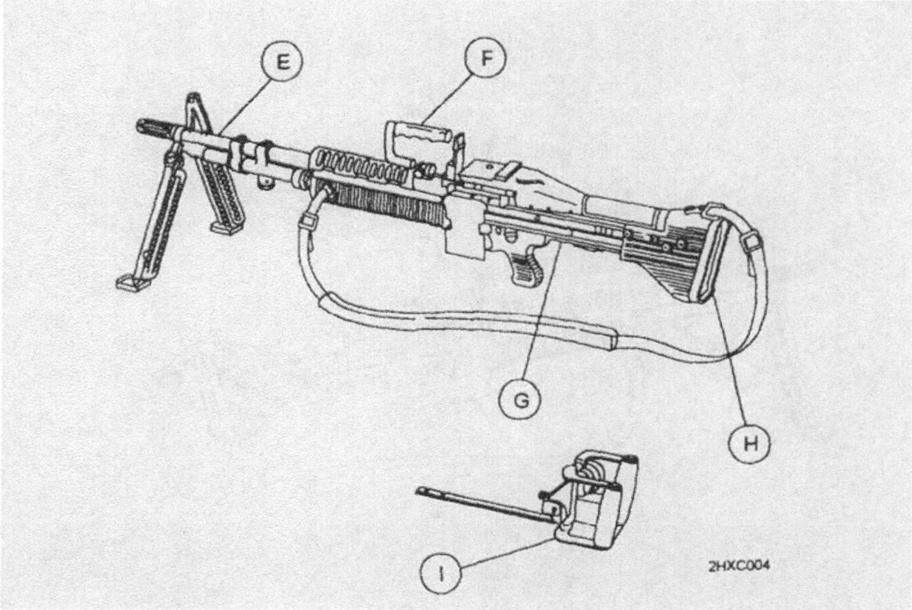

LOCATION AND DESCRIPTION OF MAJOR COMPONENTS (cont)

J **CARTRIDGE FEED TRAY AND HANGER ASSEMBLY (M60 ONLY)** - Guides cartridges for positioning and feeding. The hanger assembly supports the bandoleer.

K **CARTRIDGE TRAY ASSEMBLY (M60D ONLY)** - Guides cartridge for positioning and feeding.

L **REAR SIGHT ASSEMBLY (M60 ONLY)** - Provides a means to aim the machine gun at the target with accuracy. The sight adjusts horizontally as well as vertically.

M **REAR SIGHT ASSEMBLY (M60D ONLY)** - Provides a means to aim the machine gun in the general area of the target. The rear sight is nonadjustable.

N **FOREARM ASSEMBLY (M60 ONLY)** - Provides a hand hold when firing from the hip or from a standing or kneeling position.

O **SMALL ARMS SLING (M60 ONLY)** - Used for support during assault firing and for transport.

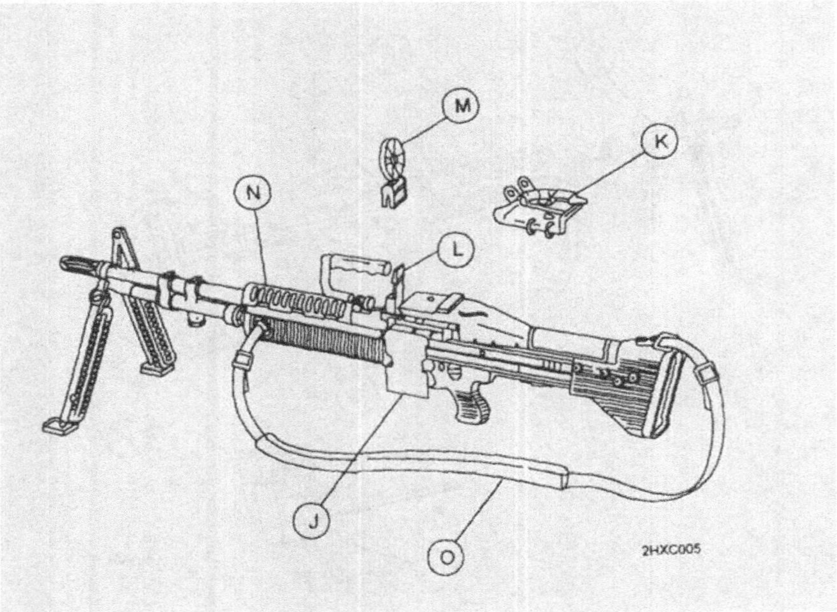

DIFFERENCES BETWEEN MODELS - Differences between the M60 and M60D machine gun are listed below.

Assembly/Component	M60	M60D
Tray and hanger assembly	X	
Cartridge feed tray assembly		X
Dust and moisture seal boot		X
Forearm assembly	X	
Grip and trigger assembly		X
Gun adapter		X
Magazine hanger assembly		X
Rear sight (adjustable)	X	
Rear sight (nonadjustable)		X
Receiver assembly	X	
Receiver assembly		X
Sear and safety housing assembly		X
Sear assembly link and spring		X
Shoulder gun stock	X	
Small arms sling	X	
Trigger mechanism grip assembly	X	
Quick release pin		X

EQUIPMENT DATA

M60 Machine Gun.

Weight	23 lb (10.43 kg)
Length	43.5 in. overall (1.1m overall)
Range:	
Maximum Effective Range	1100meters
Maximum Range	3725 meters
Tracer Burn Out	900 meters or more
Rates of fire:	
Sustained	100 rd per min (4 to 6 seconds between burst) Recommend barrel change every 10 min
Rapid	200 rd per min (2 to 3 seconds between burst) Recommend barrel change every 2 min
Cyclic	550 rd per min (approx) Recommend barrel change every min
Muzzle velocity	2800 FPS
Capacity of bandoleer	100 rds
Rifling:	
Number of lands	4
Right hand twist	One turn in 12 in (30.54 cm)
Trigger pull:	
Maximum	11.5 lb (5.2 Kg)
Minimum	6.0 lb (2.7 kg)

EQUIPMENT DATA (cont)

M60D Machine Gun

Weight	25 lb (11.33 kg)
Length	43.5 in overall (1.1m overall)
Range:	
Maximum Effective Range	1100 meters
Maximum Range	3725 meters
Tracer Burn Out	900 meters or more
Rates of fire:	
Sustained	100 rd per min (4 to 6 seconds between burst) Recommend barrel change every 10 min
Rapid	200 rd per min (2 to 3 seconds between burst) Recommend barrel change every 2 min
Cyclic	550 rd per min (approx) Recommend barrel change every min
Rifling:	
Number of lands	4
Right hand twist	One turn in 12 in (30.54 cm)
Trigger pull at sear activator:	
Maximum	20 lb (9.06 kg)
Minimum	10.5 lb (4.75 kg)

CHAPTER 2

OPERATING INSTRUCTIONS

This page intentionally left blank.

DESCRIPTION AND USE OF OPERATOR'S CONTROLS AND INDICATORS 0003 00

M60 MACHINE GUN

COCKING HANDLE. Pull back to move bolt rearward.

COVER LATCH. Holds cover closed. Pull back on lower end of latch to open.

BARREL LOCK LEVER. Locks barrel in place. To unlock barrel, push in on barrel lock lever from left side of weapon and raise lever.

BIPOD LEG PLUNGER. Depress to extend or retract bipod foot.

TRIGGER. Pull rearward to fire weapon. Release to stop firing.

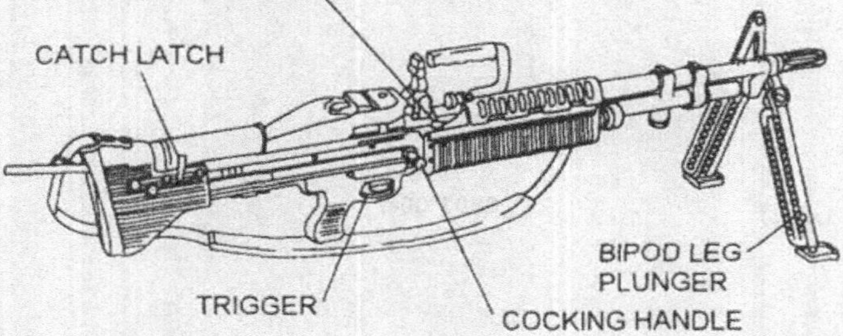

M60

M60 MACHINE GUN (cont)

SAFETY. Move safety up to 'F' (fire) and down to 'S' (safe).

REAR SIGHT. Adjustable to 1100 meters.

- Range Scale - graduated from 300 to 1100 meters.
- Scale Retaining/Adjusting Screw - locks range scale in place.
- Elevating Knob - use for minor (fine) adjustments in range.
- Windage Scale - adjust for left or right wind.
- Rear Sight Aperture - lines up with front sight for long range targets.

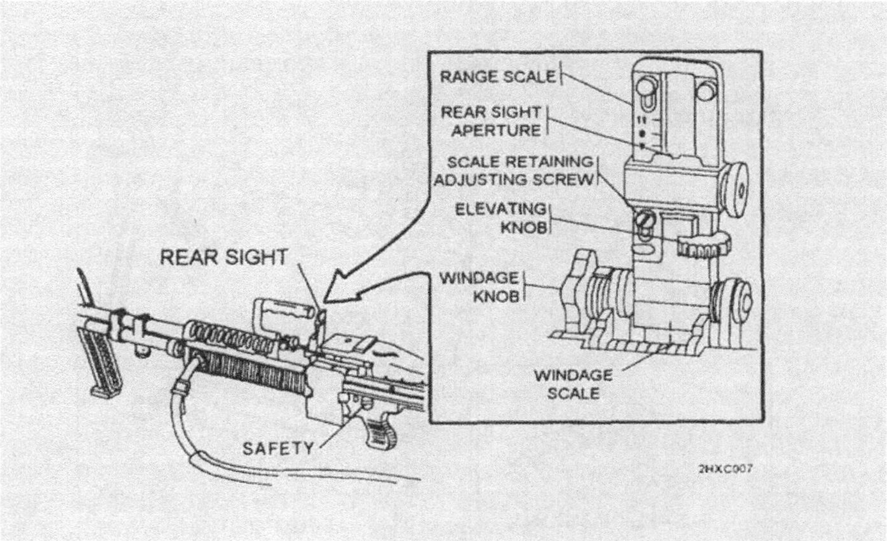

M60D MACHINE GUN

GRIP AND TRIGGER ASSEMBLY. Pull back on both triggers to fire. Release to stop firing.

SAFETY. Push to right for safe 'S', to left for fire 'F'.

REAR AND FRONT SIGHTS. Aim through the ring sight (rear sight) and front sight to line up target. Fire. Adjust range by watching tracers.

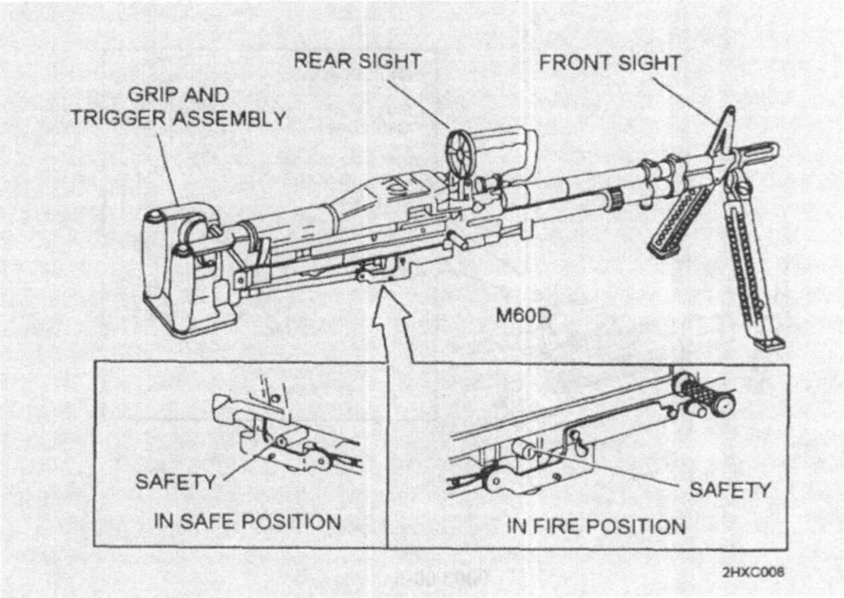

PREVENTIVE MAINTENANCE CHECKS AND SERVICES (PMCS) 0004 00

GENERAL

Always keep in mind the CAUTIONS and WARNINGS when performing your before (B) or during (D) PMCS. The numbers in the item number column shall be used for the "TM Number" column on DA Form 2404, Equipment Inspection and Maintenance Worksheet, in recording results of PMCS. After you operate, perform your after (A) PMCS.

Perform BEFORE PMCS if: (1) you are the assigned operator and the machine gun has been stored and not used for a period of 90 days or(2) you have been issued the machine gun for the first time.

WARNING

Do not interchange barrel assembly or bolt assembly from one machine gun to another. Doing so may result in injury or death.

The climate temperature in different regions will make a difference as to what constitutes a hot gun. A hot, sunny day can cause a cookoff within 50 rounds, weapon and ammunition in sun.

NOTE

PMCS is the same for both M60 and M60D machine gun except where noted.

Item No.	Interval	Item to Check/ Service	Procedure	Not Fully Mission Capable if:
			WARNING Before starting inspection, be sure to clear weapon (WP 0009 00).	
1	B	M60/M60D Machine Gun Tools and Equipment	Visually check for missing or damaged tools and equipment (WP 0034 00/0035 00) including tripod and components.	Any BII/COEI items are missing ordamaged.

Item No.	Interval	Item to Check/ Service	Procedure	Not Fully Mission Capable If:
2	B	Barrel Assemblies	Open cover, pull bolt to rear, push cocking handle forward and lock. Close cover. place on safe (S). Tilt weapon end to end to make sure gas piston slides freely (a clicking sound will be heard as the piston slides In the cylinder). Run a dry swab through the bore to remove excess lubricant or obstructions.	Gas piston will notslide. Obstruction in barrel. Gas cylinder vent, cylinder plug, and cylinder nutloose.
3	B	Trigger Mechanism Grip Assembly (M60)/ Sear and Safety Housing Assembly (M60D)	Make sure flat leaf spring is in place (WP 0018 00). Check that safety is in 'F' position. Pull bolt to rear, move safety to 'S'. Push cocking handle forward and lock and pull the trigger.	Weapon functions with safety at 'S'.

Item No.	Interval	Item to Check/ Service	Procedure	Not Fully Mission Capable If:
4	B	Cover Assembly	Lightly lubricate all moving parts with lubricant on a clean swab (item 6, WP 0038 00).	
5	B	Gun Receiver Assembly	Check for free movement of cocking handle. Check barrel lock lever. Make sure it locks in place.	Cocking handle binds or barrel is loose.
6	D	M60/M60D Machine Gun	Erratic or sluggish firing may indicate carbon buildup. Switch barrels (WP 0018 00). If firing is still erratic or sluggish, field-strip the weapon and perform the necessary maintenance to complete the tire mission.	Weapon ceases to operate.

Item No.	Interval	Item to Check/ Service	Procedure	Not Fully Mission Capable If:
7	D	Small Arms Sling (M60)	Inspect sling for damage and for missing or loose sling swivels.	Swivels loose or missing. Sling is damaged.
8	A	M60/M60D Machine Gun	Fieldstrip, clean, inspect and lubricate entire weapon (WP 0018 00).	
9	A	M60/M60D Machine Gun (Inspection)	Inspect operating rod assembly and barrel assembly for burrs, cracks and chips. Make sure the gas piston slides back and forth freely. Inspect components of trigger assembly for wear. Place safety in 'F' position, sear should move slightly. Inspect breechbolt assembly for defects (e.g., cracks or chips). Reassemble weapon and check for proper functioning.	Weapon fails to function or has damaged or missing parts.

OPERATION UNDER USUAL CONDITIONS - LOADING 0005 00

NOTE
The following procedures illustrate the M60 machine gun except where the M60 parts differ from the M60D park, then both weapons are illustrated.

1 Point weapon down range.

2 Safety should be on F.
- M60 - Move safety to F.
- M60D - Push in on F.

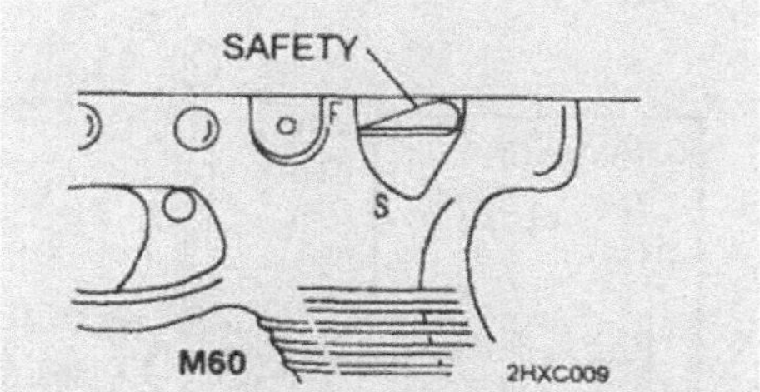

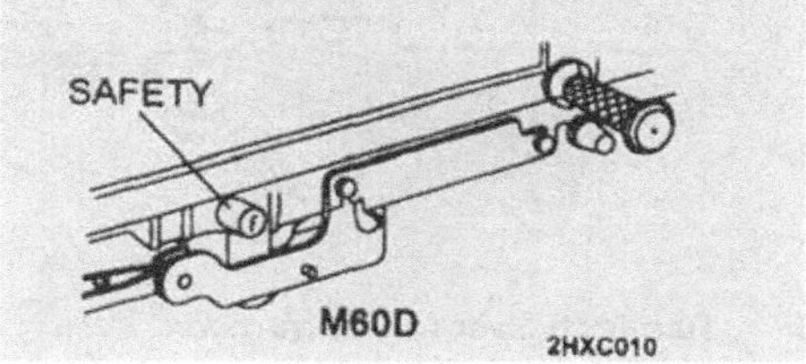

OPERATION UNDER USUAL CONDITIONS - LOADING (cont) 0005 00

3 Pull cocking handle rearward, then push it forward until it locks.

4 M60 - Move safety to S.
M60 D - Press in on S.

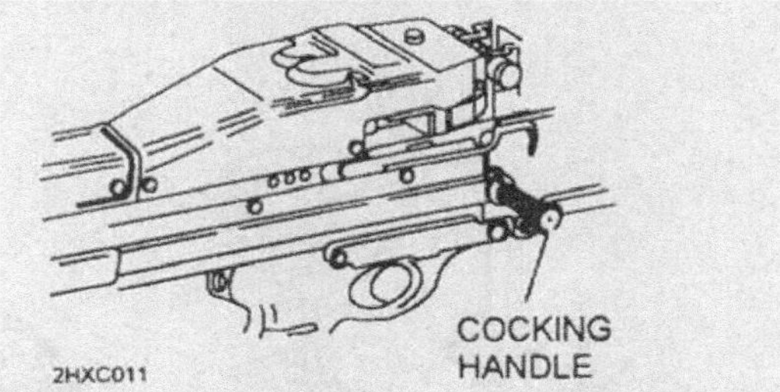

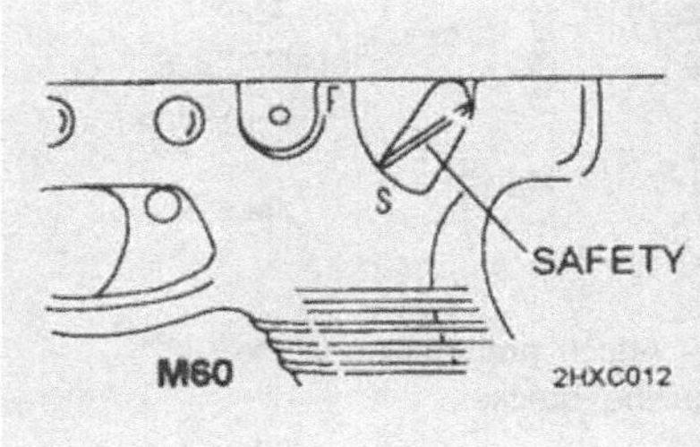

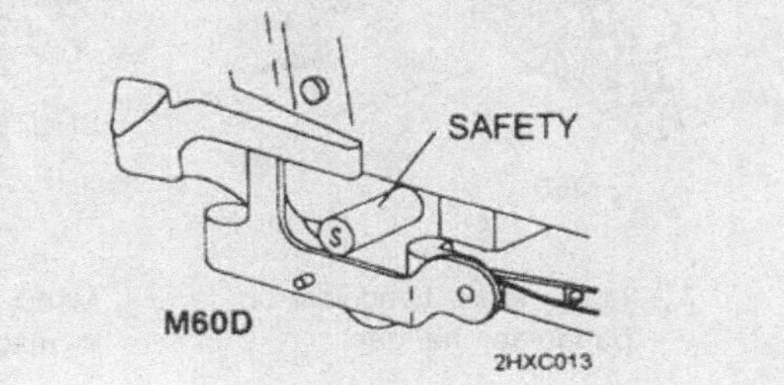

CAUTION

Bolt assembly must be rearward when opening or closing the cover.

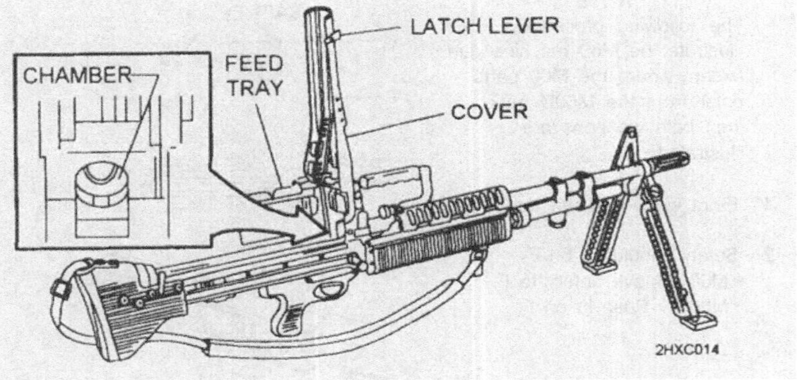

5 Turn latch lever and open cover.

6 Raise cartridge feed tray and check chamber for ammunition.

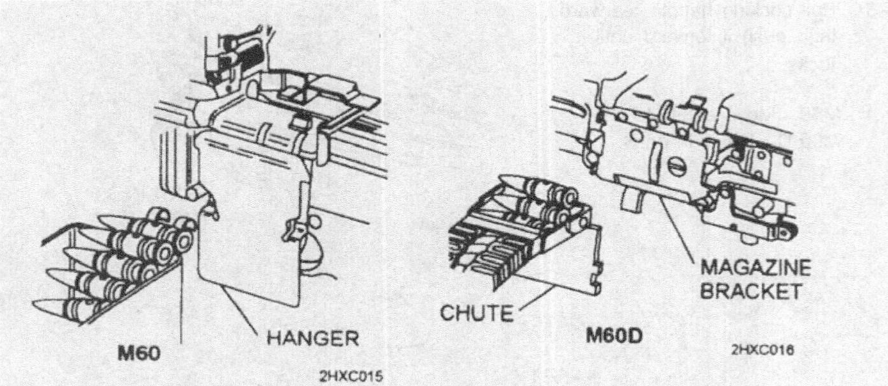

7 M60 - Place bandoleer on bandoleer hanger.

M60D - Attach ammunition chute to magazine bracket.

8 Place ammunition on feed tray, with open side of links down and with first round in the feed groove. (Make sure cartridge retainer pawl holds ammunition link on feed tray.)

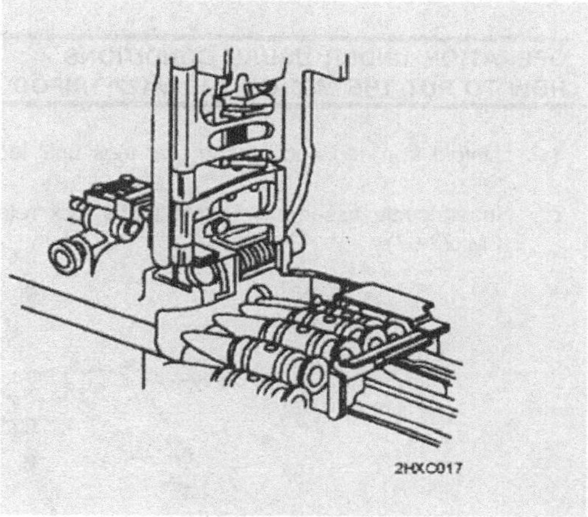

CAUTION

Bolt assembly must be rearward when opening or closing the cover.

9 Close cover.

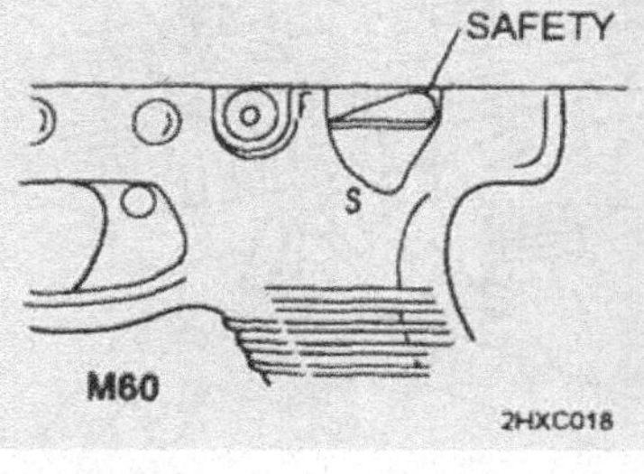

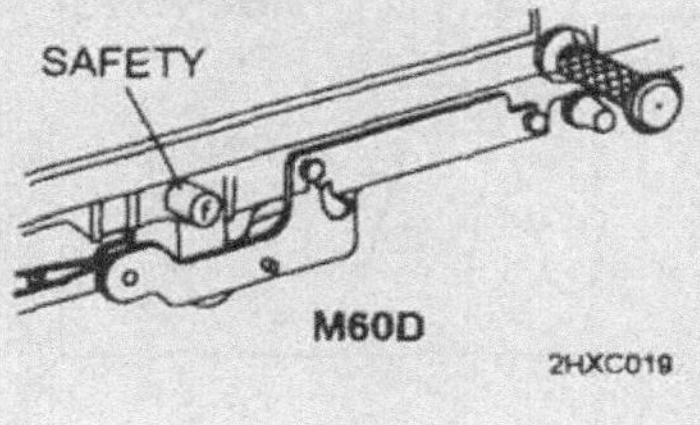

10 M60 - Move safety to 'F'.
M60D - Press in on 'F'.

11 The gun is ready to fire.

OPERATION UNDER USUAL CONDITIONS - HOW TO PUT THE M60 ON THE M122 TRIPOD 0006 00

1 Unfold front leg and spread rear legs until leg lock engages.

2 Insert pintle assembly. Rotate pintle lock release cam to lock (see FM 23-67).

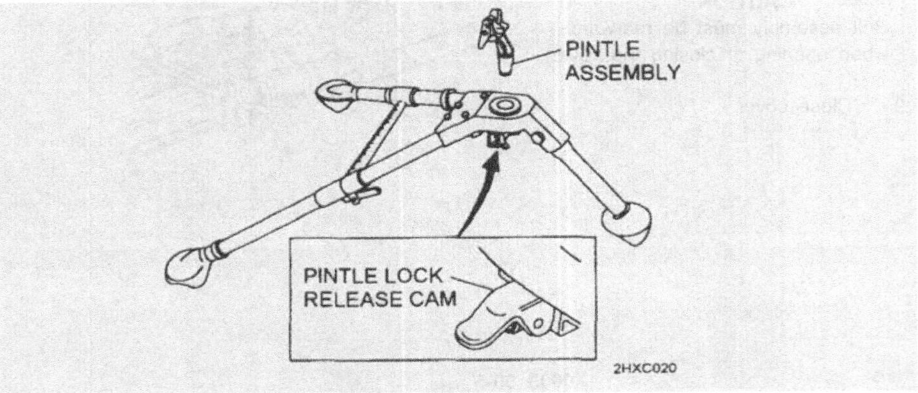

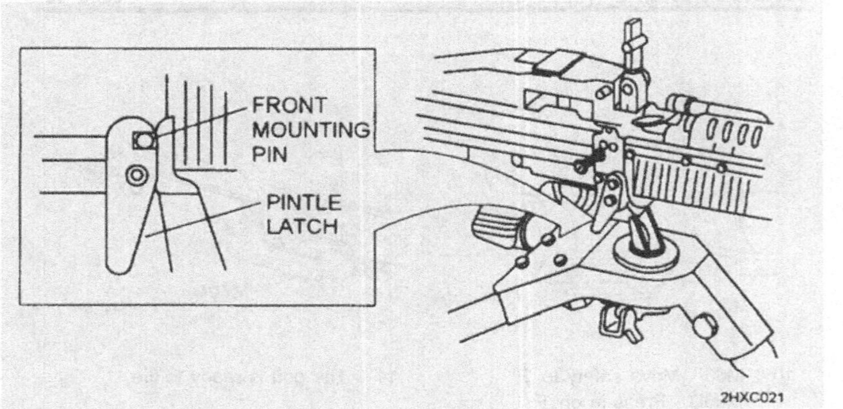

3 Place ends of machine gun's front mounting pin on top of pintle assembly.

4 Press bottom of latch to open pintle assembly. The ends of machine gun mounting pin should lock in place on pintle assembly.

5 Attach gun mount adapter to Traversing and Elevating Mechanism using straight pin and clips.

6 Center (zero) traversing hand-wheel and elevating handwheel on traversing and elevating mechanism assembly (see FM 23-67).

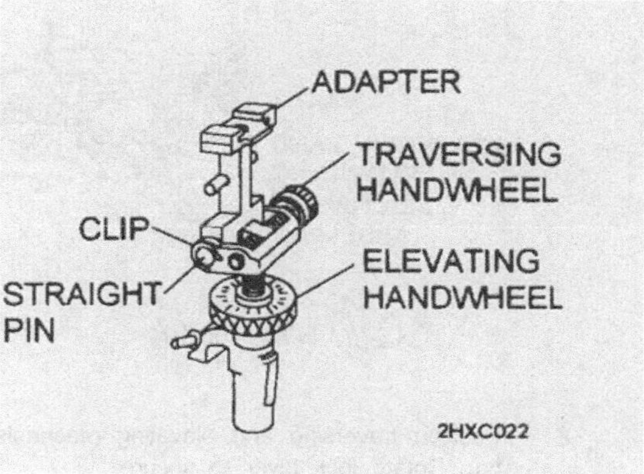

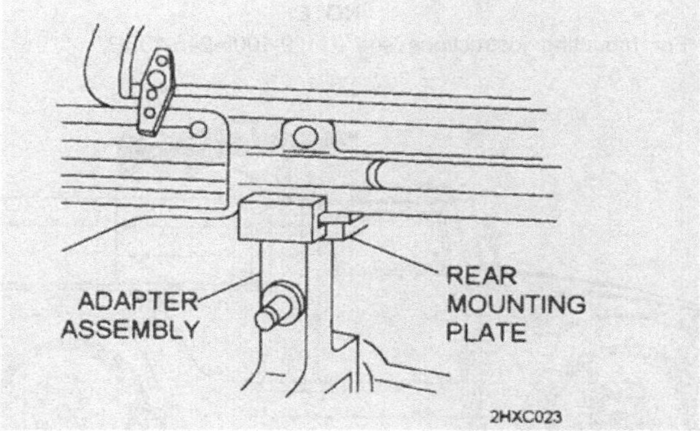

7 Slide adapter assembly on machine gun's rear mounting plate from the rear to the front.

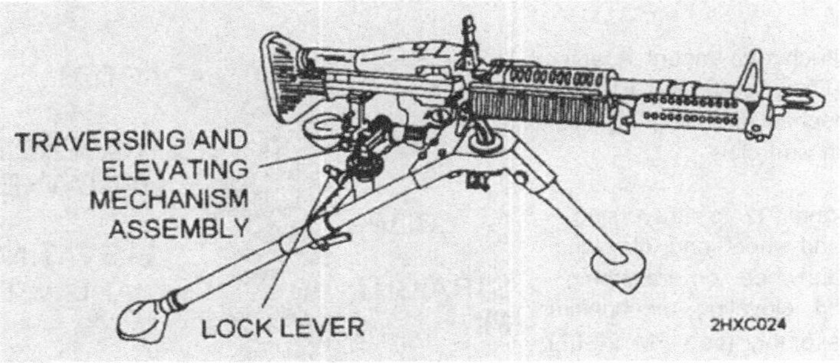

8 Position traversing and elevating mechanism assembly on traversing bar. Rotate lock lever to secure.

OPERATION UNDER USUAL CONDITIONS - HOW TO PUT THE M60 MACHINE GUN ON THE M6 PEDESTAL MOUNT/ M197 MOUNT 0007 00

NOTE

For mounting instructions see TM 9-1005-245-13&P.

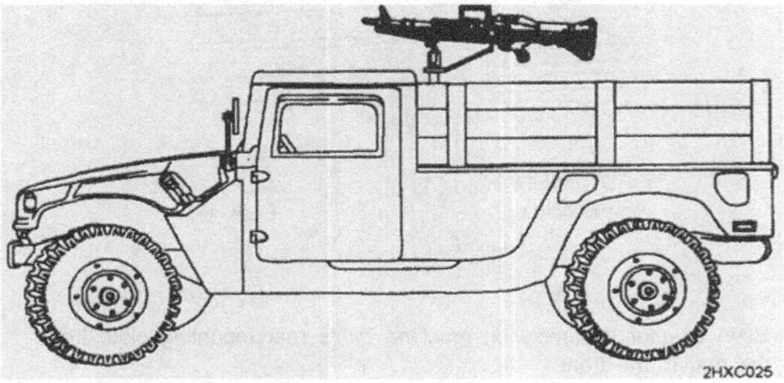

WARNING

The canvas cover above the driver and passenger seats should always be in place when firing.

Firing on-the-move is not permitted from M998/M1038/M1097.

Firing on-the-move is restricted to 5 miles per hour cross-country and 10 miles per hour on improved roads when mounted on M1025 and M1026 HMMWVs (Armament Carriers).

Ground personnel should not be within 10 meters of the vehicle when firing.

CAUTION

When the weapon is not in use/manned it should be secured in the travel lock. When not manned for extended cross-country travel weapon should be removed from the mount,

1 Ensure bipod legs are in the rearward locked position and the M197 pintle latch is rearward.

2 Press bottom of latch to open the pintle assembly. The ends of the machine gun mounting pin should lock in place on the pintle assembly.

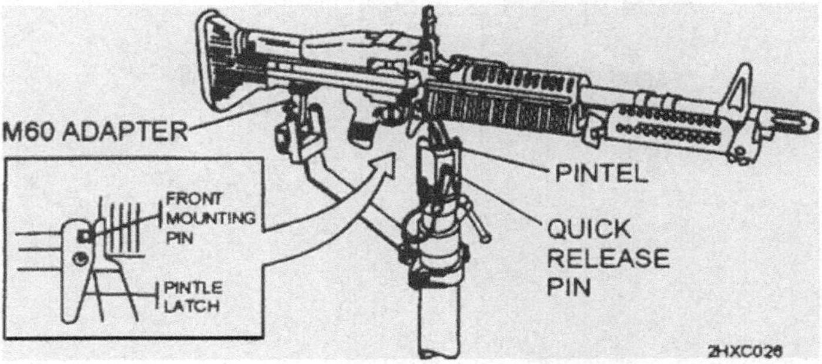

3 Place the end of the machine gun's front mounting pin on top of the pintle assembly.

4 Pivot the M60 adapter assembly into the upright position.

5 Slide the adapter assembly on the machine gun's mounting plate from the rear to the front.

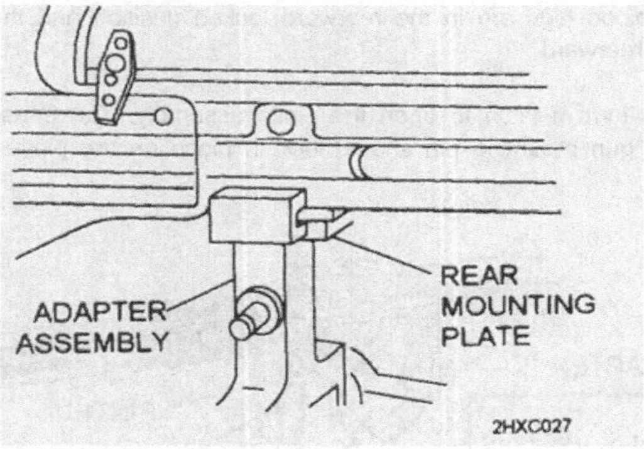

OPERATION UNDER USUAL CONDITIONS - HOW TO PUT M60D ON VEHICLES 0008 00

1 Pull quick release pin.

2 Lower gun adapter.

3 Attach M60D machine gun to aircraft mount using applicable manual.

4 Raise gun adapter.

5 Insert quick release pin

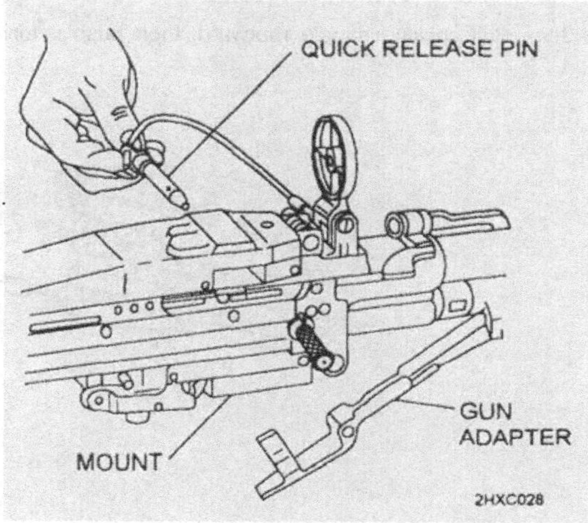

OPERATION UNDER USUAL CONDITIONS - CLEARING AND UNLOADING 0009 00

WARNING

If cover is opened on a hot cartridge (hot barrel), an open cover cookoff could occur and result in serious injury or death. Evacuate area for 15 minutes and then do REMEDIAL ACTION. Refer to WP 0011 00.

1 Point weapon down range.

2 Safety should be on 'F'.
- M60 - Move safety to 'F'.
- M60D - Press in on 'F'.

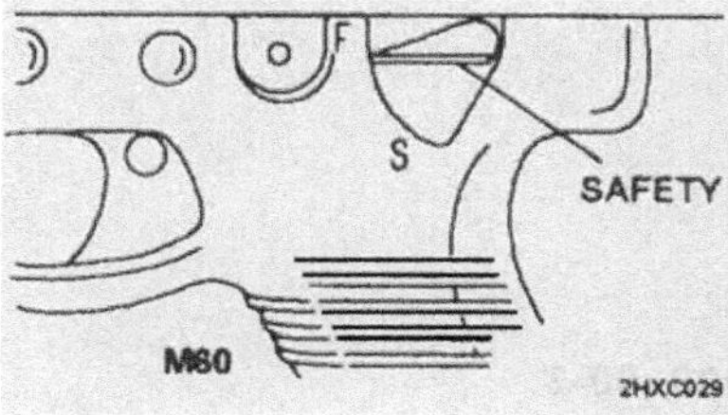

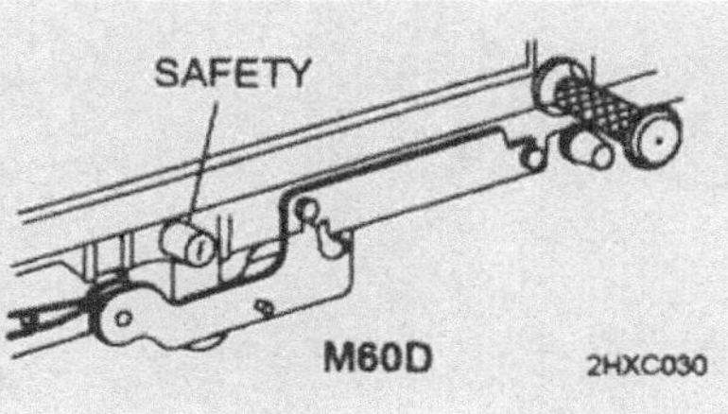

3 Pull cocking handle rearward, then push it forward until it locks.

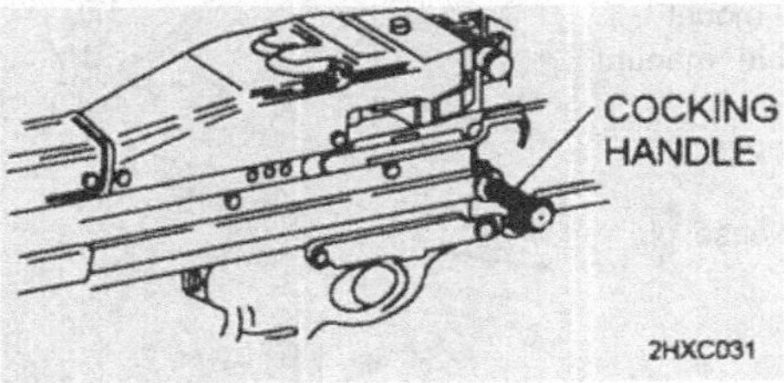

4 M60 - Move safety to 'S'.
M60D - Press in on 'S'.

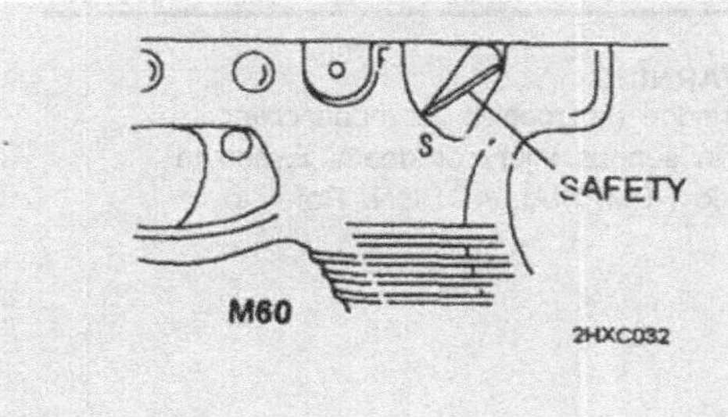

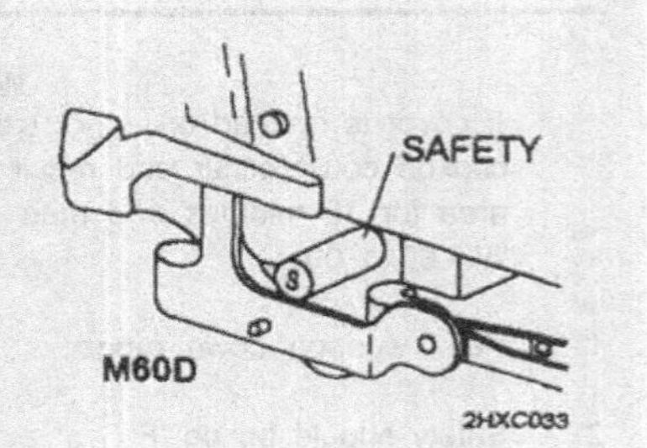

5 Turn latch lever and open cover.

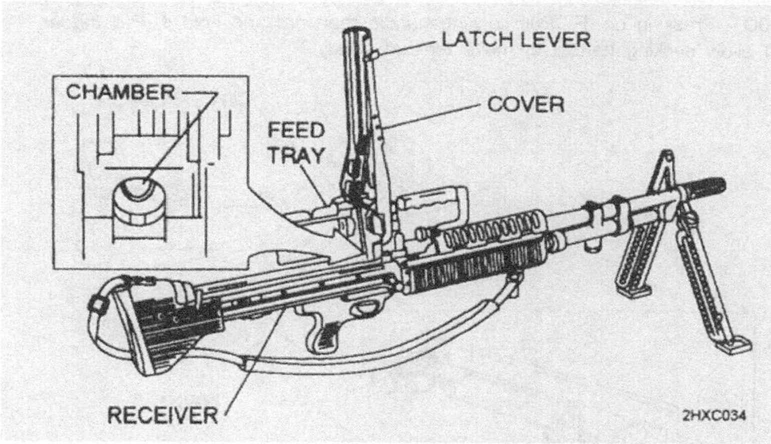

6 Remove ammunition and link belt and raise feed tray.

7 Eyeball chamber. If empty, go to step 8. If a round is chambered, see WP 0012 00 for stuck or ruptured cartridge case and WP 0017 00 for troubleshooting.

8 Lower feed tray.

9 Close cover. Cover latch should catch and hold cover closed.

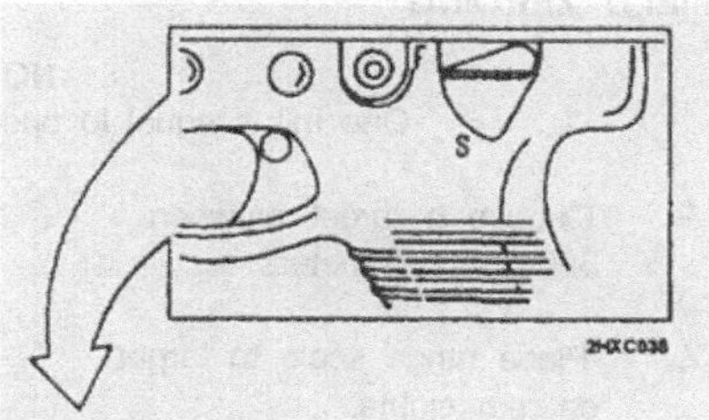

10 M60 - Move safety to 'F'. Pull cocking handle rearward and hold it. Pull trigger and allow cocking handle to move forward slowly.

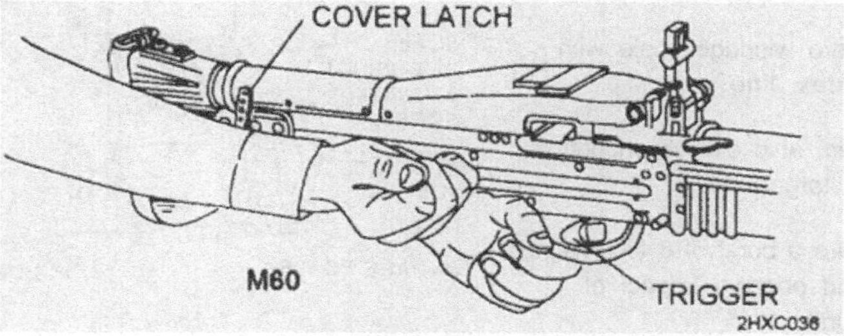

M60D - Press in on 'F'. Pull cocking handle rearward and hold it. Pull trigger and allow cocking handle to move forward slowly.

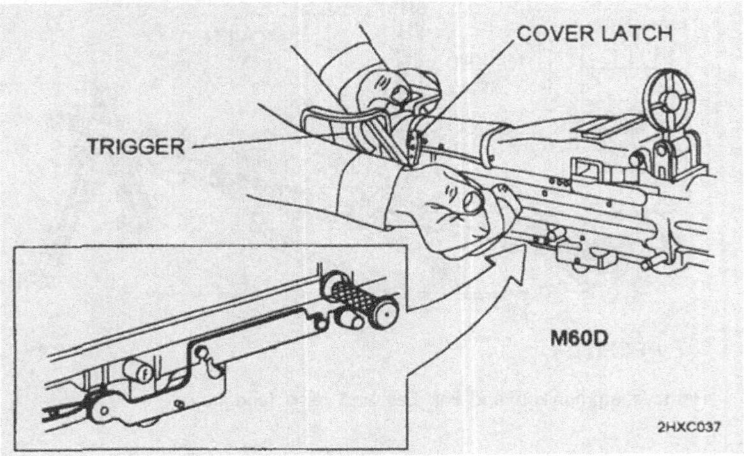

FIELD ZEROING

NOTE

One mil is equal to one meter at 1000 meters.

1 Choose a target between 300 and 700 meters.

2 Place range scale to target on rear sights.

3 Zero windage scale with index line.

4 Aim at 6 o'clock on bottom of target.

5 Fire a burst of 6 to 9 rounds and observe impact of burst.

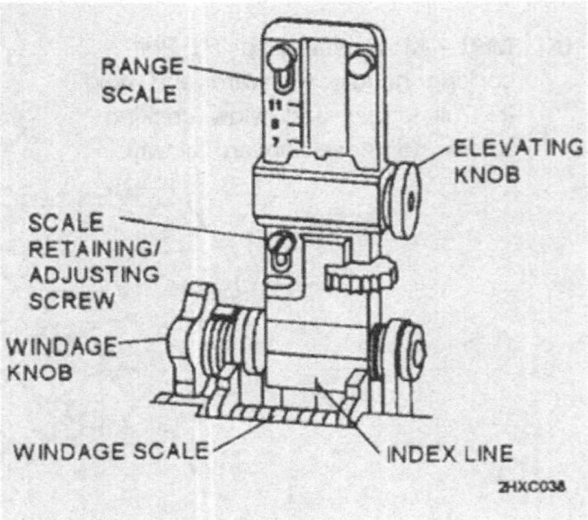

FIELD ZEROING (cont)

6 Correct for left or right wind by adjusting windage knob. Then correct range by adjusting elevating knob. Adjust until a burst of 6 to 9 rounds is on target.

7 Loosen scale retaining/adjusting screw and adjust elevation scale to reflect the range to the target.

IMMEDIATE ACTION

RUNAWAY M60 MACHINE GUN

1 If runaway occurs (machine gun won't stop firing), take the following actions to correct it quickly.

WARNING
Always keep machine gun pointed down range.

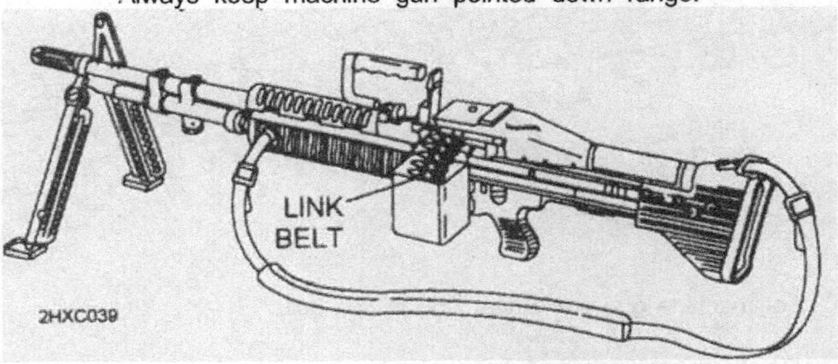

2 Break link belt (grasp belt and twist it firmly) or let machine gun fire if near end of link belt.

IMMEDIATE ACTION (cont)

RUNAWAY M60 MACHINE GUN (cont)

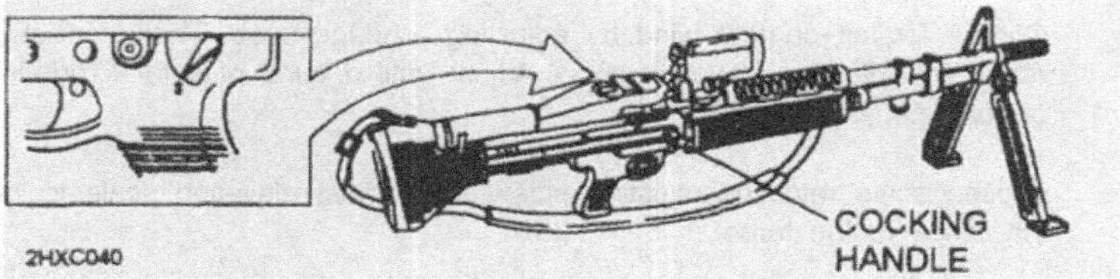

3 Pull cocking handle all the way back and hold it. Place safety to 'S' and remove link belt.

4 Clear machine gun (see WP 0009 00).

WARNING

Never reload a runaway machine gun until it has been repaired. Be sure machine gun is cleared before moving it.

5 Notify unit maintenance for repair.

RUNAWAY M60D MACHINE GUN

1 If runaway occurs (machine gun won't stop firing), take the following actions to correct it quickly.

WARNING

Always keep machine gun pointed down range.

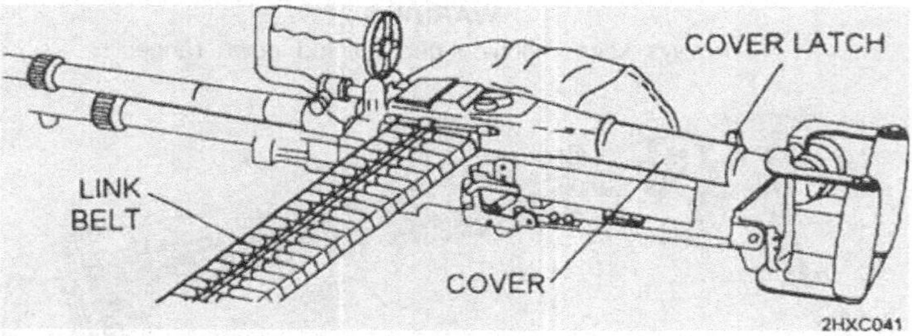

2 Let machine gun fire if near end of link belt.

3 Rotate cover latch and open cover.

IMMEDIATE ACTION (cont)

RUNAWAY M60 MACHINE GUN (cont)

4 Clear machine gun (see WP 0009 00).

WARNING

Never reload a runaway machine gun until it has been repaired. Be sure machine gun is cleared before removing it from mount.

5 Notify unit maintenance for repairs.

MISFIRE

If the weapon stops firing before you complete your mission, take IMMEDIATE ACTION within 10 seconds.

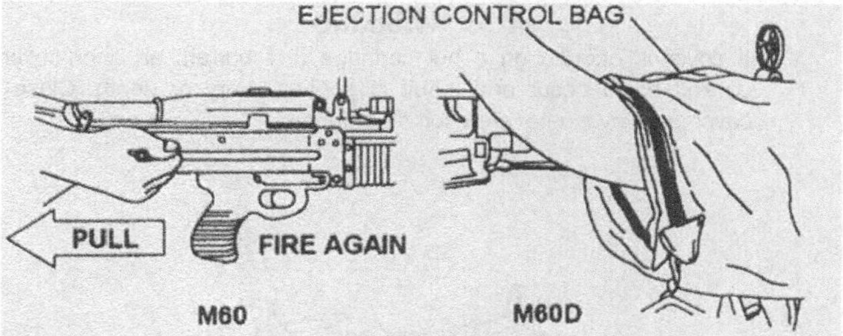

1 M60 - Pull cocking handle rearward.
M60D - Unzip ejection control bag and pull cocking handle rearward.

2 Ensure bolt is locked rearward and return cocking handle to forward locked position.

3 Check ejection port (M60 only).

MISFIRE (cont)

4 If a round or cartridge case is ejected, attempt to fire again.

WARNING

If cover is opened on a hot cartridge (hot barrel), an open-cover cookoff could occur and result in serious injury or death. Close cover and evacuate area for 15 minutes.

The climate temperature in different regions will make a difference as to what constitutes a hot gun. A hot, sunny day can cause a cookoff within 50 rounds, weapon and ammunition in sun.

5 If nothing is ejected and the barrel is hot enough to cause a cookoff (200 rounds fired within 2 minutes), watt at least 15 minutes (make sure bolt is locked rearward). Repeat steps for IMMEDIATE ACTION.

6 If IMMEDIATE ACTION fails to remove cartridge case, take REMEDIAL ACTION.

REMEDIAL ACTION

STEPS TO TAKE IF IMMEDIATE ACTION - MISFIRE - DOES NOT WORK

WARNING

If cover is opened on a hot cartridge (hot barrel), an open-cover cookoff could occur and result in serious injury or death. Close cover and evacuate area for 15 minutes.

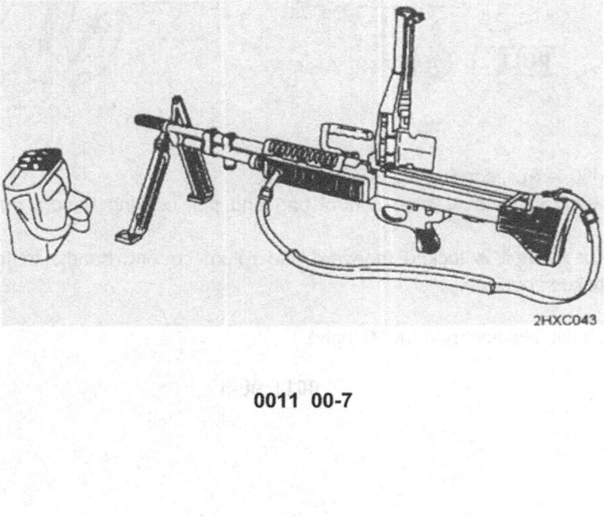

2HXC043

REMEDIAL ACTION (cont)

1 Keep weapon on target (down range/impact area). Clear weapon when barrel is cool (after 15 minutes wait) using the following steps.

2 Pull cocking handle rearward locking bolt to the rear. Return charging handle to forward locked position.

3 If a round is not ejected:
M60 - Place safety to 'S'.
M60D - Press in on 'S'.

4 Open cover, remove ammunition link belt and raise feed tray.

5 Inspect receiver, chamber, ejector, and ammunition.

6 If a round is in chamber, lower feed tray and close cover.

7 M60 - Place safety to 'F'.
M60D - Press in on 'F'.

6 Attempt to fire.

9 If a round is fired and ejected, reload and continue to fire.

10 If weapon does not fire or eject, clear and unload weapon, and notify unit maintenance for repair.

WARNING

When removing a stuck unfired cartridge, stay clear of the muzzle. Do not allow cartridge to contact any hard surface. Cartridge may fire on contact. Remove a stuck unfired cartridge using the same procedures for removing a stuck or ruptured cartridge case.

1 Pull cocking handle rearward and place safety to ‘S’.

2 Remove barrel (see WP 0018 00).

3 **STUCK CARTRIDGE CASE/ UNFIRED CARTRIDGE-** Insert cleaning rod in muzzle and tap out case/cartridge.

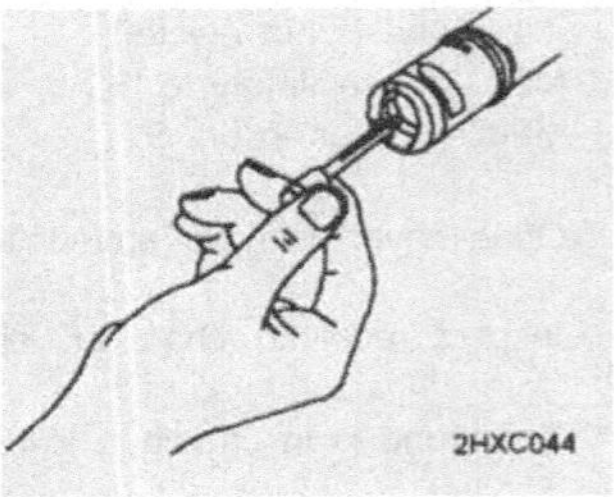

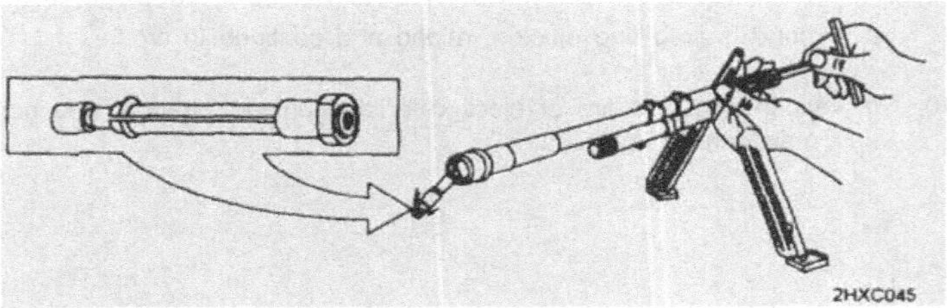

3 **RUPTURED CARTRIDGE CASE (cont)**

a. Push ruptured cartridge extractor inside ruptured case in chamber until it seats.

b. Lightly tap out with cleaning rod.

c. Remove extractor from cartridge case.

d. Unscrew extractor tool and remove cartridge case. Reassemble extractor tool.

4 Install barrel (see WP 0018 00).

5 Close cover.

6 Hold cocking handle rearward.

7 Place safety to 'F'.

8 Pull trigger and ease cocking handle forward.

9 Once weapon is cleared, do not place safety to 'S' when bolt is forward. Leave safety at 'F'.

OPERATION OF AUXILIARY EQUIPMENT 0013 00

INSTALLING M13A1 BLANK FIRING ATTACHMENT

1 Loosen lock nut and turn restrictor tube out a few times.

2 Loosen wing nut a few turns.

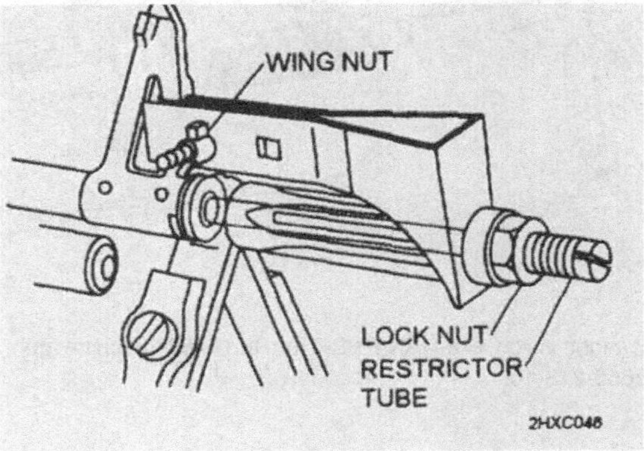

INSTALLING M13A1 BLANK FIRING ATTACHMENT (cont)

3 Slide restrictor tube in flash suppressor as far as possible,

4 Clamp blank firing attachment around front sight.

5 Tighten wing nut fingertight.

6 Screw restrictor tube in until it seats tight against muzzle end of barrel to prevent gas leakage.

7 Tighten lock nut.

NOTE

After Firing 500 Blank Rounds: Clean gun including barrel assembly, gas cylinder, gas piston, gas port, chamber bore, and Blank Firing Attachment (BFA).

NIGHT VISION SIGHT MOUNTED ON M60 MACHINE GUN

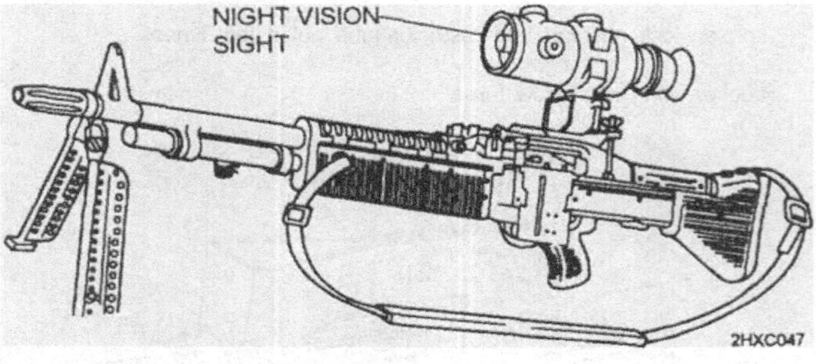

To mount night vision sight AN/PVS-4 on the M60 machine gun, refer to TM 11-5655-213-10.

PREPARATION FOR MOVEMENT 0014 00

Refer to FM 23-67.

OPERATION UNDER UNUSUAL CONDITIONS 0015 00

Under unusual conditions, clean and lubricate machine gun more often.

HOT, DUSTY, AND SANDY AREAS. Clean often. Wipe oil from exposed surfaces with clean wiping rag (item 5, WP 0039 00). Cover weapon as much as possible to keep dust and sand out of parts.

HOT, WET CLIMATE. Inspect often. Dry, clean, and lubricate lightly as necessary.

EXTREMELY COLD CLIMATE. Keep free of moisture. Lightly oil with LAW (item 3, WP 0039 00).

AFTER EXPOSURE TO WATER. Disassemble, clean, oil and reassemble as soon as possible. Make sure it's dry.

NUCLEAR, BIOLOGICAL, AND CHEMICAL (NBC) DECONTAMINATION. Decontamination procedures can be found in FM 3-5.

CHAPTER 3

MAINTENANCE INSTRUCTIONS

LUBRICATION INSTRUCTIONS 0016 00

LUBE GUIDE

Under all but the coldest arctic conditions, LSA (item 4, WP 0039 00), and CLP (item 1, WP 0039 00) are the lubricants to use on your machine gun. Remember to remove excessive oil from the bore before firing.

- Lightly Lubricate. A film of lubricant barely visible to the eye.
- Generously Lubricate. Heavy enough so lubricant can be spread with finger.

NOTE

Lubrication instructions are mandatory. Do not mix lubricants on the same machine gun. The weapon must be thoroughly cleaned during change from one lubricant to another. Dry cleaning solvent (available to unit maintenance) is recommended for cleaning during change from one lubricant to another.

LUBE GUIDE (cont)

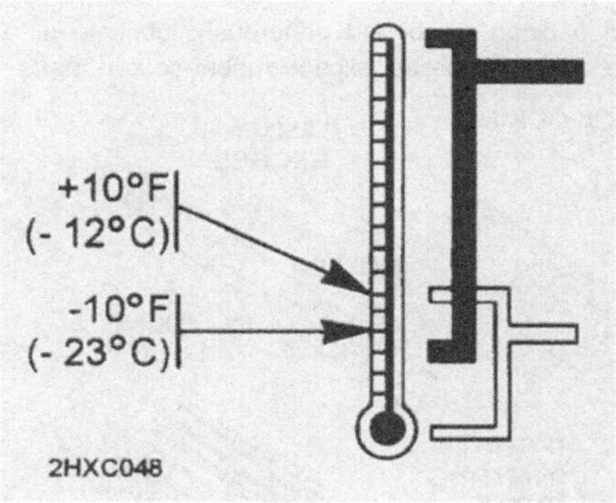

CLP - Cleaner, lubricant and preservative (item 1, WP 0039 00)

LSA - Weapons lubricating oil, semi-fluid (item 4, WP 0039 00)

LAW - Weapons lubricating oil, arctic (item 3, WP 0039 00)

Between 10°F (-12°C) and -10°F (-23°C) use CLP, LSA or LAW.
Below -10°F (-23°C) use only LAW.

WARNING

Be sure to clear weapon before disassembling, cleaning, inspecting, or storing.

1 After firing, field-strip your machine gun (WP 0018 00). Clean bore and chamber with cleaning rod, chamber brush, bore brush, and swab soaked with rifle bore cleaner (RBC) or CLP until a clean swab can be run through the bore without getting dirty.

2 Clean powder-fouled parts, except the buffer, with a wiping rag dampened with RBC or CLP.

3 Wipe dry and lube as required. Inspect and run a lightly oiled swab through the bore and chamber.

4 If your machine gun is not used, it still needs complete cleaning and lubing at least every 90 days. (Unusual conditions could shorten this time.)

LUBE GUIDE (cont)

5 External Surfaces. Put CLP on a clean swab and generously lubricate all shaded areas. Lightly lubricate all other areas except rubber-coated parts.

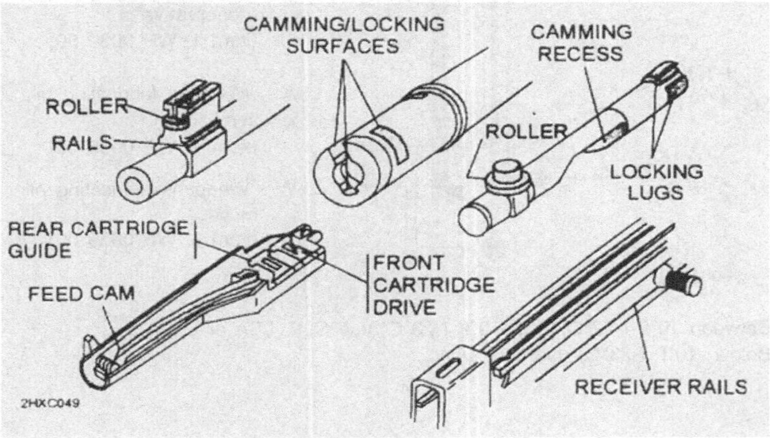

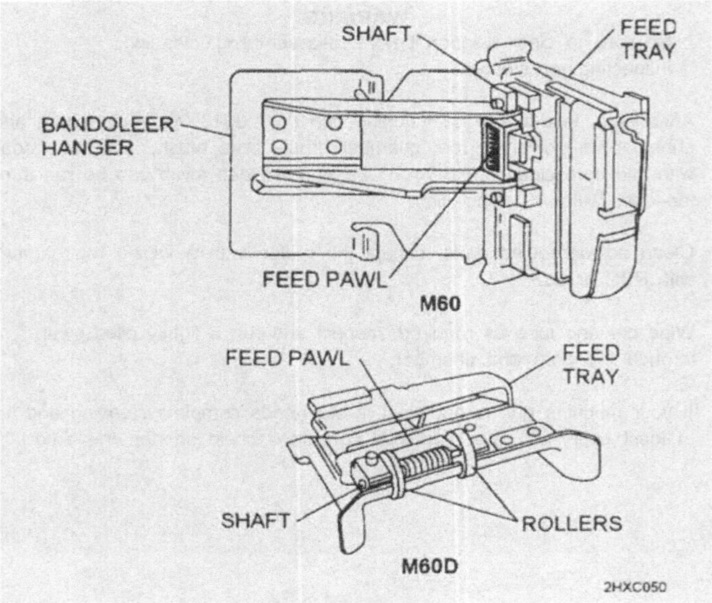

INTRODUCTION

The table lists the common malfunctions which you may find during the operation or maintenance of the machine gun or its components. You should perform the tests/inspections and corrective actions in the order listed.

This manual cannot list all malfunctions that may occur nor all tests or inspections and corrective actions. If a malfunction is not listed or is not corrected by listed corrective actions, notify your supervisor.

TROUBLESHOOTING

WARNING

Keep weapon pointed at target impact area. Never stand in front of weapon or expose body or hands to breech, ejection port, or muzzle.

MALFUNCTION	TEST OR INSPECTION	CORRECTIVE ACTION
COOKOFF*	Hot barrel	
	If you have tired 200 or more rounds within 2 minutes (rapid rate of tire), barrel will be hot enough to cause cookoff.	Cool weapon for 15 minutes.
FAILURE TO FIRE	Check ammunition.	Replace faulty ammunition.
	Check for broken or damaged firing pin.	Notify unit maintenance
	Check for weak or broken firing pin spring, guide assembly, or driving spring	Notify unit maintenance
FAILURE TO EXTRACT	Broken extractor or spring.	Notify unit maintenance

*COOKOFF - A misfire of a chambered round caused by the weapon overheating

MALFUNCTION	TEST OR INSPECTION	CORRECTIVE ACTION
FAILURE TO EXTRACT (cont)	Short recoil.	Clean gas port, piston, gas cylinder, operation rod assembly, and chamber (see WP 0023 00).
	Gas piston installed backwards	Install correctly.
	Dirty ammunition or chamber.	Clean chamber and/or use new ammunition.
SLUGGISH OPERATION	Friction from dirt, carbon, burrs or lack of lubrication.	Clean and lubricate or notify unit maintenance of any burrs.
WADDING AND UNBURNED POWDER	Manually charge weapon (if firing blanks)	Clean and lubricate or notify unit maintenance.
FAILURE TO CHAMBER	Ruptured cartridge case	Remove cartridge case (WP 00012 00).
	Carbon buildup in gas cylinder	Remove carbon.
	Carbon buildup in receiver assembly.	Remove carbon

MALFUNCTION	TEST OR INSPECTION	CORRECTIVE ACTION
FAILURE TO CHAMBER (cont)	Damaged round.	*Remove round and recharge weapon.
	Dirty chamber.	Clean or replace barrel.
FAILURE TO EJECT	Frozen, damaged or missing ejector spring.	Notify unit maintenance.
	Short recoil.	Clean gas port, piston, gas cylinder, chamber, and operating rod guide (WP 0023 00).
FAILURE TO COCK OR RUNAWAY GUN	Operating rod sear notch or sear worn.	Notify unit maintenance
	Sear plunger or spring broken or missing.	Notify unit maintenance.

*Dispose of damaged round according to range SOP

MALFUNCTION	TEST OR INSPECTION	CORRECTIVE ACTION
FAILURE TO FEED	Insufficient gas pressure (short recoil).	Clean gas system, chamber, and barrel socket (WP 0023 00).
	Improper lubrication.	Lubricate as necessary (WP 0016 00).
	Defective ammunition or links.	Use new ammunition belt.
	Ammunition belt installed wrong.	Install with open side of link down (WP 0005 00).
	Damaged or weak operating parts.	Notify unit maintenance
	Obstruction in receiver assembly.	Remove

END OF TASK

FIELD-STRIPPING MACHINE GUN 0018 00

M60 SHOULDER GUN STOCK

WARNING

Before field stripping, clear the weapon and check that the bolt is in forward position.

NOTE

The following procedures illustrate the M60 machine gun except where M60 parts differ from M60D pans, then both weapons are illustrated.

1 Lift shoulder rest (1), insert cleaning rod (2) in hole to release latch. Pull shoulder gun stock (3) from receiver.

2 Turn latch lever (4) and open cover.

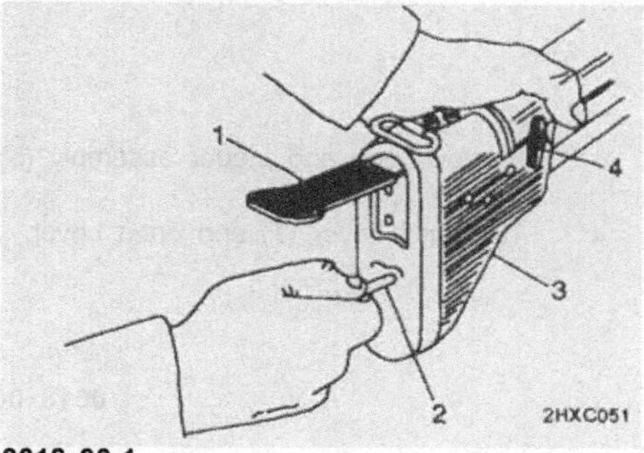

M60D SPADE AND TRIGGER GRIP ASSEMBLY

WARNING

To prevent possible bodily injury and aircraft damage, personnel should not stand the weapon on its barrel assembly when disassembling or assembling the weapon.

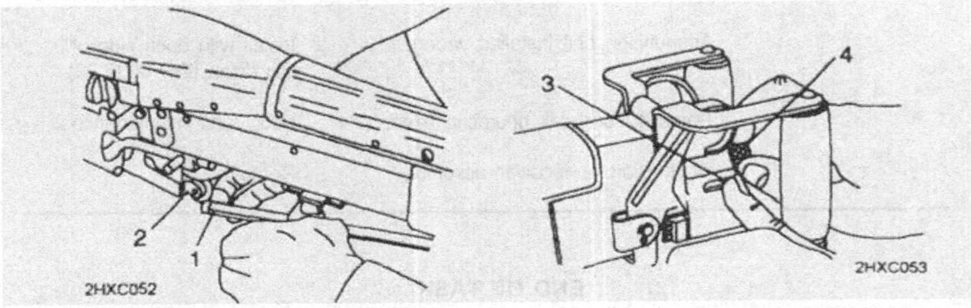

1 Unsnap sear assembly link and spring (1) from rear of sear assembly activator (2) by pressing upward with fingers.

2 Press and hold end of spring pin (3) forward while unscrewing knob and pin assembly (4).

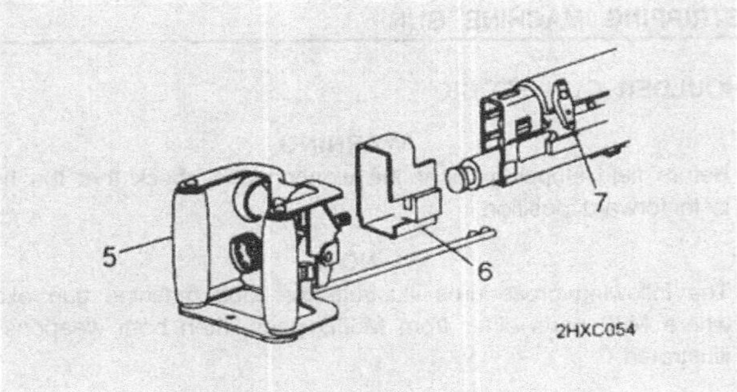

3 Remove grip and trigger assembly (5) and dust and moisture boot (6).

4 Turn latch lever (7) and open cover.

BOLT ASSEMBLY AND OPERATING ROD ASSEMBLY

1 Apply slight palm pressure to rear of hydraulic buffer assembly (1). Lift out buffer retaining yoke (2) and remove buffer assembly (1).

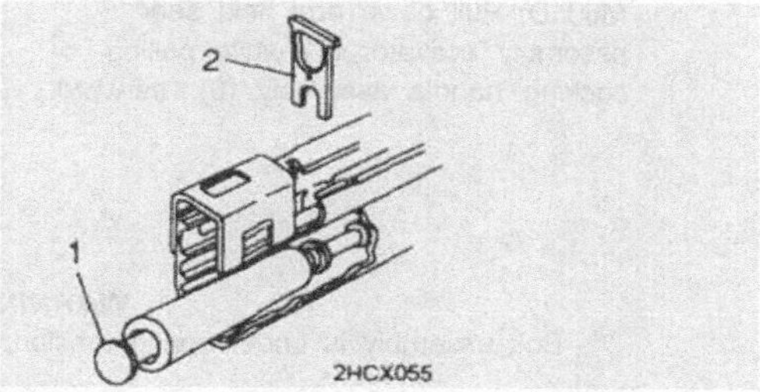

2 Remove guide assembly (3) and drive spring (4).

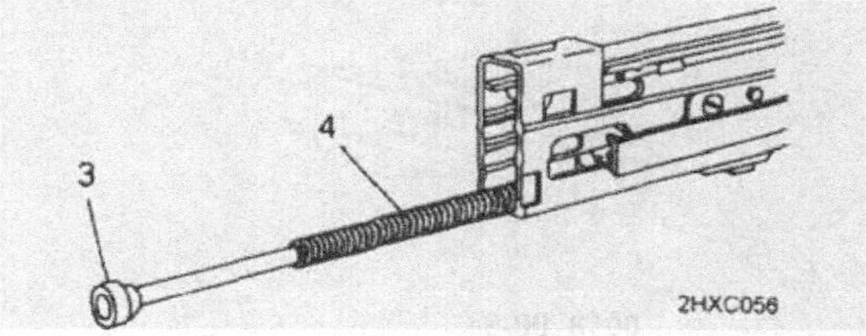

3 M60. Place safety to ‘F’. M60D. Press in on ‘F’.

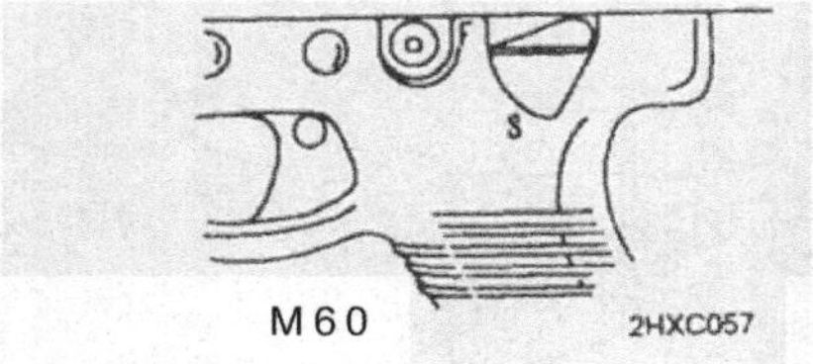

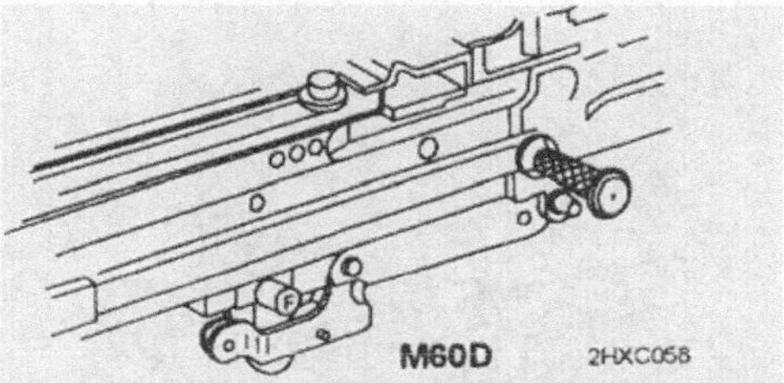

4 M60. Pull trigger (5) and hold it while pulling cocking handle assembly (6) rearward.

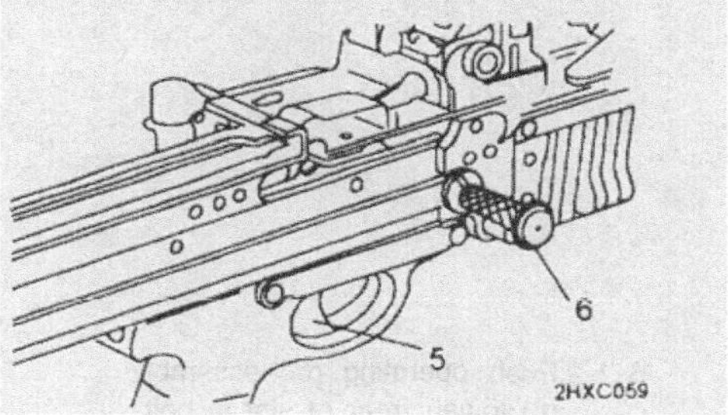

BOLT ASSEMBLY AND OPERATING ROD ASSEMBLY (cont)

M60 D. Pull down and hold sear assembly activator (7) while pulling cocking handle assembly (6) rearward.

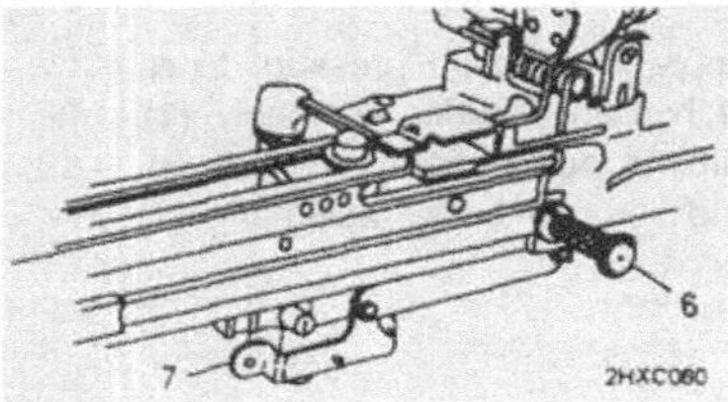

WARNING

Bolt assembly is under spring tension; it can twist and injure your hand.

5 Removing operating rod assembly (8) and bolt assembly (9) as a unit.

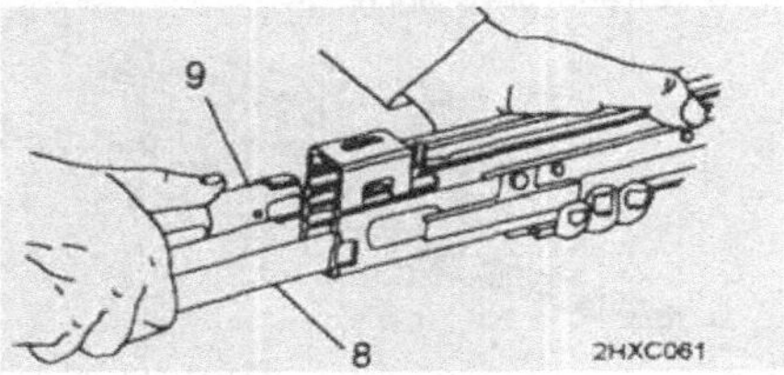

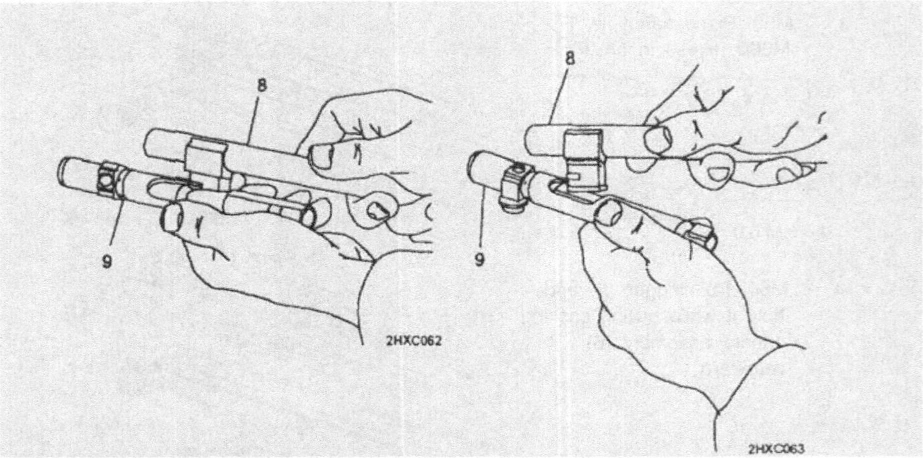

6 Push operating rod assembly (8) toward rear of slot in bolt assembly (9).

7 Separate operating rod assembly (8) from bolt assembly (9).

COVER ASSEMBLY AND M60 HANGER AND CARTRIDGE FEED TRAY ASSEMBLY OR M60D CARTRIDGE FEED TRAY ASSEMBLY

CAUTION

Do not use tip of driving spring guide assembly as a tool.

Use a cleaning rod (1) to unlatch hook of hinge pin latch (2). Remove hinge pin latch (2) and cover hinge pin (3).

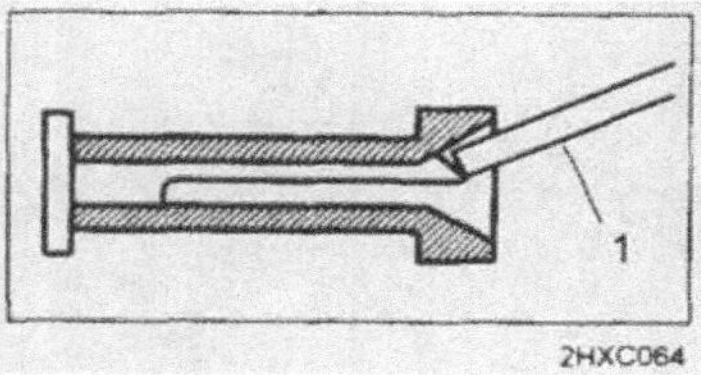

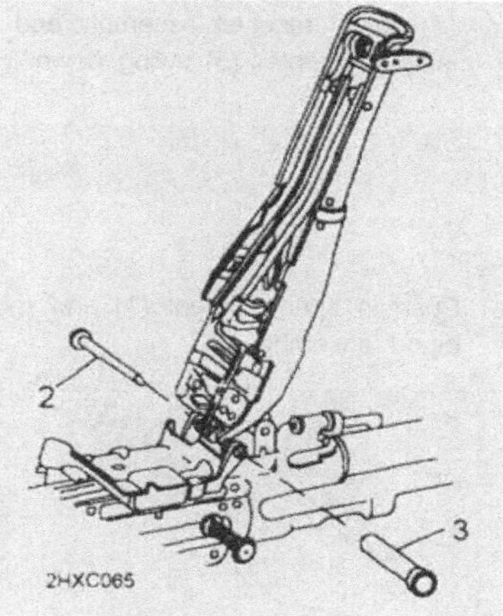

2 Remove cover assembly (4), torsion spring (5), and M60 hanger and cartridge feed tray assembly (6) or M60D cartridge feed tray assembly (7).

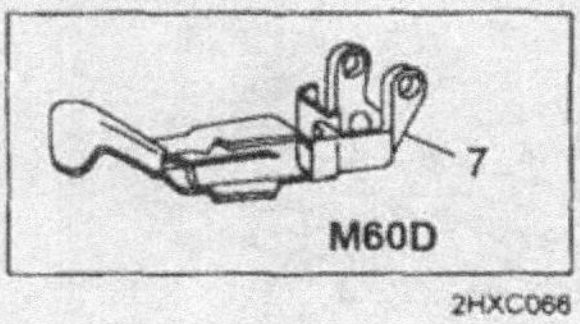

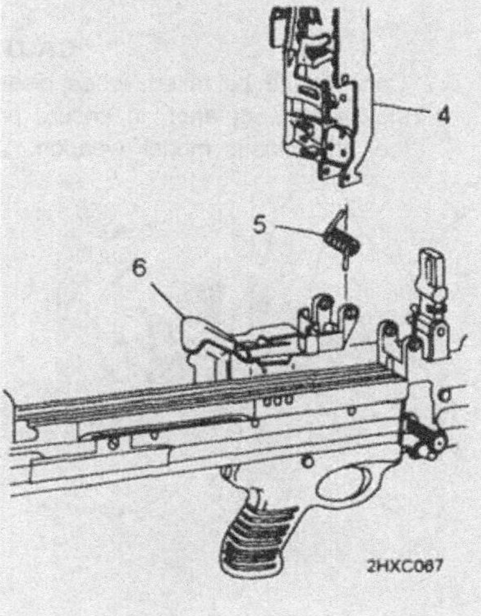

BARREL AND BIPOD ASSEMBLY

1 M60D. Pull quick release pin (4) out of receiver assembly and let gun adapter (5) swing down.

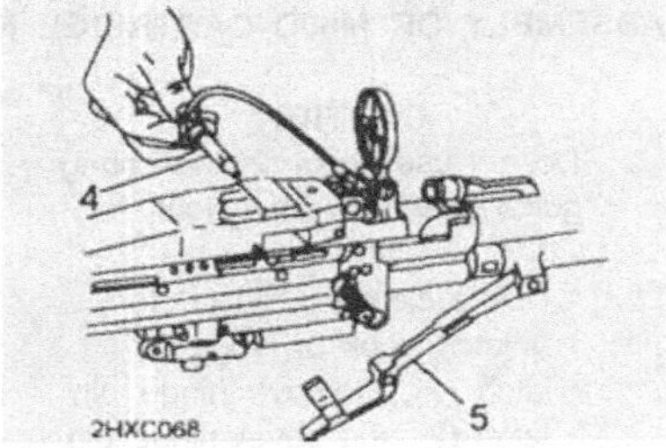

2 Push in spring detent (1) and raise barrel lock (2). Remove barrel and bipod assembly (3).

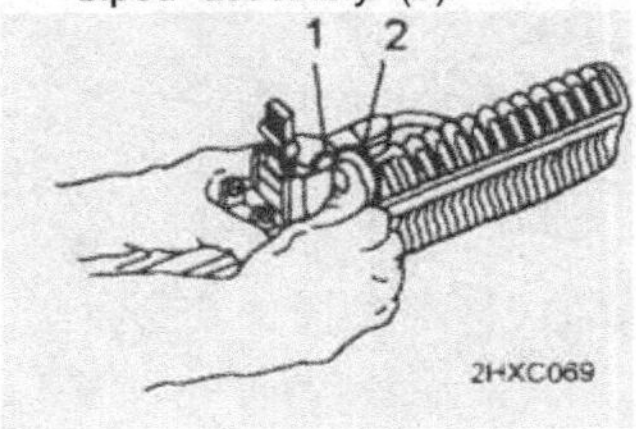

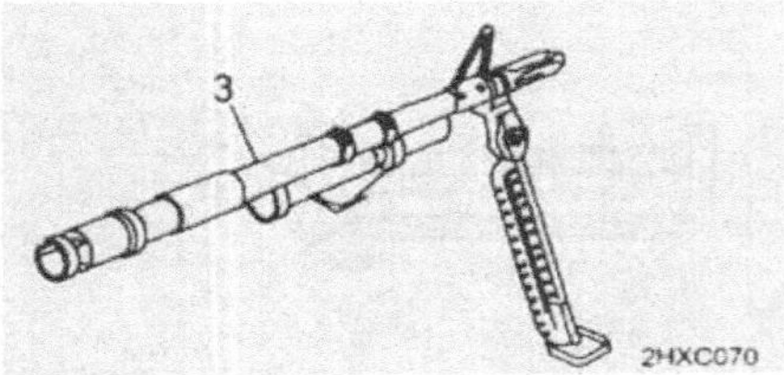

M60 TRIGGER MECHANISM GRIP ASSEMBLY OR M60D SEAR AND SAFETY HOUSING ASSEMBLY

CAUTION

Care should be taken, when disassembling M60 and M60D in the same work area, to ensure proper leaf spring remains with the appropriate model weapon.

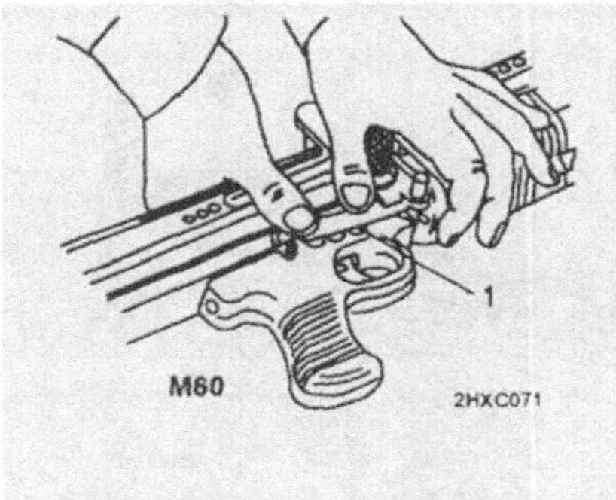

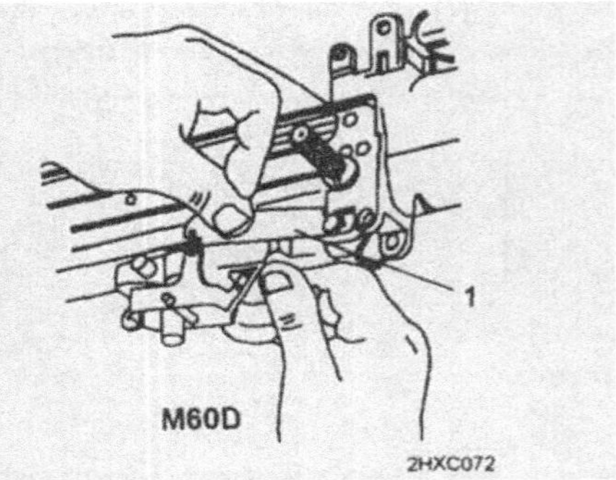

M60 and M60D. Push in and remove flat leaf spring (1).

M60 TRIGGER MECHANISM GRIP ASSEMBLY OR M60D SEAR AND SAFETY HOUSING ASSEMBLY (cont)

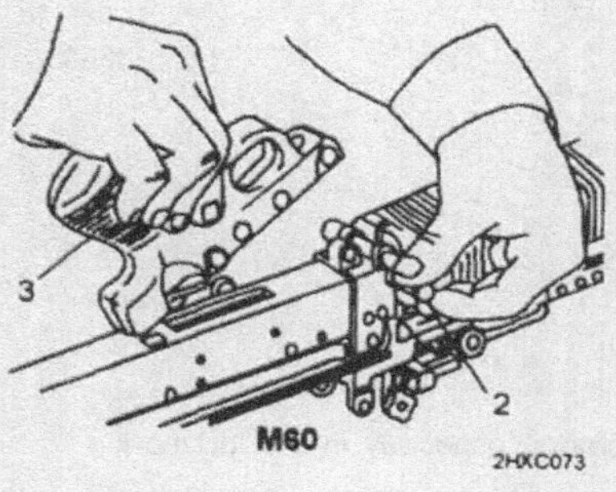

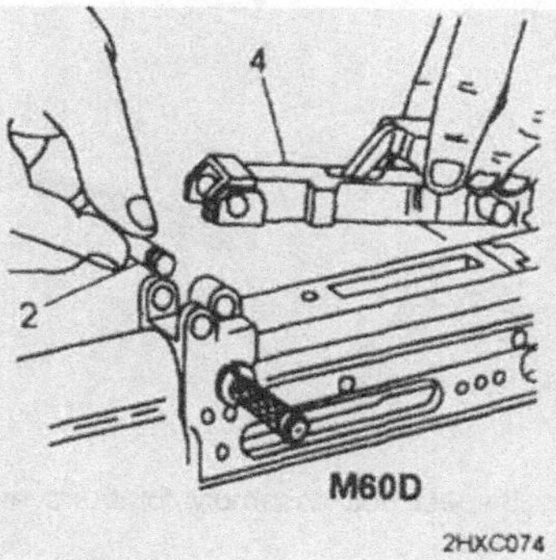

2 M60. Push out front pin (2) and slide trigger mechanism grip assembly (3) slightly forward and pull out to remove.

M60D. Push out front pin (2) and slide sear and safety housing (4) slightly forward and pull out to remove.

FOREARM ASSEMBLY

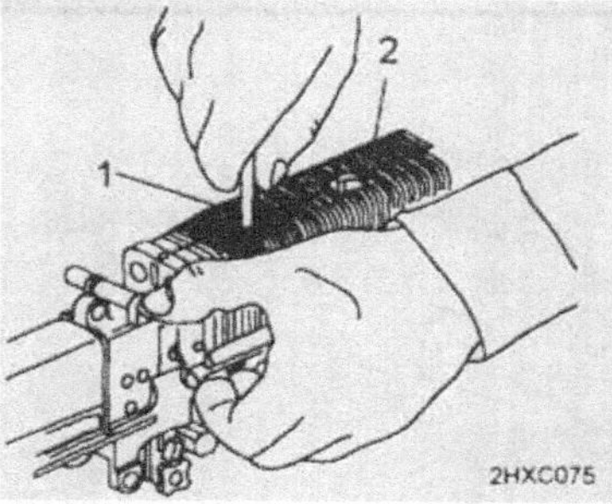

1 Insert cleaning rod (1) or combination wrench through opening in forearm assembly (2) and push down on spring.

CAUTION

Be careful not to damage internal ribs of forearm assembly.

2 Lift and gently slide forearm assembly (2) from receiver.

END OF TASK

INSPECT AND CLEAN

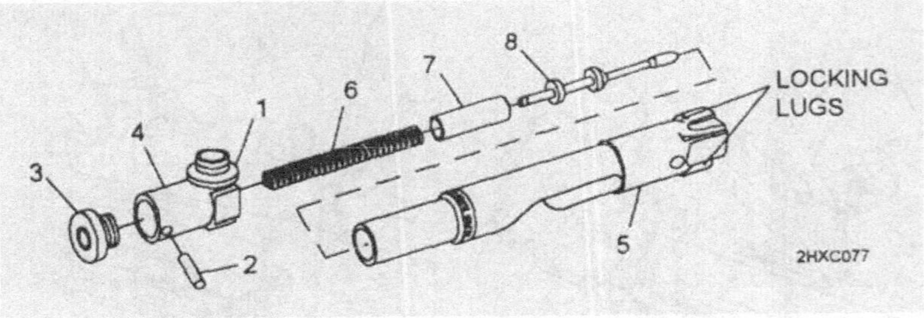

1 Inspect bolt assembly for burrs and cracks, especially in the locking lug area.

2 Make sure roller (1) operates freely and is not cracked.

INSPECT AND CLEAN (cont)

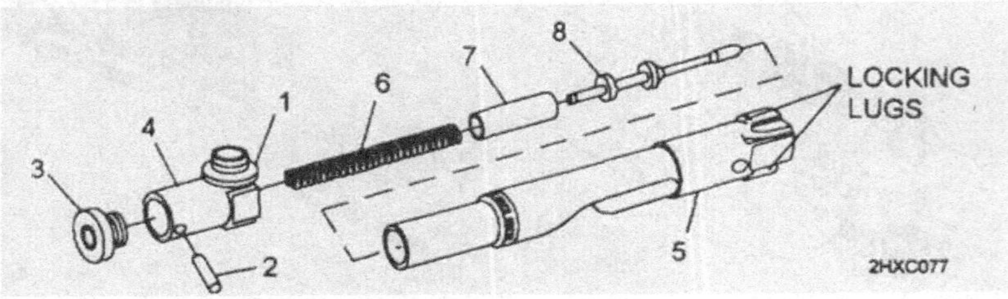

NOTE

Rotate cam actuator assembly so holes line up with headless straight pin. Use reamer of combination wrench to remove or replace headless straight pin.

3 Remove headless straight pin (2) and unscrew bolt plug assembly (3). Remove cam actuator assembly (4) from breech bolt (5). Separate spring (6), guide (7). and firing pin (8).

4 Make sure spring (6) is not kinked.

5 inspect threads on bolt plug assembly (3) and in breech bolt (5) for damage.

6 Clean powder fouling, corrosion, and dirt from all parts with CLP/RBC (items 1 and 2,WP 0039 00).

7 Drop firing pin (8) into breech bolt (5). Assemble and drop guide (7) and spring (6) into breech bolt (5).

8 Assemble cam actuator assembly (4) on breech bolt with roller to front. Screw in bolt plug assembly (3) and align three holes. Insert headless straight pin (2). Make sure cam actuator assembly (4) rotates freely when headless straight pin (2) is fully seated.

END **OF** TASK

MAINTENANCE OF OPERATING ROD ASSEMBLY **0020 00**

INSPECT AND CLEAN

1 Inspect operating rod (1), yoke (2), sear notch (3), and pins (4) for burrs, cracks, and chips.

2 Make sure roller (5) operates freely.

3 Inspect driving spring (6) for tension, kinks, breaks, and wear.

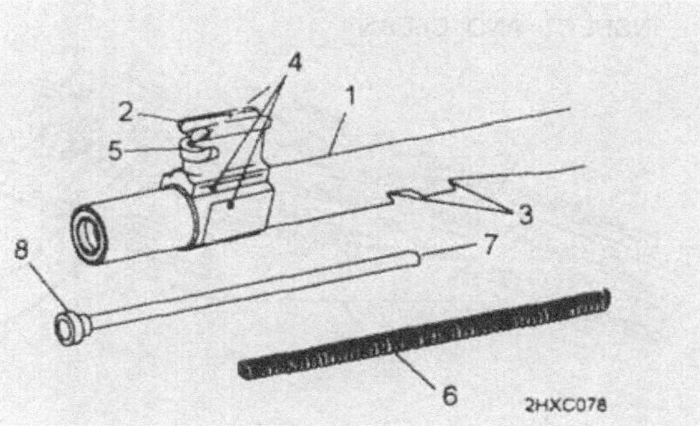

4 Check guide assembly (7) to make sure stop (8) is tight. If loose, notify unit maintenance.

5 Clean powder fouling, corrosion, and dirt from all parts with CLP/RBC (items land 2, WP 0039 00).

INSPECT AND CLEAN (cont)

6 Clean fluted carbon remover (9),

NOTE

Use fluted carbon remover (9) to clean inside of receiver (see WP 0025 00).

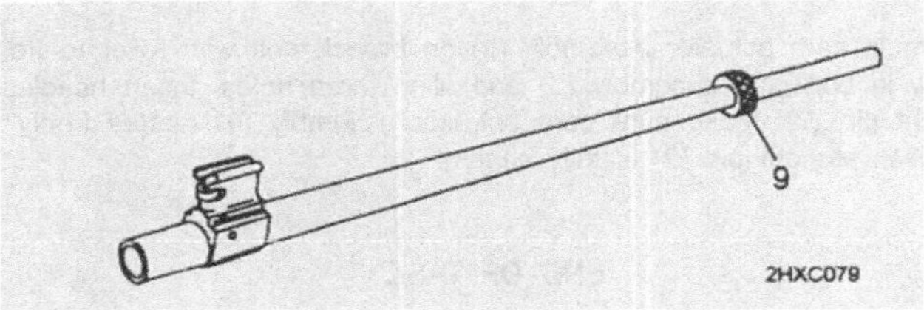

END OF TASK

MAINTENANCE OF COVER ASSEMBLY (M60 AND M60D) 0021 00

INSPECT AND CLEAN

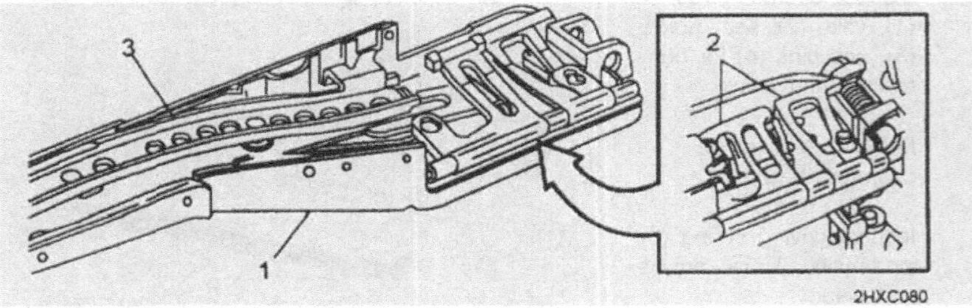

1 Inspect cover assembly (1) for bent, burred, cracked, loose, or missing parts.

2 Check spring action of front and rear cartridge guides (2), feed pawls (beneath the cartridge guides), and feed cam assembly (3).

MAINTENANCE OF COVER ASSEMBLY (M60 AND M60D) (cont) 0021 00

INSPECT AND CLEAN (cont)

3 Make sure feed cam assembly is secure and operates freely.

4 Clean powder fouling, corrosion, and dirt from all areas with CLP/RBC (items 1 and 2, WP 0039 00).

END OF TASK

MAINTENANCE OF M60 HANGER AND CARTRIDGE FEED TRAY ASSEMBLY OR M60D CARTRIDGE FEED TRAY ASSEMBLY 0022 00

INSPECT AND CLEAN

1 Inspect for burred, cracked, missing, or loose parts.

2 Check that feed tray (1) fits on receiver and that cartridge retaining pawl (2) works properly.

3 M60D - Make sure rollers (3) do not bind.

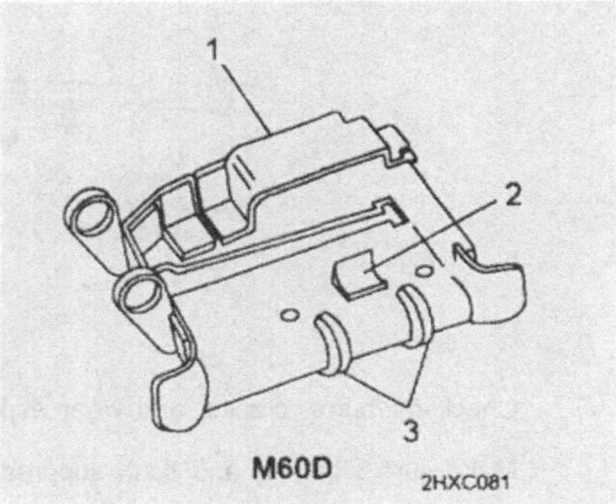

MAINTENANCE OF M60 HANGER AND CARTRIDGE FEED TRAY ASSEMBLY OR M60D CARTRIDGE FEED TRAY ASSEMBLY (cont) 0022 00

INSPECT AND CLEAN (cont)

4 M60 - Inspect for obstructions in hanger (4).

5 Clean corrosion and dirt from ail areas with CLP/RBC (items 1 and 2, WP 0039 00).

6 Remove CLP/RBC from top surface of feed tray (1) if gun is to be fired.

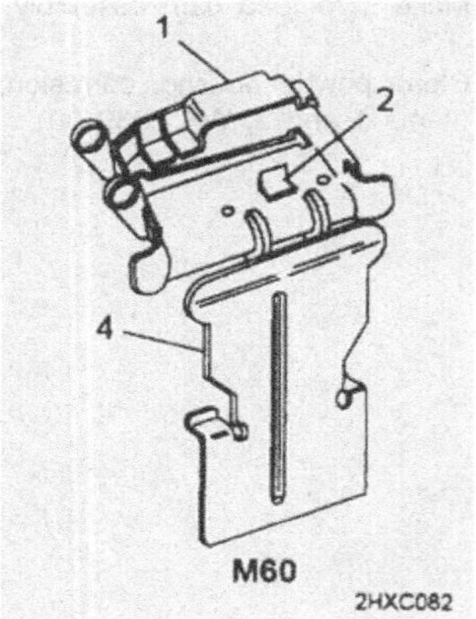

END OF TASK

MAINTENANCE OF BARREL AND BIPOD ASSEMBLY (INCLUDING GAS SYSTEM) 0023 00

BIPOD ASSEMBLY

INSPECT AND CLEAN

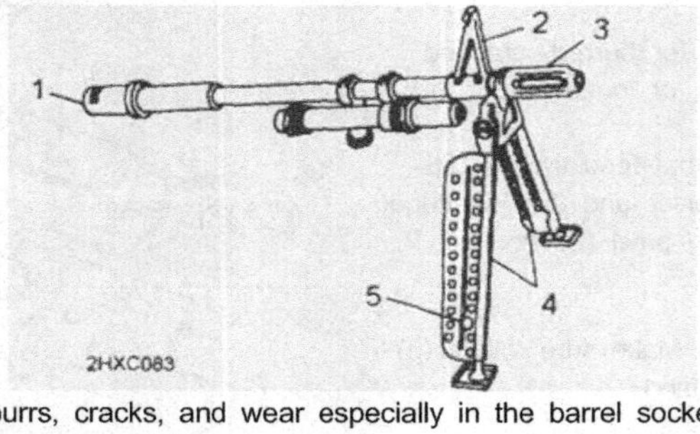

1 Check for burrs, cracks, and wear especially in the barrel socket area (1).

2 Make sure sight (2) and flash suppressor (3) are tight.

3 Make sure bipod (4) works, its legs are straight, and connections are tight.

4 Make sure bipod plungers (5) operate smoothly.

BIPOD ASSEMBLY (cont)

INSPECT AND CLEAN (cont)

WARNING

Barrels issued for a specific gun will not be changed from gun to gun. Each barrel and bolt assembly should be tagged during cleaning with the serial number of the receiver (gun) and should remain together as initially assigned.

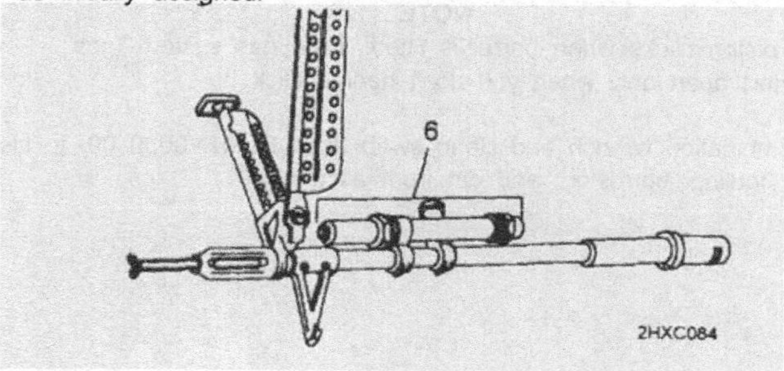

0023 00-2

5 Turn barrel upside down for cleaning. This will keep gas piston and inside of gas cylinder (6) dry.

6 Run a cleaning rod with swab (item, 6 WP 0039 00) saturated with CLP/ RBC (items 1 and 2,WP 0039 00) through bore. Leave bore wet.

GAS SYSTEM

INSPECT AND CLEAN

CAUTION

No abrasive materials will be used to clean the gas piston or inside the gas clinder.

NOTE

Gas piston clicks when barrel is tilted. Take gas system (gas cylinder) apart only when you don't hear a click.

1 Use combination wrench and clean swab (item 6, WP 0039 00) to clean powder fouling, corrosion, and dirt from all parts.

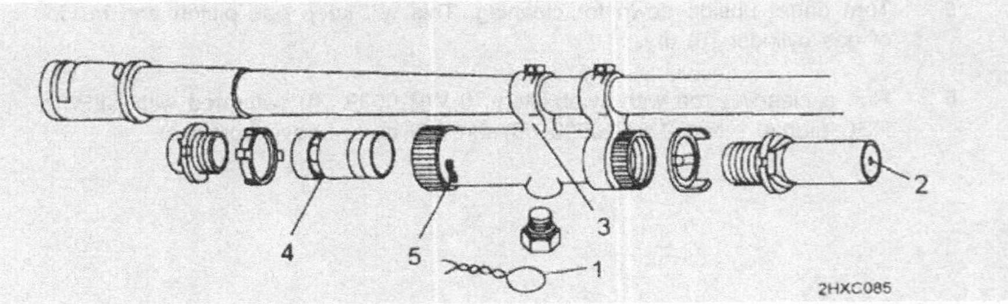

2 Remove safety wire (1) (notify unit maintenance).

3 Disassemble gas cylinder with combination wrench

4 Clean extension vent hole (2) with safety wire.

5 Clean gas port (3), gas piston hole (4), and gas cylinder vent holes (5) with reamer portion of wrench.

GAS SYSTEM (cont)

INSPECT AND CLEAN (cont)

CAUTION

Insert reamer carefully to prevent damage.

6 Bottom the reamer in the hole in the gas port (3) to make sure you have removed all carbon. Run small arms cleaning brush (bore) through bore to remove carbon deposits.

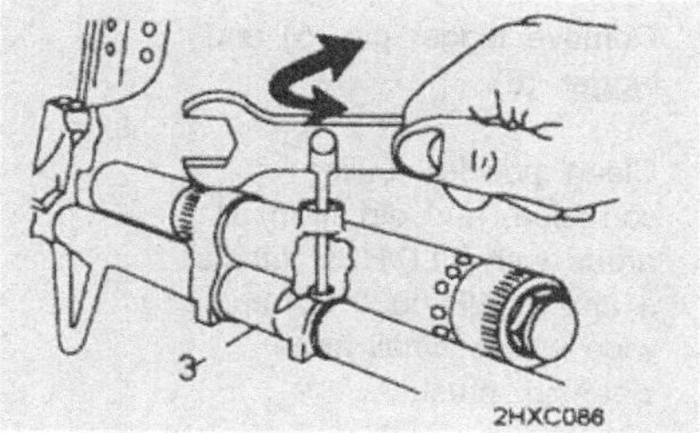

CAUTION

Make sure you do not install piston backward. Do not get CLP/RBC in gas system.

7 Reassemble gas system (gas cylinder), making sure gas cylinder vent, plug and nut are tight.

8 Notify unit maintenance to replace safety wire (1) promptly on completion of mission.

9 When safety wire is not available, check gas system periodically for tightness.

END OF TASK

INSPECT AND CLEAN

1 Hold down on sear (1) and remove sear pin (2).

2 Remove sear (11, sear plunger (3), and spring (4).

3 Remove trigger pin (5) and trigger (6).

4 Clean powder fouling, corrosion, and dirt from all areas with CLP/RBC (items 1 and 2, WP 0039 00) and wipe with a small arms cleaning brush.

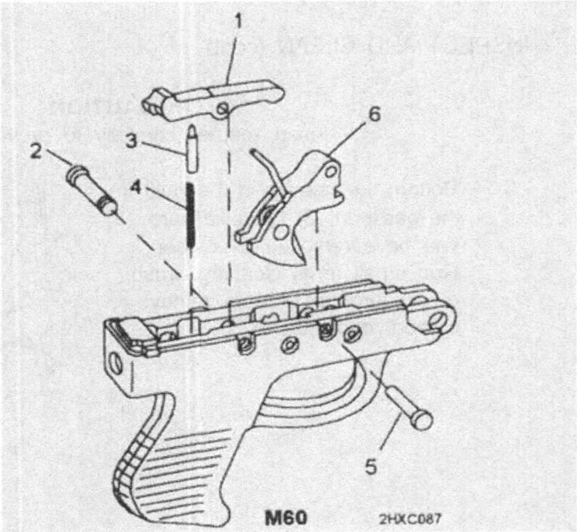

INSPECT AND CLEAN (cont)

5 Inspect sear (1) for chips, cracks and signs of wear.

6 Inspect sear plunger (3) and spring (4) for signs of wear.

7 Position trigger (6) in the grip so that the trigger spring is under the channel. Align holes and insert trigger pin (5) through the right side of grip and secure trigger (6).

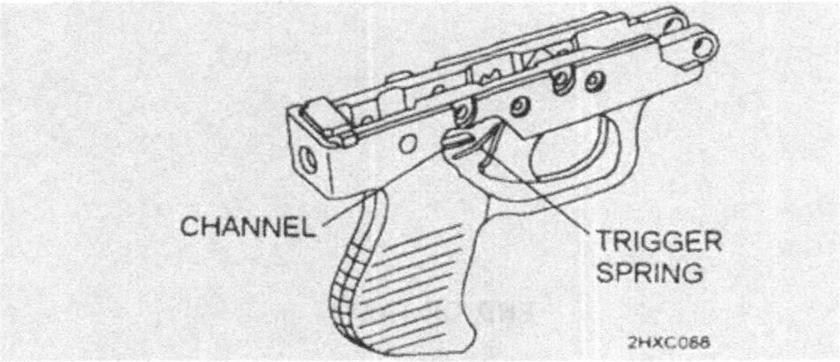

8 Insert spring (4) and sear plunger (3) into hole. Position sear (1) and apply slight pressure to align holes. Insert sear pin (2) through left side of grip and secure sear.

9 Check safety while pulling trigger back:
- 'F' - Sear moves freely.
- 'S' - Sear moves slightly.

END OF TASK

MAINTENANCE OF M60D SEAR AND SAFETY HOUSING ASSEMBLY 0025 00

INSPECT AND CLEAN

1 Hold down top of sear (1) and remove sear pin (2) sear assembly activator (3) and sleeve bushing (4).

2 Lift out sear (1) sear plunger (5), and spring (6).

3 Clean powder fouling, corrosion, and dirt from all parts with CLP/RBC (items 1 and 2, WP 0039 00) with a small arms cleaning brush.

4 Check sear (1) for chips, cracks, and signs of wear.

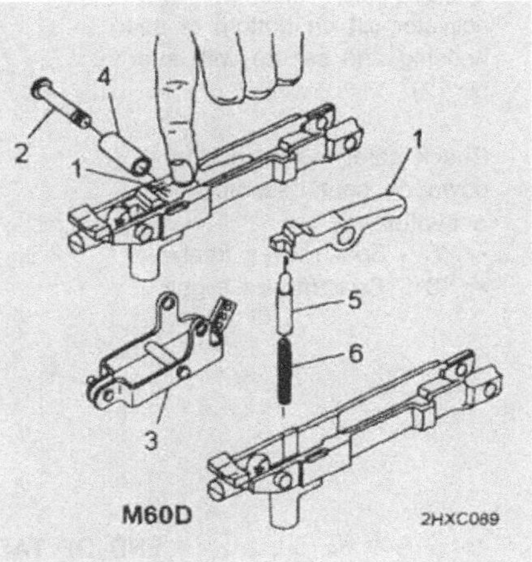

INSPECT AND CLEAN (cont)

5 Check sear plunger (5) and spring (6) for signs of wear.

6 Insert spring (6) and sear plunger (5) into hole. Position sear (1) into housing.

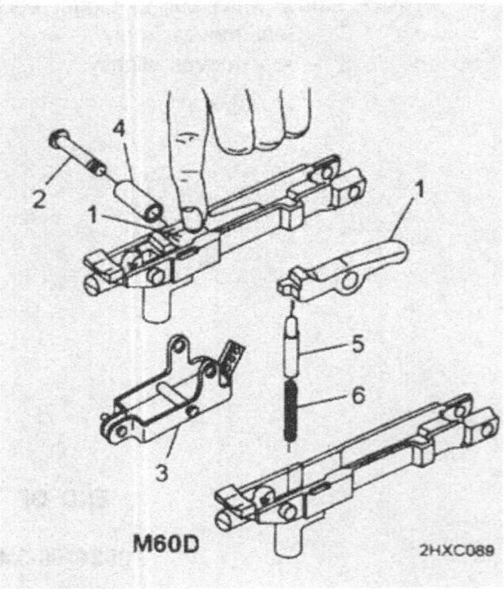

7 Apply slight pressure to align holes and insert sleeve bushing (4).

8 Position sear assembly activator (3) on bottom of sear housing and secure with sear pin (2).

9 Check safety while pulling down on sear assembly activator.

- 'F' - Sear moves freely.
- 'S' - Sear moves slightly.

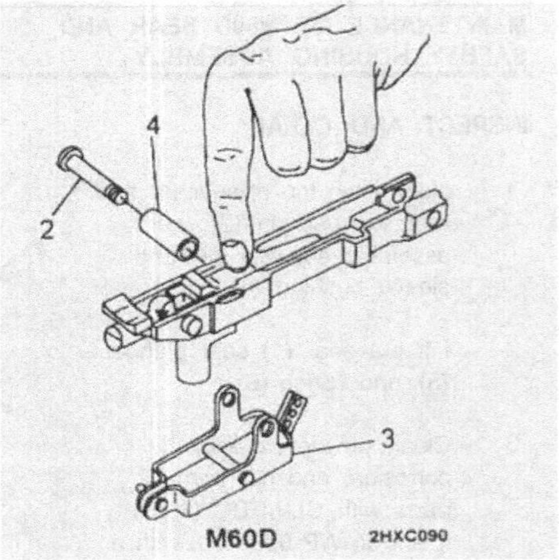

END OF TASK

INSPECT AND CLEAN

1 Clean powder fouling, corrosion, and dirt from all areas with CLP/RBC (items 1 and 2, WP 0039 00) and small arms brushes.

2 When weapon is disassembled, use fluted end of operating rod assembly to remove carbon from operating rod guide of receiver (WP 0020 00).

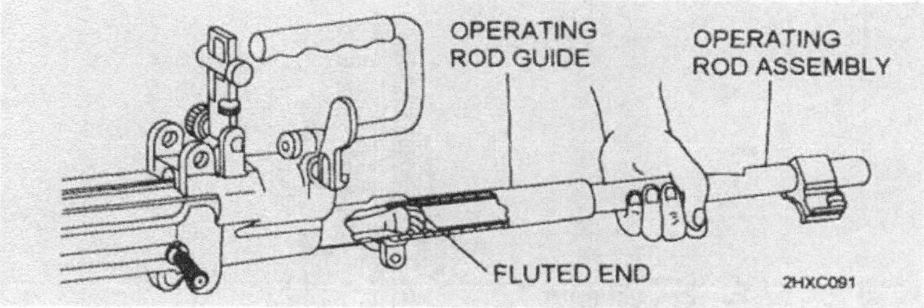

INSPECT AND CLEAN (cont)

3 Check receiver rails for burrs and wear.

4 Check that cocking handle moves freely.

END OF TASK

M60 FOREARM ASSEMBLY

CAUTION

When installing forearm assembly, be careful not to damage its internal ribs.

1 Slightly tip forearm assembly and slide it onto receiver.

2 Press in on bottom of forearm assembly to latch.

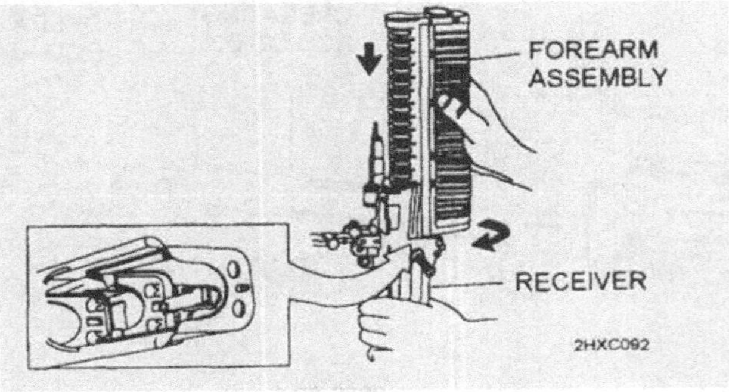

M60 TRIGGER MECHANISM GRIP ASSEMBLY OR M60D SEAR AND SAFETY HOUSING ASSEMBLY

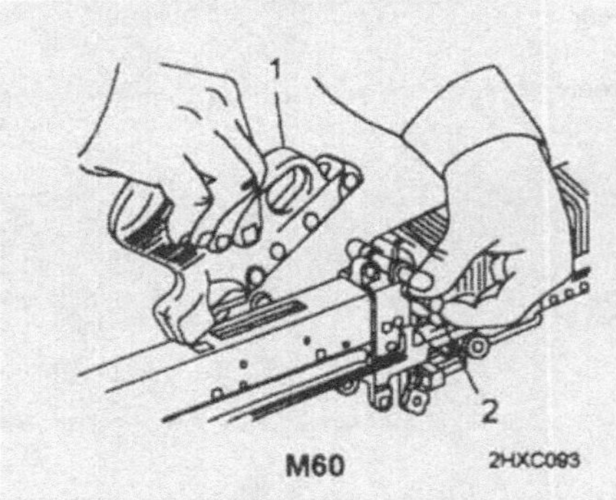

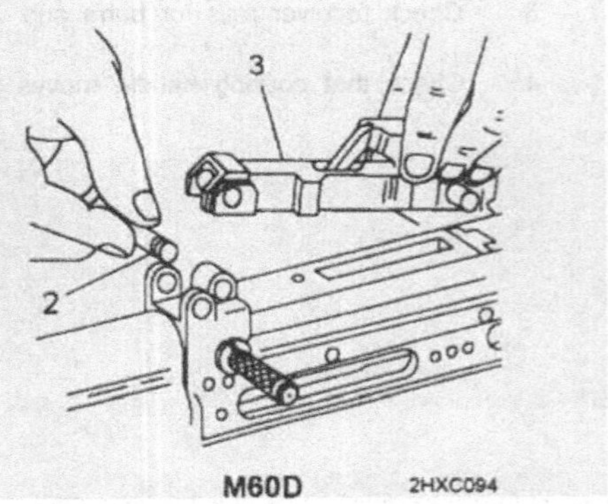

1 M60. Position trigger mechanism assembly (1) on bottom of receiver, align holes, and install front pin (2) from the left side.

M60D. Position sear and safety housing assembly (3) on bottom of receiver, align holes and install front pin (2) from the left side.

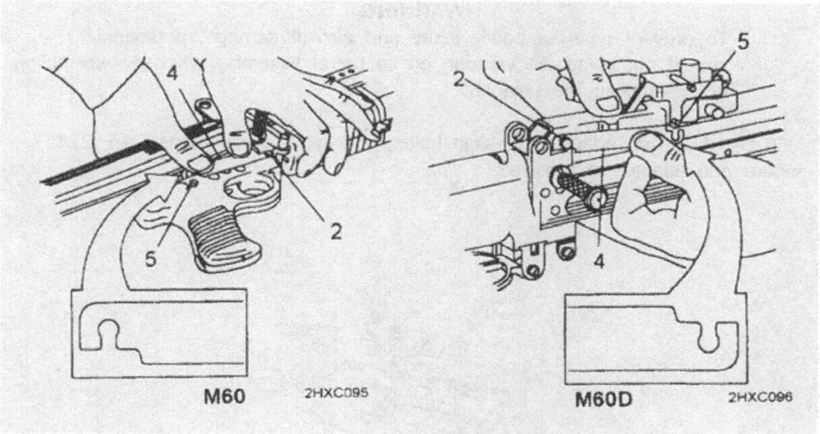

WARNING

Do not interchange leaf springs from one model weapon to another.

2 M60 and M60D. Slide slotted end of flat leaf spring (4) on front pin (2). The hooked end of flat leaf spring should bend outward. Push down and slide hooked end of flat leaf spring onto grooved pin (5).

BARREL AND BIPOD ASSEMBLY

M60 and M60D. Make sure barrel lock (3) is up (unlocked). Install barrel and bipod assembly (4). Push barrel lock (3) down to lock.

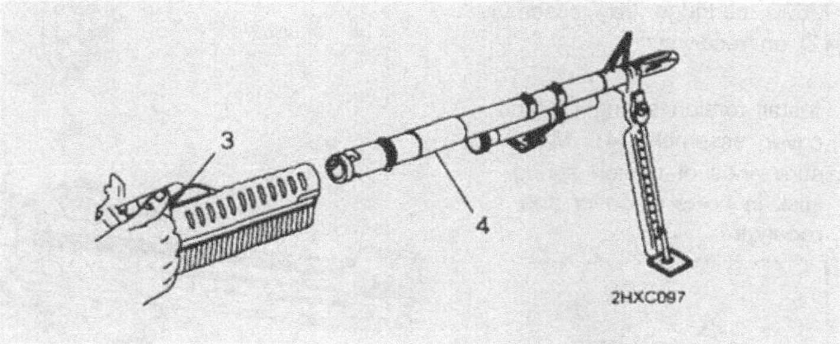

WARNING

To prevent possible bodily injury and aircraft damage, personnel should not stand the weapon on its barrel assembly when disassembling or assembling the weapon.

M60 D. Raise gun adapter (1), align holes, and insert quick release pin (2) to secure gun adapter to receiver.

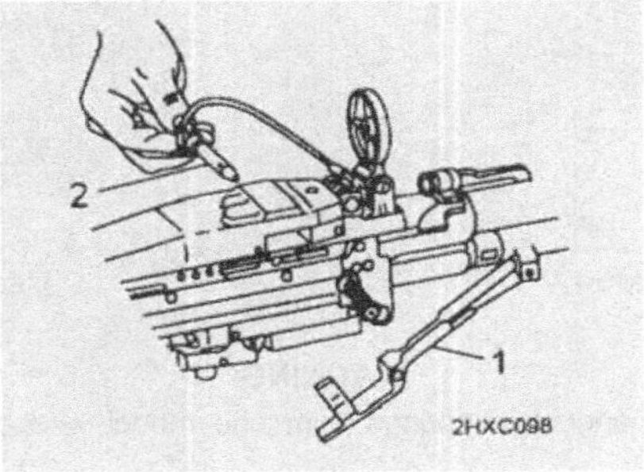

COVER ASSEMBLY AND M60 HANGER AND CARTRIDGE FEED TRAY ASSEMBLY OR M60D CARTRIDGE FEED TRAY ASSEMBLY

1 M60 or M60D. Position M60 hanger and cartridge feed tray assembly (1) or M60D cartridge tray assembly (2) on receiver.

2 Install torsion spring (3) and cover assembly (4). Make sure ends of torsion spring stick in holes of cover and receiver.

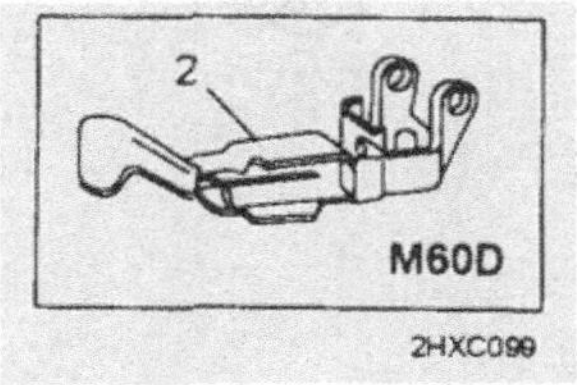

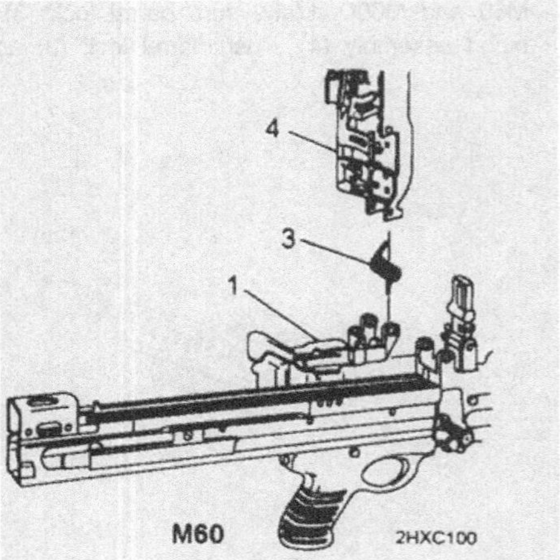

NOTE

Insert the hinge pin from one side and the hinge pin latch from the opposite side.

3 Apply slight pressure to line up torsion spring (3) and insert cover hinge pin (5). Install hinge pin latch (6) through cover hinge pin until it locks. Keep cover assembly open until machine gun is reassembled.

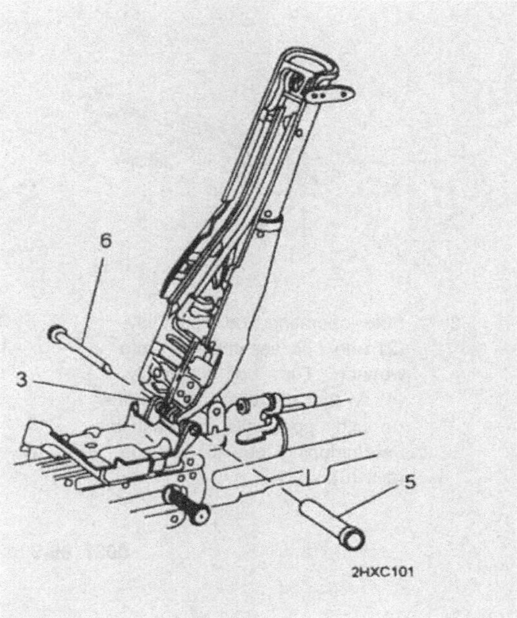

BOLT ASSEMBLY AND OPERATING ROD ASSEMBLY

1 Place yoke (1) of operating rod assembly (2) against spool of firing pin (3) in slot of bolt assembly (4) and push spool in direction of arrow.

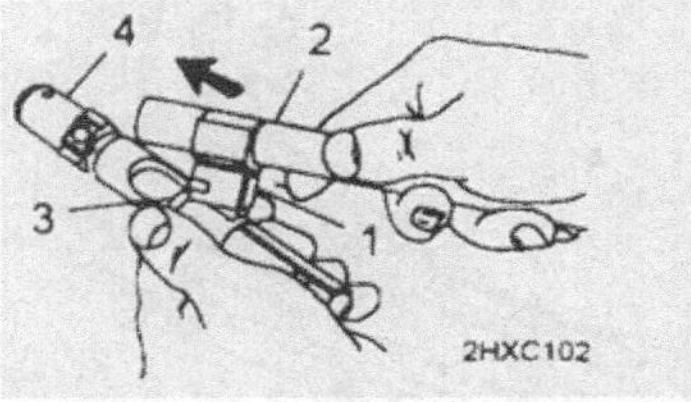

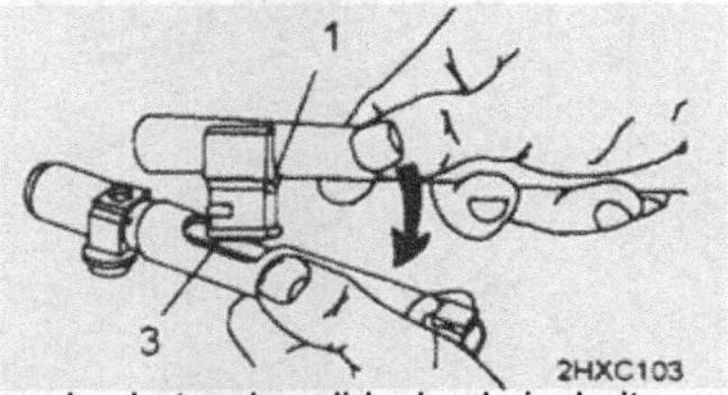

2 Seat yoke (1) between spools of firing pin. Let yoke slide back in bolt assembly (4).

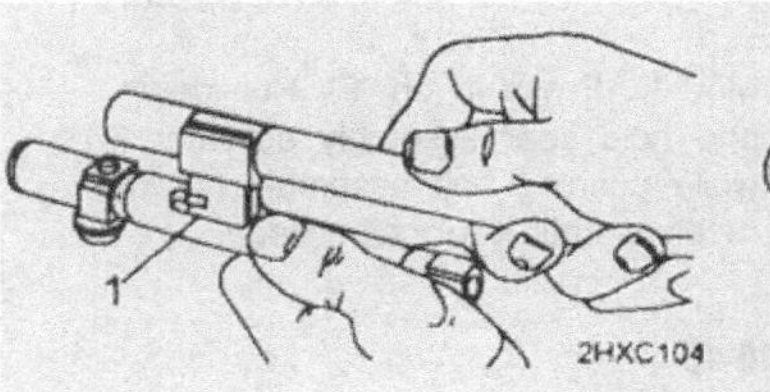

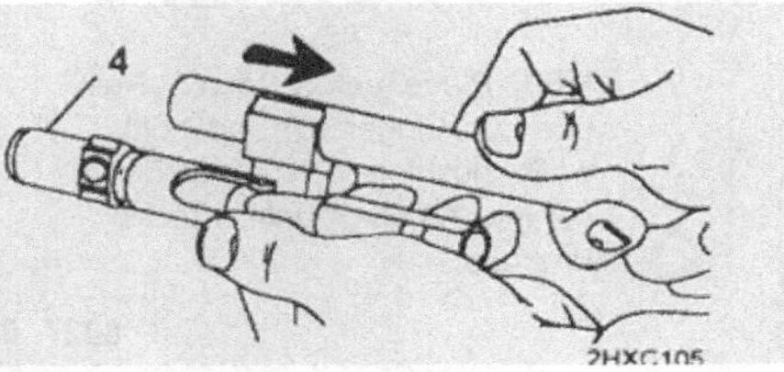

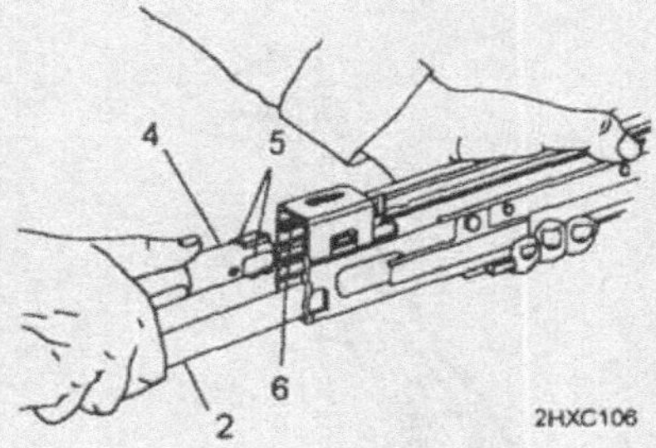

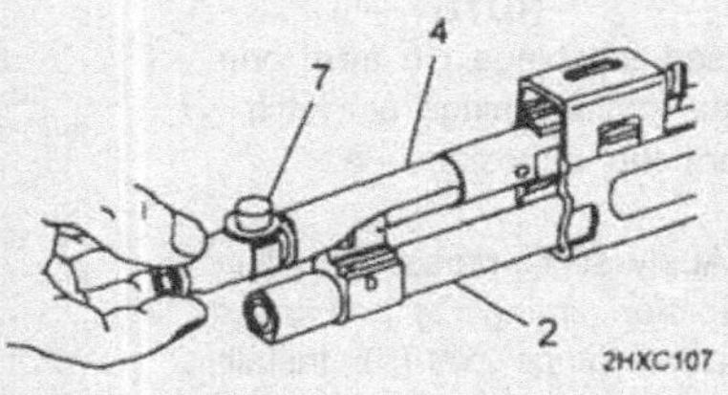

3 Start operating rod assembly (2) with bolt assembly (4) into weapon. Turn bolt assembly (4) so that stripping lugs (5) line up with upper rails (6). Push in until stripping lugs (5) engage rails (6).

4 Turn roller (7) straight up and push bolt assembly (4) and operating rod assembly (2) as a unit into weapon.

BOLT ASSEMBLY AND OPERATING ROD ASSEMBLY (cont)

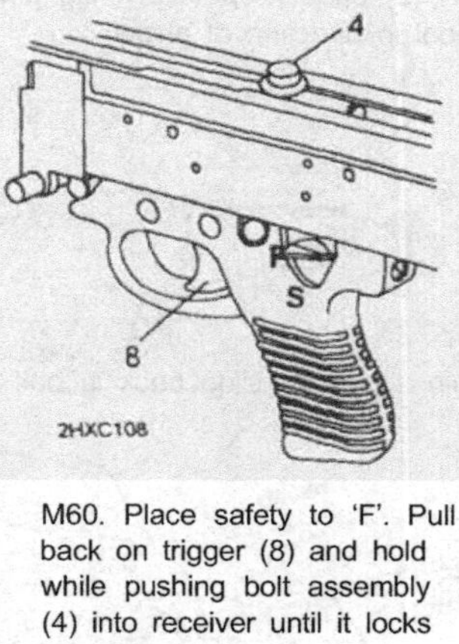

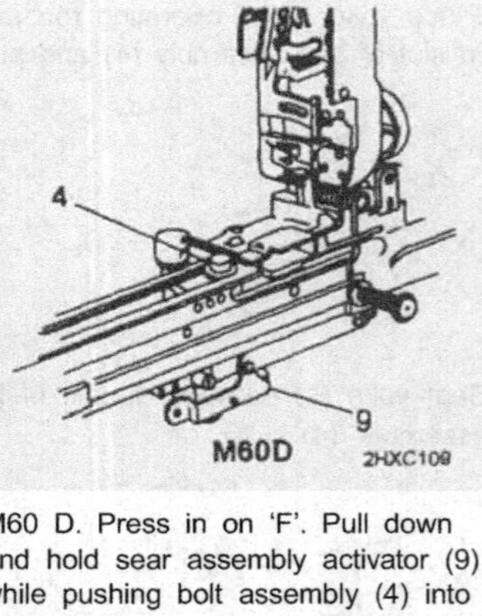

5 M60. Place safety to 'F'. Pull back on trigger (8) and hold while pushing bolt assembly (4) into receiver until it locks in place.

M60 D. Press in on 'F'. Pull down and hold sear assembly activator (9) while pushing bolt assembly (4) into receiver until it locks in place.

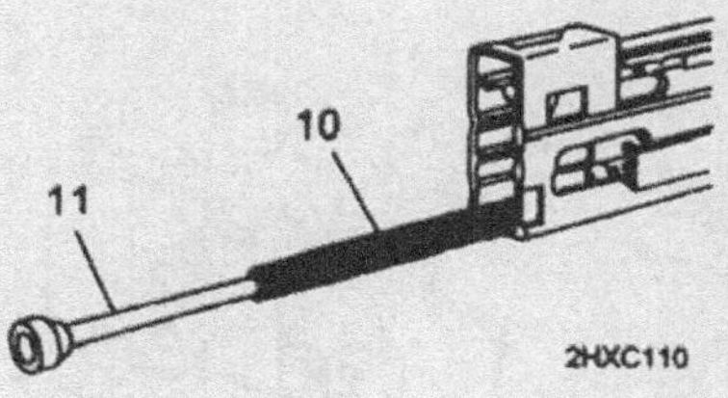

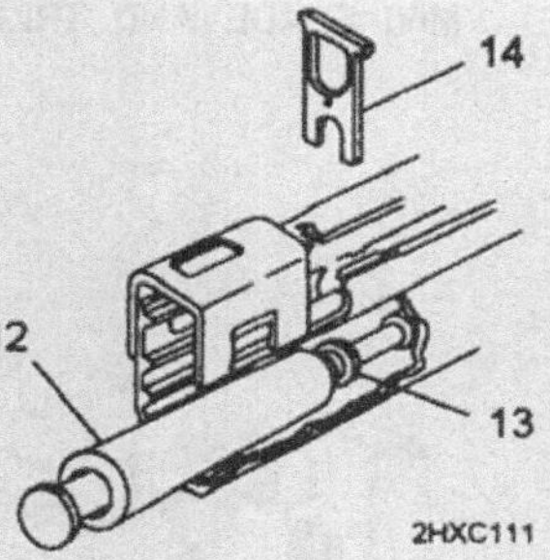

6 Install drive spring (10) and guide assembly (11) in rear of receiver.

7 Insert buffer assembly (12) against end of guide assembly and push until groove (13) on buffer assembly lines up with slot for buffer retaining yoke (14).

6 Insert buffer retaining yoke(14).

M60 SHOULDER GUN STOCK

1 Position shoulder gun stock on rear of receiver and push until it snaps in place.

2 Pull cocking handle rearward to lock bolt assembly to rear. Place safety to 'S'.

3 Close cover, move safety to 'F' position.

4 Hold charging handle, pull trigger, and ease charging handle forward.

M60 SPADE AND TRIGGER GRIP ASSEMBLY

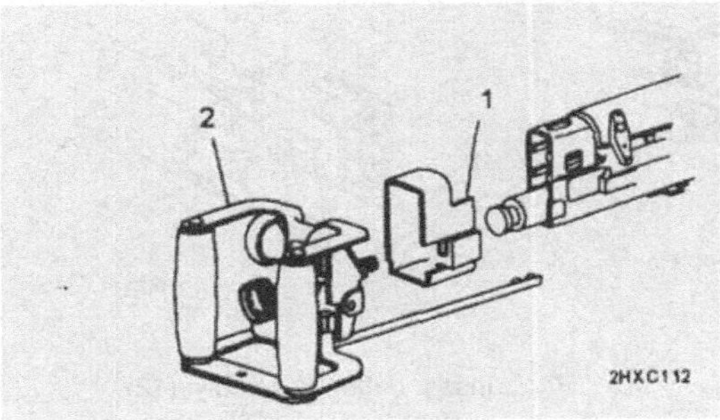

1 Install dust and moisture boot (1) and position grip and trigger assembly (2) on receiver.

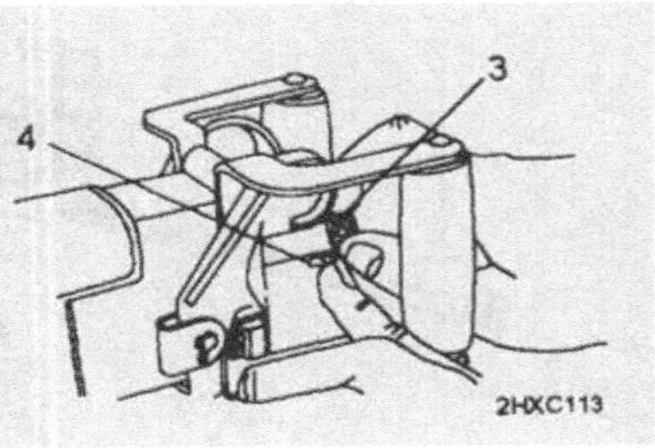

2 Turn knob and pin assembly (3) until clicking is heard. Then press and hold spring pin (4) and turn knob (3) until tight.

3 Release spring pin (4) and turn knob and pin assembly (3) until it locks.

M60 SPADE AND TRIGGER GRIP ASSEMBLY (cont)

4 Pull cocking handle rearward to lock bolt assembly to rear. Press in on ‘S’.

5 Close cover (5).

6 Snap end of sear assembly link and spring (6) on end of sear assembly activator (7).

7 Press safety in on ‘F’, hold cocking handle, pull trigger, and ease **bolt** forward.

END OF TASK

MAINTENANCE CHECKLIST 0028 00

Check the cleanliness and overall condition of the following:

- Barrel assemblies. Chamber bore and flash suppressor, gas cylinder, gas port, and gas piston.
- Receiver assembly. Operating rod guide, rails, and rear bridge rivets.
- Bolt assembly.
- Front and rear sights.
- M60 Shoulder gun stock.
- Bipod assembly.
- M60D Sear and safety housing assembly.
- M60 Trigger mechanism grip assembly.
- M60D Spade and trigger grip assembly.
- Cocking handle assembly.
- Cover assembly and M60 hanger and cartridge feed tray assembly.
- Cover assembly and M60D cartridge feed tray assembly.
- Forearm assembly.
- Blank firing attachment (BFA). Clean after firing 500 rounds or after each day's firing, whichever comes first. Remember BFA causes carbon fouling.
- Small arms sling.

0028 00-1/2 blank

FUNCTION CHECK 0029 00

Function check shall be performed after any maintenance or disassembly of the weapon.

Bolt assembly rearward? Good. Safety in 'S'? Good. Gun cleared? Good.

1 Load machine gun with dummy ammunition (WP 0005 00).

2 Pull trigger. Nothing should happen.

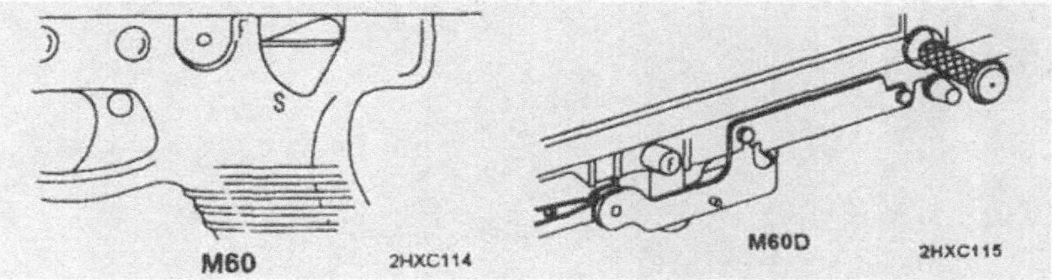

3 M60. Place safety to 'F'.
M60D. Press in on 'F'.

4 Pull trigger. Bolt assembly should move forward, strip and chamber a dummy round, and lock.

5 Pull cocking handle back. Dummy round should eject, and bolt assembly should lock to rear.

WARNING

After unloading dummy round, be sure the barrel is cleared.

6 M60. Place safety to 'S'.
M60D. Press in on 'S'.

7 Unload dummy ammunition and clear weapon (see WP 0009 00).

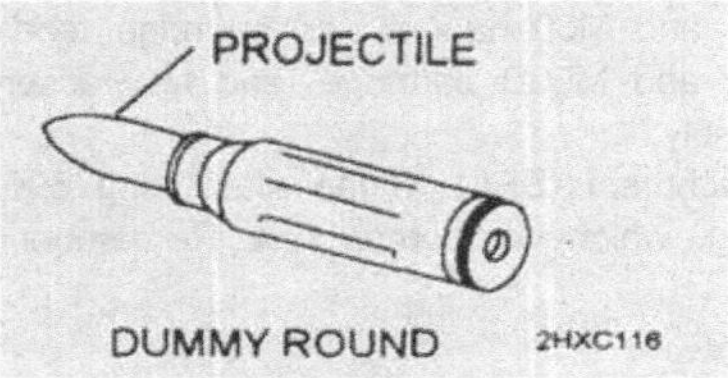

8 Be sure that projectile stays on dummy round.

9 Close **cover** and latch.

10 M60. Place safety to 'F'. Pull cocking handle, pull trigger, and ease bolt assembly forward.
M60D. Press in on 'F'. Pull cocking handle, pull trigger, and ease bolt assembly forward.

END OF TASK

CHAPTER 4

MAINTENANCE OF AUXILIARY EQUIPMENT

MAINTENANCE OF AUXILIARY EQUIPMENT 0030 00

MACHINE GUN M122 TRIPOD MOUNT. Refer to TM 9-1005-245-13&P.

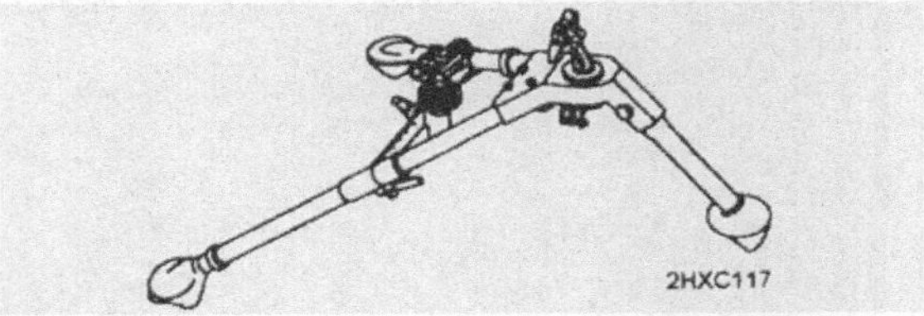

NIGHT VISION SIGHT AN/PVS-4. Refer to TM 11-5855-213-10 for maintenance.

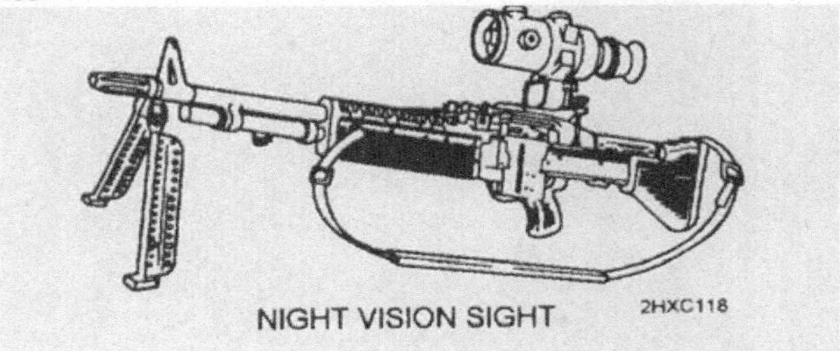

MOUNT PEDESTAL, MACHINE GUN, M6 AND MOUNT, MACHINE GUN M197. Refer to TM 9-1005-245-13&P

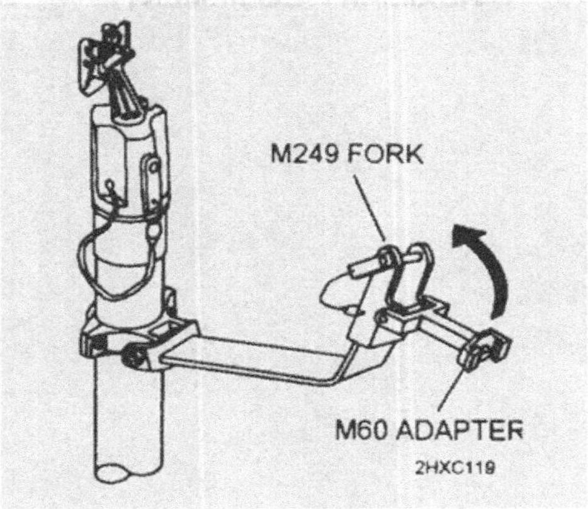

CHAPTER 5

AMMUNITION

WARNING

This is the only ammunition authorized for use in your machine gun. If it s not shown, it is not authorized.

AMMUNITION WHICH FAILS TO FIRE. Dispose of any ammunition, which fails to fire, according to authorized procedures.

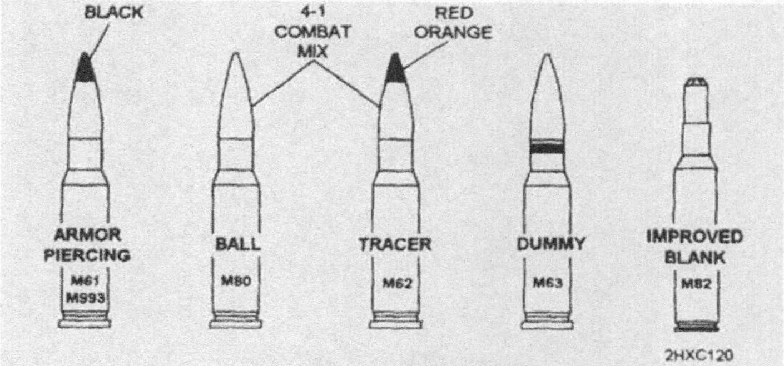

7.62-MM NATO AMMUNITION

CARE, HANDLING, AND PRESERVATION.

a. Do not open ammunition containers until the ammunition is to be used. Ammunition removed from the airtight containers, particularly in damp climates, is likely to corrode.

b. Protect ammunition from mud, dirt, and water. If the ammunition gets wet or dirty, wipe it off prior to use. Wipe off light corrosion as soon as it is discovered. Heavily corroded cartridges or cartridges which have dented cases or loose projectiles should not be fired.

c. Do not expose the ammunition to the direct rays of the sun. If the powder is hot, excessive pressure may be developed when the gun is fired.

d. Do not oil or grease ammunition. Dust and other abrasives collecting on oiled or greased ammunition will damage the operating parts of the gun, and oil on cartridges will produce excessive chamber pressure.

SCOPE

This appendix lists all forms, field manuals, and technical manuals referenced in this manual.

CONSOLIDATED TABLE OF ALLOWANCES

CTA 8-100	Army Medical Department Expendable/ Durable Items
CTA 50-970	Expendable/Durable Items (Except: Medical, Class V, Repair Parts, and Heraldic Items)

FIELD MANUALS

FM 3-87	Nuclear, Biological and Chemical (NBC) Reconnaissance and Decontamination Operations (How to Fight)
FM 21-11	First Aid for Sholdiers
FM 21-40	NBC (Nuclear, Biological and Chemical) Defense
FM 23-67	Machine-gun 7.62-MM, M60

FIRING TABLE

FT 7.62-A-2	Machine Gun, 7.62-MM: M60 on Mount, Machine Gun: 7.62-MM, M122 and Machine Gun, 7.62-MM: M73 on Tank Combat, Full-Tracked: 105-MM Gun, M60 Series and Rifle, 7.62-MM: M14: Firing Cartridge, 7.62-MM, Ball, NATO, M59; Cartridge, 7.62-MM, Ball, NATO, M80; Cartridge, 7.62-MM: AP, NATO, M61/M993 and Cartridge, 7.62-MM; Tracer, NATO, M62

FORMS

DA Form 2028	Recommended Changes to Publications and Blank Forms
SF 368	Quality Deficiency Report

TECHNICAL MANUALS

TM 3-220	Chemical, Biological, and Radiological (CBR) Decontamination
TM 11-5855-213-10	Operator's Manual for Night Vision Sight, Individual Served Weapon, AN/PVS-4
TM 9-1005-245-13&P	Machine Gun Mounts and Combinations for Tactical/Armored Vehicles and Ground Mounting
TM 9-4933-273-12&P	Boresighting Equipment, Weapon Small Arms, M30

MISCELLANEOUS PUBLICATIONS

DA PAM 738-750	The Army Maintenance Management System (TAMMS)

COMPONENTS OF END ITEM

AND

BASIC ISSUE ITEMS LISTS

INTRODUCTION 0033 00

SCOPE

These work packages lists COEI and BII for the M60/M60D machine gun to help you inventory items required for safe and efficient operation of the equipment.

GENERAL

The Components of End Item and Basic Issue Items Lists are divided into the following lists:

Components of End Item (COEI). This list is for informational purposes only and is not authority to requisition replacements. These items are part of the M60/M60D machine gun. As part of the end item, these items must be with the end item whenever it is issued or transferred between property accounts. Items of COEI are removed and separately packaged for transportation or shipment only when necessary. Illustrations are furnished to help you find and identify the items.

GENERAL (cont)

Basic Issue Items (BII). These essential items required to place the M60/M60D machine gun in operation, operate it, and to do emergency repairs. Although shipped separately packaged, BII must be with the M60/M60D machine gun during operation and whenever it is transferred between property accounts. Listing these items is your authority to request/requisition them for replacement based on authorization of the end item by the TOE/MTOE. Illustrations are furnished to help you find and identify the items.

EXPLANATION OF COLUMNS IN THE COEI LIST AND BII LIST

Column (1) - Illus Number. Gives you the number of the item illustrated.

Column (2) - National Stock Number (NSN). Identifies the stock number of the item to be used for requisitioning purposes.

Column (3) - Description, CAGEC, and Part Number. Identifies the Federal name (in all capitol letters) followed by a minimum description when needed. The stowage location of COEI and BII is also included in the column. The last line below the description is the CAGEC (Commercial and Government Entity Code) (in parentheses) and the part number.

Column (4) - Usable On Code. When applicable, gives you a code if the item you need is not the same for different models of equipment.

Code	Used On
023	Model M60
G79	Model M60D

Column (5) - Unit of Measure (U/M). Indicates the physical measurement or count of the item as issued per the National Stock Number shown in column (2).

Column (6) - Qty Rqr. Indicates the quantity required.

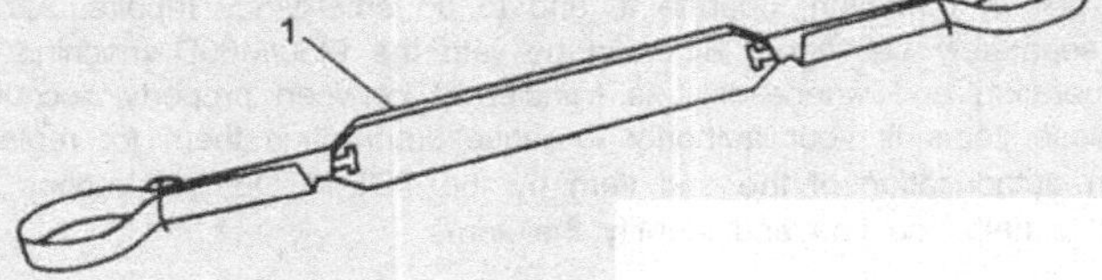

(1) ILLUS NO.	(2) NATIONAL STOCK NUMBER	(3) DESCRIPTION, CAGEC. AND PART NUMBER	(4) USABLE ON CODE	(5) U/M	(6) QTY RQR
1	1002-00-312-7177	SLING, SMALL ARMS (19204) 12003483	023	EA	1

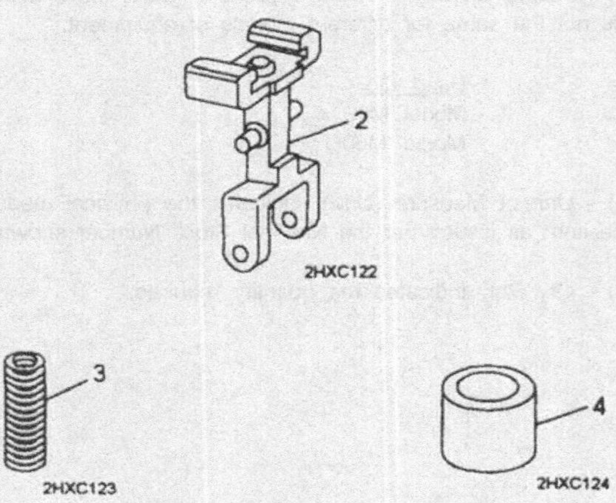

(1) ILLUS NO.	(2) NATIONAL STOCK NUMBER	(3) DESCRIPTION, CAGEC, AND PART NUMBER	(4) USABLE ON CODE	(5) U/M	(6) QTY RQR
2	1005-00-772-0194	ADAPTER ASSEMBLY (19204) 7792991	023	EA	1
3	5360-00-772-0193	HELICAL SPRING (19204) 7792987	023	EA	1
4	1005-00-772-0192	STUD, PLUNGER (19204) 7792986	023	EA	1

BASIC ISSUE ITEMS **0035 00**

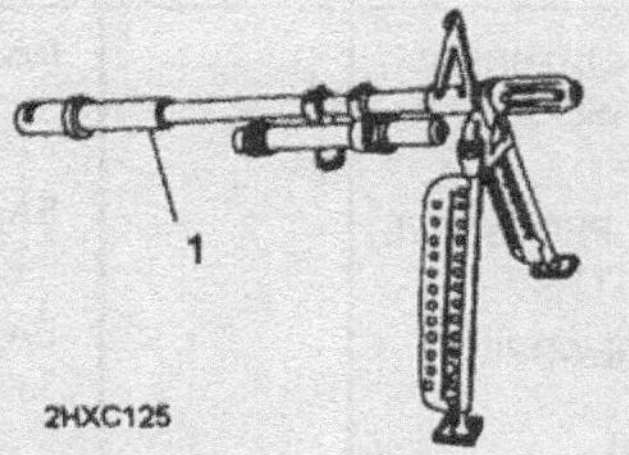

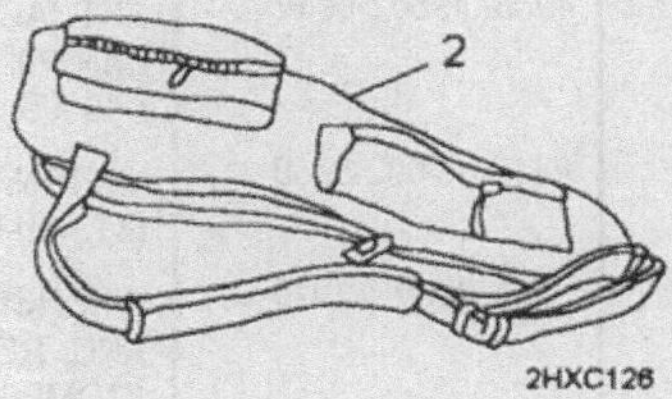

(1) ILLUS NO.	(2) NATIONAL STOCK NUMBER	(3) DESCRIPTION, CAGEC, AND PART NUMBER	(4) USABLE ON CODE	(5) U/M	(6) QTY RQR
1	1005-00-608-5001	BARREL AND BIPOD ASSEMBLY (19204) 7269027		EA	1
2	1005-00-791-5420	CASE, CARRYING, BARREL ASSEMBLY AND EQUIPMENT (19205) 7791009		EA	1

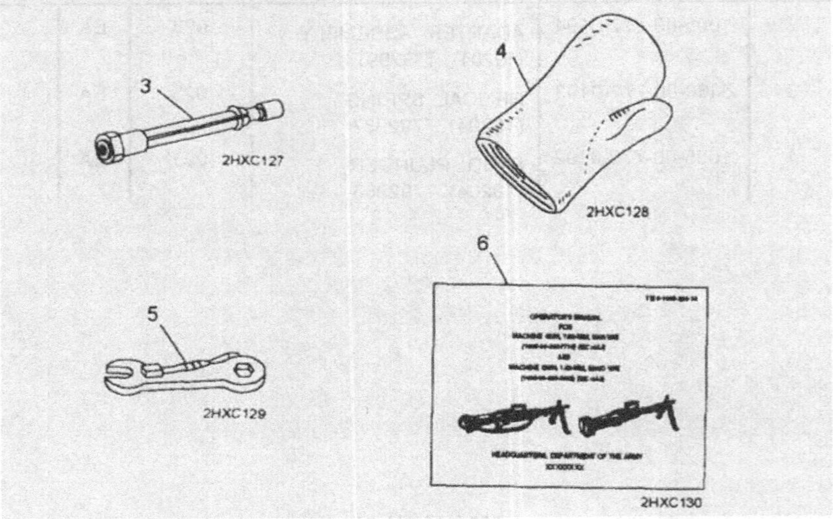

(1) ILLUS NO.	(2) NATIONAL STOCK NUMBER	(3) DESCRIPTION, CAGEC. AND PART NUMBER	(4) USABLE ON CODE	(5) U/M	(6) QTY RQR
3	4933-00-652-9950	EXTRACTOR, RUPTURED CARTRIDGE CASE (19205) 7790352		EA	1
4	8415-01-092-0039	MITTEN, HEAT PROTECTIVE (81349) MIL-M-11199		EA	1
5	5120-00-461-1075	WRENCH, SCREWDRIVER AND REAMER (COMBINATION) (19204) 8448458		EA	1
6		TM 9-1005-224-10		EA	1

ADDITIONAL AUTHORIZED LISTS

INTRODUCTION 0036 00

SCOPE

This work package lists additional items you are authorized for the support of the M60/M60D machine Gun.

GENERAL

This list identifies items that do not have to accompany the M60/M60D machine gun and that do not have to be turned in with it. These items are all authorized to you by CTA, MTOE, TDA, or JTA.

EXPLANATION OF COLUMNS IN AAL

Column (1) - National Stock Number (NSN). Identifies the stock number of the item to be used for requisitioning purposes.

Column (2) - Description, Commercial and Government Entity Code (CAGE), and Part Number (P/N). identifies the Federal item name (in all capital letters) followed by a minimum description when needed. The last line below the description is the CAGEC (in parentheses) and the part number.

EXPLANATION OF COLUMNS IN AAL (cont)

Column (3) - Usable On Code. When applicable, gives you a code if the item you need is not the same for different models of equipment.

Code	Used On
023	Model M60
G79	Model M60D

Column (4) - Unit of Measure (U/M). Indicates the physical measurement or count of the item as issued per the National Stock Number shown in column (1).

Column (5) - Qty Recm. Indicates the quantity recommended.

(1) NATIONAL STOCK NUMBER	(2) DESCRIPTION, CAGEC. AND PART NUMBER	(3) USABLE ON CODE	(4) U/M	(5) QTY RECM
	MTOE AUTHORIZED ITEMS			
5855-00-629-5334	SIGHT, NIGHT VISION AN/PVS-4 (60063) SMD850300-1	023	EA	1
1005-00-710-5599	M122 MACHINE GUN TRIPOD (19205) 7790723		EA	1
	CTA AUTHORIZED ITEMS			
1005-00-556-4174	BRUSH, CLEANING, SMALL ARMS: (BORE) (19204) 5564174		EA	2
1005-00-690-3115	BRUSH, CLEANING, SMALL ARMS: (CHAMBER) (19204) 7790452		EA	1
1005-00-350-4100	BRUSH, CLEANING, SMALL ARMS: (RECEIVER) (19204) 8848466		EA	1

(1) NATIONAL STOCK NUMBER	(2) DESCRIPTION, CAQEC, AND PART NUMBER	(3) USABLE ON CODE	(4) U/M	(5) QTY RECM
	CTA AUTHORIZED ITEMS (cont)			
1005-00-140-3515	FIRING ATTACHMENT, BLANK (19204) 12002981		EA	1
1005-00-793-6761	HANDLE ASSEMBLY: CLEANING ROD (19204) 7266115		EA	1
1005-00-726-6109	ROD, SECTIONS, CLEANING, SMALL ARMS (19204) 7266109		EA	5
	SWAB HOLDER SECTION, SMALL ARMS CLEANING ROD (19204) 7266110		EA	1

EXPENDABLE/DURABLE SUPPLIES AND MATERIALS LIST

SCOPE

This work package lists expendable and durable items that you will need to operate and maintain the M60/M60D machine gun. This list is for information only and is not authority to requisition the listed items. These items are authorized to you by CTA 50-970, Expendable/Durable Items (Except Medical, Class V Repair Parts, and Heraldic Items), or CTA 8-100, Army Medical Department Expendable/Durable Items.

EXPLANATION OF COLUMNS IN THE EXPENDABLE/DURABLE ITEMS LIST

Column (1) - Item Number. This number is assigned to the entry in the list and is referenced in the narrative instructions to identify the item (e.g., Use wiping rag, (item 5, WP 0034 00).)

Column (2) - Level, This column identifies the lowest level ofmaintenance that requires the listed item.

C - Operator/Crew

EXPLANATION OF COLUMNS IN THE EXPENDABLE/DURABLE ITEMS LIST (cont)

Column (3) - National Stock Number (NSN). This is the NSN assigned to the item which you can use to requisition it.

Column (4) - Item Name, Descprition, Commercial and Government Entity Code (CAGEC), and Part Number (P/N). This column provides the other information you need to identify the item.

Column (5) - Unit of Measure (U/M). This code shows the physical measurement or count of an item, such as gallon, dozen, gross, etc.

EXPENDABLE/DURABLE SUPPLIES AND MATERIALS LIST

(1) ITEM NUMBER	(2) LEVEL	(3) NATIONAL STOCK NUMBER	(4) DESCRIPTION, CAGEC, AND PART NUMBER	(5) U/M
1	C		Cleaner, Lubricant and Preservative: grade 2 (CLP)	
		9150-01-102-1473	1/2 oz bottle	EA
		9150-01-079-6124	4 oz bottle	EA
			(8 1349) MIL-L-63460	
2	C		Cleaning Compound, Solvent: Rifle Bore Cleaner (RBC) (81349) MIL-C- 372	
		6850-00-224-6656	2 oz can	EA
		6850-00-224-6657	8 oz can	EA
3	C		Lubricating Oil, Weapons (LAW) (81349) MIL-L-14107	
		9150-00-292-9689	1 qt can	EA

EXPENDABLE/DURABLE SUPPLIES AND MATERIALS LIST (cont)

(1) ITEM NUMBER	(2) LEVEL	(3) NATIONAL STOCK NUMBER	(4) DESCRIPTION, CAGEC, AND PART NUMBER	(5) U/M
4	C		Lubricating Oil, Weapons: semi-fluid (LSA) (19204) 8436793	
		9150-00-889-3522	4 oz bottle	EA
5	C		Rag, Wiping: (cotton) (58536) A-A-531	
		7920-00-205-1711	50 lb bdl	LB
6	C		Swab, Small Arms Cleaning: (cotton) (19204) 5019316	
		1005-00-288-3565	(1000 per pkg)	PG

ALPHABETICAL INDEX 0040 00

Subject **WP Sequence No.**

Subject **WP Sequence No.**

Subject	WP Sequence No.

O (cont)

P

Subject	WP Sequence No.

R

S

T

W

TECHNICAL MANUAL
ARMY NO. 9-1005-319-23&P
AIR FORCE TO 11W3-5-5-42

***ARMY TM 9-1005-319-23&P**
AIR FORCE TO 11W3-5-5-42

DEPARTMENTS OF THE ARMY
AND AIR FORCE
Washington, DC, 1 May 1991

Technical Manual for 5.56MM,

M16A2, M4 and M4A1 Carbine

DISTRIBUTION STATEMENT. APPROVED FOR PUBLIC RELEASE; DISTRIBUTION IS UNLIMITED.

REPORTING ERRORS AND RECOMMENDING IMPROVEMENTS

You can help improve this manual. If you find any mistakes or if you know of a way to improve the procedures please let us know.

Army users mail your letter, DA Form 2028 (recommended Changes to Publications and Blank Forms), or DA Form 2028-2 located in the back of this manual direct to: Director, Armament and Chemical Acquisition and Logistics Activity, ATTN: AMSTA-AC-NML, Rock Island, IL 61299-7630.

Air Force users submit AFTO Form 22, Technical Order System Publication Improvement Report and Reply, to: WR-ALC/MMDET, Robins AFB, GA 31098-5609.

A reply will be furnished to you.

*This manual supersedes ARMY TM 9-1005-319-23&P dated 28 August 1987, including all changes.

WARNING

ALL WARNINGS in this technical manual pertain to both the rifle and the carbines unless otherwise specified.

Before starting an inspection, be sure to clear the rifle. Do not pull the trigger until the rifle has been cleared. Inspect the chamber to ensure that it is empty and no ammunition is in position to be chambered.

Do not keep live ammunition near work area.

To avoid injury to your eyes, use care when removing and installing spring-loaded parts.

All M1 6A2 rifles and M4 carbines must be inspected and gaged at least once annually for safety and serviceability. Initial gaging is required 1 year from receipt of the weapons. Air Force users refer to inspection requirements in Air Force Manual (AFM) 36-2227, Volume 1.

All Army Reserve and Army National Guard M16A2 rifles and M4 carbines must be inspected and gaged at least once every 2 years, after the initial inspection/gaging procedures have been accomplished. This initial gaging procedure is required 1 year from receipt of the weapons. This 2 year interval may be maintained unless preventive maintenance checks and services (PMCS) or other physical evidence indicates that an individual unit's M16A2 rifles and M4 carbines require inspection/ gaging at a more frequent interval. If it is determined that a yearly inspection is necessary for an individual unit, only that unit will be affected. This will not affect other units in regard to the interval of inspection.

It is recommended that training units inspect/gage all rifles and carbines at the end of each training cycle. Training units will inspect/gage all rifles and carbines at least once annually.

Below direct support maintenance, **DO NOT** interchange bolt assemblies from one rifle/carbine to another. Doing so may result in injury to, or death of, personnel.

Bolt cam pin must be installed or rifle/carbine will blow up while firing the first round. If the bolt cam pin is not installed, injury to or death of, personnel may result.

Dry cleaning solvent is flammable and toxic and should be used in a well ventilated area. The use of rubber gloves is necessary to protect the skin when washing rifle parts.

When using solid film lubricant or dichloromethane, be sure the area is well ventilated.

When using carbon removing compound (item 8, app D), avoid skin contact. If carbon removing compound comes In contact with the skin, wash thoroughly with running water. using a good lanolin base cream after exposure to the compound is helpful. Using gloves and protective equipment is required.

The lock plate prevents the selector lever from being placed in BURST and will be installed at the discretion of the unit commander. It is mandatory for use in civil disturbance (riot control).

Only blank cartridge M200 is to be used when the blank firing attachment is attached to the rifle/carbine.

WARNING (CONT)

Do not fire blank ammunition at a representative enemy at distances of less than 20 feet (6.10 Om). The unburned propellant grains can cause injury within this distance.

For further information on safety, care, and handling of ammunition: Army and Air Force users refer to M16A2 Rifle Operator's Manual.

For additional first aid data, see Field Manual (FM) 21-11.

HOW TO USE THIS MANUAL

Read this manual carefully before performing required maintenance. This manual will be referred to for Inspection/Maintenance and Repair procedures.

GENERAL

There are several things you need to know to use this manual efficiently.

1. All references in the manual are to pages only. Reference to maintenance procedures is to the page where the respective initial setup appears.

2. Illustrations for the maintenance procedures show only those parts affected by the operation being performed.

3. Whenever the male gender is mentioned in the manual (i.e., crewman, repairman), it also pertains to females.

4. When the term "evacuate to support maintenance" is used, the entire rifle must be evacuated.

5. When a procedure is common to M16A2 rifle, and M4/M4A1 carbine, **ONLY** the M16A2 configuration will be depicted. If a procedure is not common to both weapons, the procedure will be incorporated.

6. When the word rifle is referenced in text, it will reference the rifle and the carbines.

INDEXES

This manual is organized to help you find the information you need quickly. There are several useful indexes.

1. Table of Contents. Lists in order all chapters, sections, and appendixes. Gives page references.

2. Nomenclature Cross-References List.

3. Chapter Overviews. Summarize material covered in the chapter. Are located at the beginning of each chapter.

4. Symptom Index. Located just before the troubleshooting table in each maintenance chapter. Lists, in alphabetical order, parts of the rifle with possible malfunctions. References pages of the troubleshooting table.

5. Alphabetical Index. Located at the end of the manual. An extensive subject index for everything in the manual. Gives page references.

MAINTENANCE PROCEDURES

There are two maintenance chapters:

Army personnel use chapter two for unit maintenance procedures and chapter three for direct support maintenance procedures.

Air Force personnel: Only Air Force Specialty Code 753XX Combat Arms Training and Maintenance (CATM) specialists, technicians, and gunsmiths are authorized to perform maintenance procedures contained in this manual.

Each maintenance task has an initial setup containing a list of the following things you will need in order to do your maintenance task.

1. **Tools and Special Tools**. For standard and special tools, see appendixes B and C. Army users are to use the Tool Set, Gage Set, and/or Shop Set listed in the initial setup.

2. **Materials/Parts**. Lists expendable materials and 100 percent replaceable parts. Each material or part is followed by a part number or appendix reference.

3. **References**. Lists other publications containing necessary information.

4. **Equipment Condition**. Lists conditions to be met before starting the procedure. The reference on the left of the condition is a page reference to instructions for setting up the condition.

5. **General Safety Instructions**. Lists safety instructions to follow before performing maintenance procedures.

EXTERNAL VIEW OF 5.56 MM RIFLE M16A2

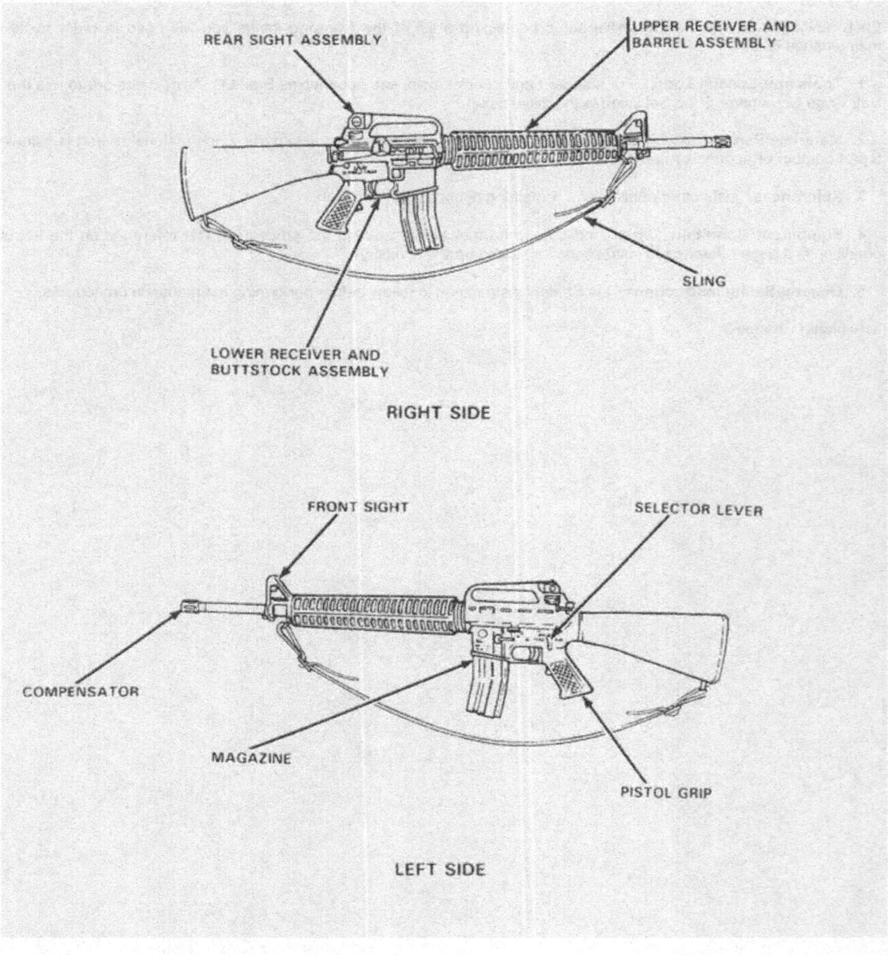

EXTERNAL VIEW OF 5.56MM CARBINE, M4/M4A1

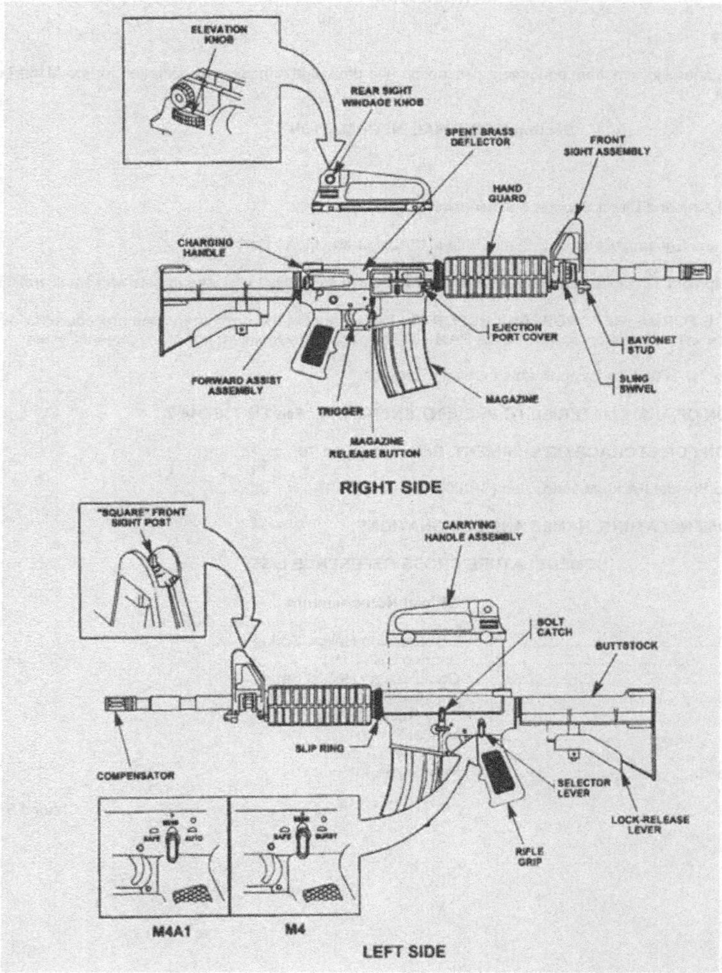

CHAPTER 1
INTRODUCTION

CHAPTER OVERVIEW

This chapter contains general information, equipment description and data, and principles of operation for the M16A2 rifle and M4/M4A1 carbines.

Section I. GENERAL INFORMATION

1-1. SCOPE.

a. Type of Manual: Unit and Direct Support Maintenance.

b. Model Number and Equipment Name: 5.56mm Rifle M16A2, M4 and M4A1 Carbines.

c. Purpose of Equipment. Provides personnel an offensive/defensive capability to engage targets with small arms fire.

1-2. MAINTENANCE FORMS, RECORDS, AND REPORTS. Department of the Army forms and procedures used for equipment maintenance will be those prescribed by DA PAM 738-750, The Army Maintenance Management System.

Air Force users refer to TO 11W-1-10 for applicable forms and records.

1-3. DESTRUCTION OF ARMY MATERIEL TO PREVENT ENEMY USE. See TM 750-244-7.

1-4. PREPARATION FOR STORAGE OR SHIPMENT. Refer to page 2-70.

Air Force users refer to Special Package Instruction (SPI) 00-856-6885.

1-5. OFFICIAL NOMENCLATURE, NAMES AND DESIGNATIONS.

NOMENCLATURE CROSS-REFERENCE LIST

Common Name	Official Nomenclature
Action Spring	Compression Helical Spring
Ball Bearing	Bearing Ball
Bolt Catch Spring	Compression Helical Spring
Bolt Carrier Key Tool	Machine Key
Burst Disconnector	Lock-Release Lever
Cam Clutch Spring	Helical Spring
Carbine	M4/M4A1 Carbine
Charging Handle Assembly	Handle Assembly

1-5. OFFICIAL NOMENCLATURE, NAMES AND DESIGNATIONS (CONT).

NOMENCLATURE CROSS-REFERENCE LIST

Common Name	Official Nomenclature
Disconnector Springs	Compression Helical Spring
Ejector Spring	Helical Spring
Extractor Spring Assembly	Spring Assembly
Hammer Spring	Torsion Helical Spring
Lower Receiver Extension	Spring Receiver Holder
Magazine	Cartridge Magazine
Magazine Catch Spring	Compression Helical Spring
Peel Washer	Shim
Pistol Grip	Rifle Grip
Pivot Pin Detent	Takedown Pin Detent
Rifle	Rifle, 5.56mm, M16A2
Rifle Barrel Assembly	Barrel Assembly
Selector Lever	Fire Control Selector
Semiautomatic Disconnector	Lock-Release Lever
Sling	Small Arms Sling
Trigger Spring	Torsion Helical Spring
Upper Receiver	Upper Cartridge Receiver

1-6. REPORTING EQUIPMENT IMPROVEMENT RECOMMENDATIONS (EIR). If your M16A2 rifle needs improvement, let us know. Send us an EIR. You, the user, are the only one who can tell us what you don't like about your equipment. Let us know why you don't like the design.

Army users submit SF 368 (Product Quality Deficiency Report) to: Commander, U.S. Army Armament Research, Development and Engineering Center, ATTN: AMSTA-AR-QAW (R)/Customer Feedback Center, Rock Island, IL 61299-7300.

Air Force users submit Materiel Deficiency Report (MDR) to: DIR MAT MGT ROBINS AFB GA//MMIBTC// and Product Quality Deficiency Report to: DIR MAT MGT ROBINS AFB GA//MMQA// IAW Technical Order 00-35D-54.

A reply will be sent to you.

1-7. CORROSION PREVENTION AND CONTROL (CPC). CPC of Army materiel is a continuing concern. It is important that any corrosion problems with this item be reported so that the problem can be corrected and improvements can be made to prevent the problem in future items,

While corrosion is typically associated with rusting of metals, it can also include deterioration of other materials such as rubber and plastic. Unusual cracking, softening, swelling, or breaking of these materials may be a corrosion problem.

If a corrosion problem is identified, it can be reported using Standard Form 368, Product Quality Deficiency Report. Use of key words such as "corrosion", "rust", " deterioration", or "cracking" will assure that the information is identified as a CPC problem.

Army users submit Product Quality Deficiency Report (SF 368) to:

Commander
U.S. Army Armament Research, Development and Engineering Center
ATTN: AMSTA-AR-QAW (R)
Rock Island, IL 61299-7300

Air Force users submit Materiel Deficiency Report (MDR) to:

DIR MAT MGT
ATTN: MMIBTC
Robins AFB, GA

and Product Quality Deficiency Report to:

DIR MAT MGT
ATTN: MMQA
Robins AFB, GA

Section II. EQUIPMENT DESCRIPTION AND DATA

1-8. EQUIPMENT CHARACTERISTICS, CAPABILITIES, AND FEATURES.

a. Characteristics.

(1) Light weight
(2) Air-cooled
(3) Gas-operated
(4) Magazine-fed
(5) Semiautomatic or burst fire

b. Capabilities. Provides personnel an offensive/defensive capability to engage targets with direct small arms fire.

c. Features.

(1) Receivers are made of light-weight aluminum alloys; however, the safety , durability, and function of the rifles are in no way reduced. The portability and logistical values are greatly increased, particularly when air transport is used.

(2) The bolt locking action is one of the mechanical features of the rifle. The bolt assembly and barrel extension contain locking lugs which engage and lock the bolt assembly firmly in the barrel extension. The initial force of the explosion of the cartridge is absorbed by the barrel, barrel extension, and bolt assembly.

1-8. EQUIPMENT CHARACTERISTICS, CAPABILITIES, AND FEATURES (CONT).

(3) The trigger guard is easily adaptable to winter operations. A spring-loaded retaining pin is depressed to allow ready access to the trigger when wearing arctic mittens.

(4) The ejection port cover prevents dirt or sand from getting into the ejection port. The ejection port cover must be closed during periods when firing is not anticipated. It opens automatically by the forward or rearward movement of the bolt carrier assembly.

1-9. LOCATION AND DESCRIPTION OF MAJOR COMPONENTS.

(A) MAGAZINE. 30 cartridge capacity.

(B) SLING. The sling is adjustable and provides a means to carry the weapon.

(C) BOLT CARRIER ASSEMBLY. Carries bolt assembly to chamber and fires the weapon. Contains the firing pin, cartridge extractor, bolt assembly, cartridge ejector, and bolt cam pin.

(D) CHARGING HANDLE ASSEMBLY. Provides a means of charging the weapon.

(E) M16A2 UPPER RECEIVER AND BARREL ASSEMBLY. Upper receiver contains rear sight assembly, ejection port, ejection port cover, and a housing for the key and bolt carrier assembly and bolt assembly. Rifle barrel assembly is air-cooled, contains compensator and front sight assembly, and holds the two handguard assemblies and the sling swivel.

(F) LOWER RECEIVER AND BUTTSTOCK ASSEMBLY. Lower receiver contains the trigger assembly, sear, hammer assembly, selector lever, rifle grip, bolt catch, and buttstock assembly. The buttstock assembly houses the action spring, buffer assembly, and extension assembly.

(G) M4/M4A1 CARRYING HANDLE. Provides a means of carrying carbine.

(H) **M4/M4A1 UPPER RECEIVER AND BARREL ASSEMBLY**. Upper receiver contains, ejection port, ejection port cover, a housing for key and bolt carrier assembly and bolt assembly, and mounting surface for the carrying handle assembly. Carbine barrel assembly is air-cooled, contains compensator and front sight assembly, and holds the two handguard assemblies and the sling swivel.

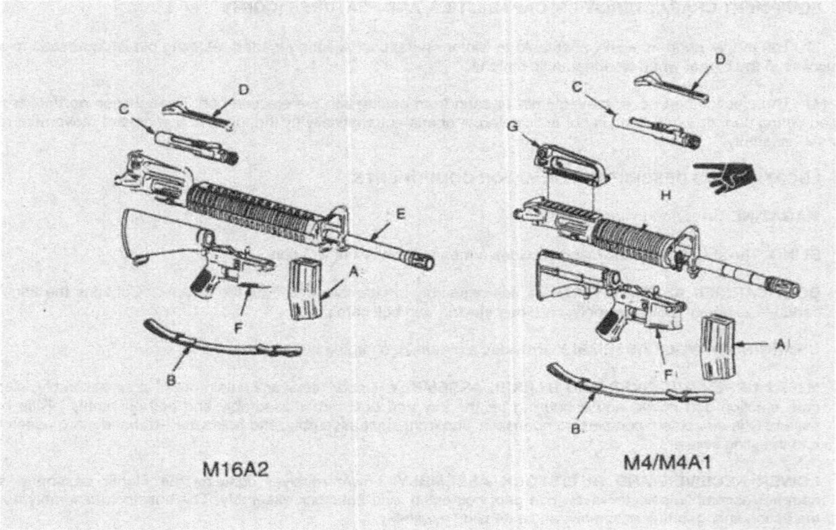
D
C
E
A
F
B
M16A2
D
C
G
H
A
F
B
M4/M4A1

1-10. EQUIPMENT DATA.

	US CUSTOMARY	METRIC
Weight:		
Carbine, M4/M4A1 without magazine and sling	6 lb 7 oz	2.91 kg
Rifle, MI 6A2 without magazine and sling	7 lb 8 oz	3.40 kg
Sling, adjustable	4 oz	0.11 kg
Empty magazine	4 oz	0.11 kg
Loaded magazine	1 lb 1 oz	0.48 kg
Carbine, M4/M4A1 w/sling and loaded magazine	7 lb 12 oz	3.51 kg
Rifle M16A2 w/sling and loaded magazine	8 lb 13 oz	4.00 kg
Bayonet-Knife M7	10.5 oz	0.30 kg
Scabbard M10 o	5 oz	0.14 kg
Length:		
Carbine with compensator, buttstock extended	33.0 in	83.82 cm
Carbine with compensator, buttstock closed	29.75 in	75.57 cm
Rifle with compensator	39.63 in	100.66 cm
Barrel (Carbine)	14.5 in	36.83 cm
Barrel (Rifle)	20 in	50.8 cm
Barrel with compensator (Carbine)	15.5 in	39.37 cm
Barrel with compensator (Rifle)	21 in	53.34 cm
Mechanical features:		
Rifling	right-hand twist 6 grooves, 1 turn in 7 inches (17.78 cm)	
Method of operation	direct gas	
Type of breech mechanism	rotating bolt	
Method of feeding	magazine	
Cooling	air	
Trigger pull (M16A2 & M4)	5.5 to 9.5 lb	2.49 to 4.31kg
Trigger pull (M4A1)	5.5 to 8.5 lb	2.49 to 3.86 kg
Ammunition:		
Caliber	223	5.56mm
Type	ball, blank, dummy, and tracer	
Firing characteristics:		
Muzzle velocity (Carbine) (approximate)	2,970 fps	905.85 mps
Muzzle velocity (Rifle) (approximate)	3,100 fps	945.5 mps
Chamber pressure	52,000 psi	358,540 kPa
Cyclic rate of fire (Carbine) (approximate)	700-970 rds/m	
Cyclic rate of fire (Rifle) (approximate)	700-900 rds/m	

1-10. EQUIPMENT DATA (CONT).

	US CUSTOMARY	METRIC
Maximum rate of fire:		
Semiautomatic	45 rds/m	
Burst	90 rds/m	
Sustained rate of fire	12/15 rds/m	
Maximum range	3,938 yards	Approximately 3,600 meters
Maximum effective range:		
Individual/point targets (Carbine)	547 yards	500 meters
Individual/point targets (Rifle)	602 yards	550 meters
Area targets (Carbine)	650 yards	600 meters
Area targets (Rifle)	875 yards	800 meters

Section III. PRINCIPLES OF OPERATION

1-11. GENERAL. The 5.56mm M16A2 and M4/M4A1 carbine:

a. Is gas-operated. It fires in either the semiautomatic or burst mode.

b. Has positive locking of the bolt. Firing pin is part of the bolt carrier assembly and cannot strike the primer until the bolt assembly is fully locked.

Section III. PRINCIPLES OF OPERATION (CONT).

1-12. PRINCIPLES OF OPERATION.

(A) **MAGAZINE**. Holds cartridges ready for feeding and provides a guide for positioning cartridges for stripping. Provides quick reload capabilities for sustained firing.

(B) **SLING**. Provides the means for carrying the weapon.

(C) **BOLT CARRIER ASSEMBLY**. Provides stripping, chambering, locking, firing, extraction, and ejection of cartridges using the drive springs and projectile propelling gases for power.

(D) **CHARGING HANDLE ASSEMBLY**. Provides initial charging of the weapon. The handle latch locks the charging handle assembly in the forward position during sustained fire to prevent injury to the operator.

(E) **M16A2 UPPER RECEIVER AND BARREL ASSEMBLY**. Provides support for the bolt carrier assembly. The barrel chambers the cartridge for firing and directs the projectile.

(F) **LOWER RECEIVER AND BUTTSTOCK ASSEMBLY**. Provides firing control for the rifle and carbine. M16A2 ONLY provides storage for basic cleaning materials.

(G) **M4/M4A1 CARRYING HANDLE ASSEMBLY**. Provides a means of carrying the carbine, contains rear sight assembly, and can be removed for mounting various optics.

(H) **M4/M4A1 UPPER RECEIVER AND BARREL ASSEMBLY**. Provides support for the bolt carrier assembly. The barrel chambers the cartridge for firing and directs the projectile.

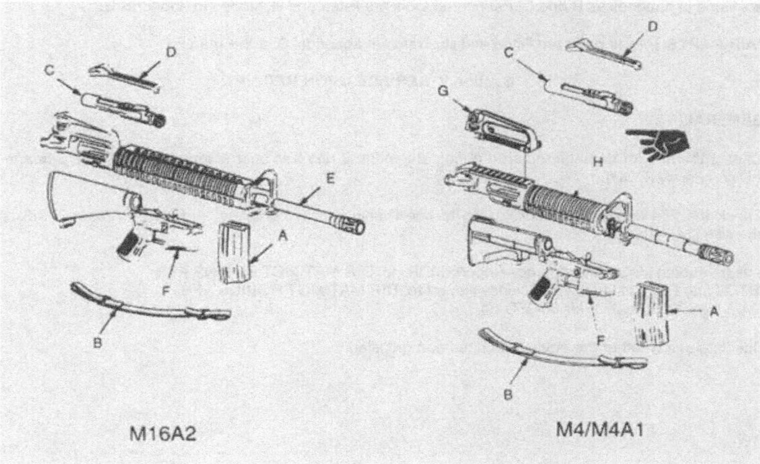

CHAPTER 2
UNIT MAINTENANCE INSTRUCTIONS

CHAPTER OVERVIEW

This chapter provides information and Instructions to help keep the rifle in good repair and contains the following sections:

a. Repair Parts, Special Tools, TMDE, and Support Equipment
b. Service Upon Receipt
c. Preventive Maintenance Checks and Services (PMCS)
d. Troubleshooting
e. Maintenance Procedures

Section I. REPAIR PARTS, SPECIAL TOOLS, TMDE, AND SUPPORT EQUIPMENT

2-1. COMMON TOOLS AND EQUIPMENT. For authorized common tools and equipment, refer to the Modified Table of Organization and Equipment (MTOE) applicable to your unit.

Air Force users must maintain the following common tools:

3-ounce soft-brass hammer
Vise
Flat tip screwdriver
Punch
Tweezers/round nose pliers
Hammer
Needle nose pliers

2-2. SPECIAL TOOLS, TMDE, AND SUPPORT EQUIPMENT. Special tools required for unit support are listed in appendixes B and C. Fabricated tools are listed and illustrated in appendix E.

2-3. REPAIR PARTS. Repair parts are listed and illustrated in appendix C of this manual.

Section II. SERVICE UPON RECEIPT

2-4. GENERAL.

a. Inspect the rifle for damage Incurred during shipment. If rifle has been damaged, report the damage on SF 364, Report of Discrepancy (ROD).

b. Check the rifle against the packing slip to see If shipment is complete. Army users report all discrepancies in accordance with DA PAM 738-750.

Air Force users submit Materiel Deficiency Report (MDR) to: DIR MAT MGT ROBINS AFB GA//MMIBTC// and Product Quality Deficiency Report to: DIR MAT MGT ROBINS AFB GA//MMQA//. IAW Technical Order 00-35D-54.

c. Check to see whether the equipment has been modified.

2-5. SERVICE UPON RECEIPT OF MATERIEL.

WARNING

Before starting an inspection, be sure to clear the rifle. Do not actuate the trigger before clearing the rifle. Inspect the chamber to make sure it is empty and free of obstructions. Check to see there are no obstructions in the barrel and no ammunition is in position to be chambered.

SERVICE UPON RECEIPT

LOCATION	ITEM	ACTION	REMARKS
1. Container	a. M16A2 rifle or M4/M4A1 carbine	a. Remove rifle from containers.	
		b. Inspect the equipment for damage incurred during shipment	If the equipment has been damaged, report the damage on SF Form 364, Report of Discrepancy (ROD).
		c. Check the equipment against the packing list to see if the shipment is complete	Report all discrepancies in accordance with the instructions of DA PAM 738-750.
	b. Basic issue items	Check for missing items	Refer to TM 9-1005-319-10 (operator's manual).
2. M16A2 rifle or M4/M4A1 carbine	a. Barrel assembly	If volatile corrosion inhibitor (VCI) is in barrel, remove and discard.	
	b. All parts	a. Field-strip rifle and inspect for missing, damaged, and rusted or corroded parts.	Refer to operator's manual.
		b. Clean and lubricate	Refer to operator's manual.
		c. Reassemble	Refer to operator's manual.
		d. Function check	Refer to page 2-69.

SERVICE UPON RECEIPT (CONT)

LOCATION	ITEM	ACTION	REMARKS
		e. Check to see whether the equipment has been modified	Refer to DA PAM 25-30.
	c. Magazine	Check for positive retention and functioning of bolt catch	Refer to operator's manual.

Section III. PREVENTIVE MAINTENANCE CHECKS AND SERVICES (PMCS)

2-6. GENERAL This section contains the procedures and instructions necessary to perform unit preventive maintenance checks and services. These services are to be performed by unit maintenance personnel with the assistance of the operator where practical.

2-7. PREVENTIVE MAINTENANCE CHECKS AND SERVICES

WARNING

Before starting an inspection, be sure to clear the rifle. Do not keep live ammunition near the work area.

a. General. The PMCS procedures are contained in the table following. They are arranged in logical sequence requiring a minimum amount of time and motion on the part of the persons performing them and are arranged so that there will be minimum interference between persons performing checks simultaneously on the same end item.

b. Item No. Column. Checks and services are numbered in disassembly sequence. This column shall be used as a source of item numbers for the "TM Number" column on DA Form 2404, Equipment Inspection and Maintenance Worksheet, in recording results of PMCS.

c. Interval Column. This column gives the designated interval when each check is to be performed.

d. Item To Be Checked Or Serviced Column. This column lists the items to be checked or serviced.

e. Procedure Column. This column contains a brief description of the procedure by which the check is to be performed. It contains all the information required to accomplish the checks and services. Information marked SH Indicates a specific equipment shortcoming and the procedure needed to correct the shortcoming.

NOTE

For the purpose of this technical manual, the following definition is supplied. This definition is not intended to apply to any other document

Shortcoming (SH): A fault that requires maintenance or supply action on a piece of equipment, but does not render equipment Not Mission Capable

f. Not Fully Mission Capable If: Column. This column contains a brief statement of the condition (e.g., malfunction, shortage) that would cause the covered equipment to be less than fully ready to perform its assigned mission.

2-7. PREVENTION MAINTENANCE CHECKS AND SERVICES (CONT).

PREVENTIVE MAINTENANCE CHECKS AND SERVICES FOR M16A2 RIFLE (CONT)

Item No.	Interval	Item to be Checked or Serviced	Procedure	Not Fully Mission Capable if:

WARNING

Before starting an inspection, be sure to clear the weapon. Do not pull the trigger until the weapon has been cleared. Inspect the chamber to ensure that it is empty and no ammunition is in position to be chambered. Do not keep live ammunition near work area.

NOTE

An inactive weapon is a weapon which has been stored in an arms room for a period of 90 days without use. The weapon may or may not have been assigned to an individual.

Inactive weapons shall receive quarterly PMCS unless inspection reveals more frequent servicing is necessary.

Normal cleaning (PMCS) of an inactive weapon will be performed every 90 days. Should the unit armorer detect corrosion on a weapon prior to the end of the 90-day period, the PMCS should be performed immediately.

Solid Film Lubricant (SFL) is the authorized touch up for the M16A2 Rifle and M4/M4A1 Carbine and may be used on up to one third of the exterior finish of the weapon.

FOR ARMY CONUS USE ONLY AND AIR FORCE TRAINING WEAPONS ONLY: Solid Film Lubricant may be used as a touch up without limitation on the upper receiver and barrel assembly. This is to say that units which **DO NOT** fall under the category of Divisional Combat Units or rapid deployment type units may have up to 100 percent of the exterior surface of the upper receiver and barrel assembly protected with SFL.. Prior to application of SFL, the surface must be thoroughly cleaned and inspected for corrosion and/or damage. If corroded or damaged, the part must be repaired or replaced prior to application of SFL. Continued use under combat conditions would result in an unprotected surface when the SFL wears off. This would result in a large light reflecting surface and accelerated deterioration of the unprotected surface. Therefore, Divisional Combat Units and units which fall under the definition of Rapid Deployment type must adhere to the limitation of **NOT** over one third of their exterior surface covered by SFL.

When determining mission capability, deadline if it is a deficiency.

PREVENTIVE MAINTENANCE CHECKS AND SERVICES FOR M16A2 RIFLE (CONT)

Item No.	Interval	Item to be Checked or Serviced	Procedure	Not Fully Mission Capable if:
1	Quarterly	Magazine (serviceability check)	Disassemble as In TM 9-1005-319-10 (operator's manual). Inspect tube (1) for bulges, dents, or damaged feeder lips (2). Inspect spring (3) and follower (4) for kinks or damage. SH-Replace the magazine if any of these conditions exist. Reassemble magazine and check for binding during operation of follower (4). SH-Replace the magazine if the follower binds. 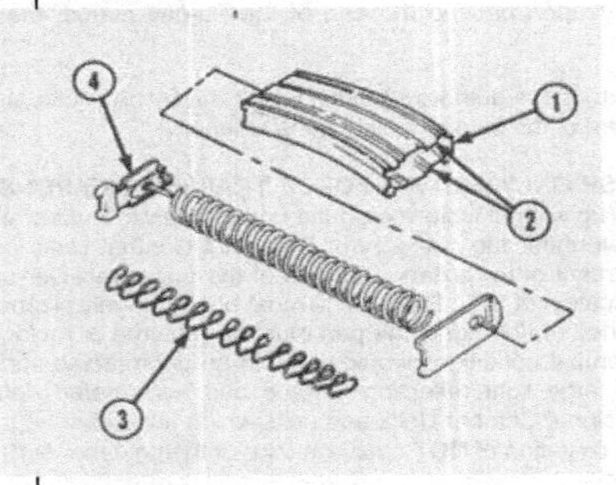	A magazine is not available fu; use with the rifle.
2	Quarterly	Charging handle assembly and selector lever	**WARNING** If the rifle falls any of the following selector lever tests, evacuate It to support maintenance. Continued use of the rifle could result In injury to, or death of, personnel.	

2-7. PREVENTIVE MAINTENANCE CHECKS AND SERVICES (CONT).

PREVENTIVE MAINTENANCE CHECKS AND SERVICES FOR M16A2 RIFLE (CONT)

Item No.	Interval	Item to be Checked or Serviced	Procedure	Not Fully Mission Capable if:
2	Quarterly (cont)	Charging handle assembly and selector lever (cont)	Pull charging handle (1) to rear. Check that chamber is clear. Let bolt carrier assembly (2) close. Leave hammer in cocked position. Do not pull trigger	Charging handle does not lock in place when in the forward position.
			Place selector lever (3) in SAFE position Pull trigger. Place selector lever (3) in SEMI position.	Hammer falls.
			NOTE For the purpose of the following test, "SLOW" is defined as 1/4 to 1/2 the normal rate of trigger release.	
			Pull trigger	Hammer does not fall.
			Hold trigger to the rear, charge weapon, and release the trigger with a slow, smooth motion, without hesitations or stops, until the trigger is fully forward (an audible click should be heard).	Hammer falls.
			Repeat the above SEMI position test five times	The weapon malfunctions during any of these five tests.

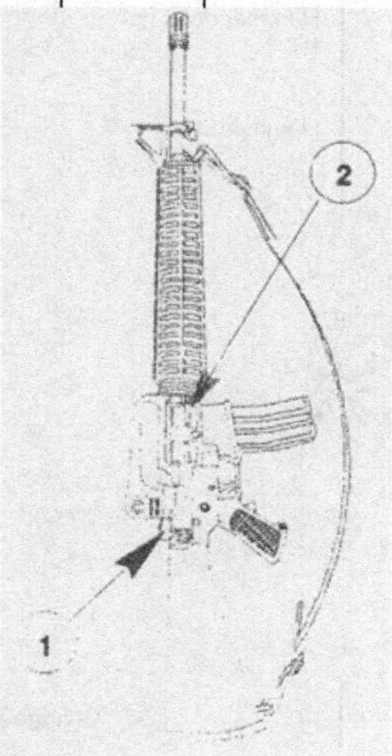

PREVENTIVE MAINTENANCE CHECKS AND SERVICES FOR M16A2 RIFLE (CONT)

Item No.	Interval	Item to be Checked or Serviced	Procedure	Not Fully Mission Capable if:
			M16A2 and M4 ONLY	
2	Quarterly (cont)	Charging handle assembly and selector lever (cont)	Place selector lever (3) in BURST position. Charge weapon and squeeze trigger.	Hammer does not fall.
			Hold trigger to the rear, pull the charging handle to the rear and release it three times. Release trigger. Squeeze trigger.	Hammer does not fall.
			NOTE The burst disconnector should have held the hammer to the rear when it engaged the deep notch of the burst cam.	
			M4A1 ONLY	
			Place selector lever (3) in AUTO position Charge carbine and squeeze trigger Hammer should fall.	Hammer does not fall.
			Hold trigger to the rear, charge carbine, and release trigger. Squeeze trigger. Hammer should not fall.	Hammer falls.

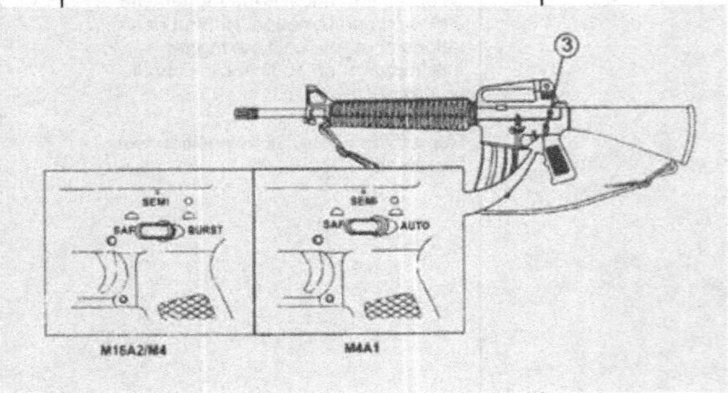

2-7. PREVENTIVE MAINTENANCE CHECKS AND SERVICES (CONT).

PREVENTIVE MAINTENANCE CHECKS AND SERVICES FOR M16A2 RIFLE (CONT)

Item No.	Interval	Item to be Checked or Serviced	Procedure	Not Fully Mission Capable if:
2	Quarterly (cont)	Charging handle assembly and selector lever (cont)	**NOTE** Automatic sear should have released hammer while holding trigger in the squeezed position before releasing and resqueezing the trigger.	
			All weapons With hammer in forward position, using moderate finger/thumb pressure attempt to place the selector lever (3) in SAFE position	Moderate finger/ thumb pressure moves selector lever to SAFE position.

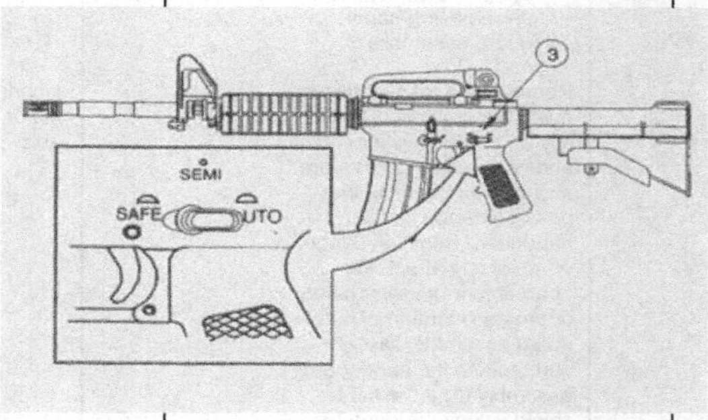

2-7. PREVENTIVE MAINTENANCE CHECKS AND SERVICES (CONT).

PREVENTIVE MAINTENANCE CHECKS AND SERVICES FOR M16A2 RIFLE (CONT)

Item No.	Interval	Item to be Checked or Serviced	Procedure	Not Fully Mission Capable if:
3	Quarterly	Upper receiver and barrel assembly (handguard assemblies) 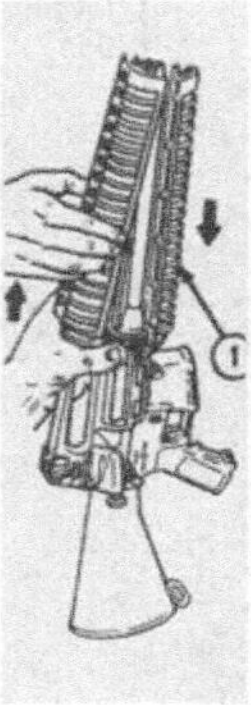	**CAUTION** Do not use screwdriver or any other tool when removing the handguard assemblies, doing so may damage the handguard assemblies and/or slip. **NOTE** Refer to operator's manual for "buddy system" procedure on removing handguard assemblies. Remove and Inspect handguard assemblies (1) internally and externally for cracks and/or damage. Cracks are acceptable providing they do not extend into the handguard retaining flange, or adversely affect rifle operation or operator safety or proper retention of handguard assembly. Discard and replace the handguard assembly (1) If the heatshield is loose enough to rattle when Installed on rifle.	Handguard missing or unserviceable.

PREVENTIVE MAINTENANCE CHECKS AND SERVICES FOR M16A2 RIFLE (CONT)

Item No.	Interval	Item to be Checked or Serviced	Procedure	Not Fully Mission Capable if:
4	Quarterly	Upper receiver and barrel assembly (serviceability check) 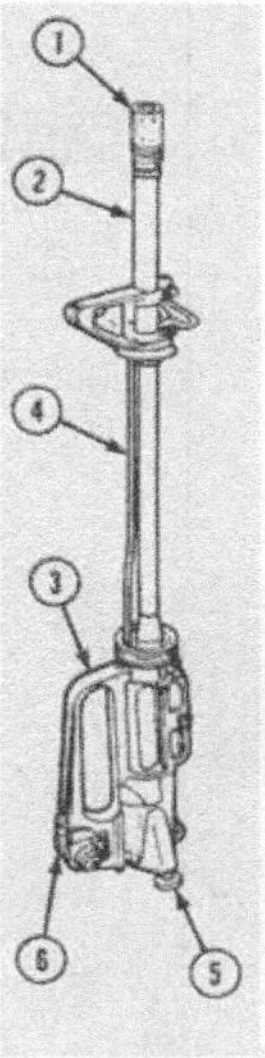	**WARNING** Dry cleaning solvent is flammable and toxic and should be used in a well ventilated area The use of rubber gloves is necessary to protect the skin when washing rifle parts **CAUTION** Damage may occur if excessive force is used to release takedown pin or pivot pin. Use hand pressure ONLY. Release takedown pins and open and separate receivers Hand check compensator (1) for looseness on barrel (2), then hand check barrel (2) for looseness on upper receiver (3). Check center slot of compensator for alignment (p 2-50). If compensator or barrel is loose, evacuate to support maintenance Check gas tube (4), forward assist assembly (5), and rear sight assembly (6) for damage, The rear sight spring should retain the rear sight assembly (6) In either position with firmness SH - If damaged, evacuate to support maintenance.	Compensator or barrel is loose

PREVENTIVE MAINTENANCE CHECKS AND SERVICES FOR M16A2 RIFLE (CONT)

Item No.	Interval	Item to be Checked or Serviced	Procedure	Not Fully Mission Capable if:
4	Quarterly (cont)	Upper receiver and barrel assembly (serviceability check) (cont) 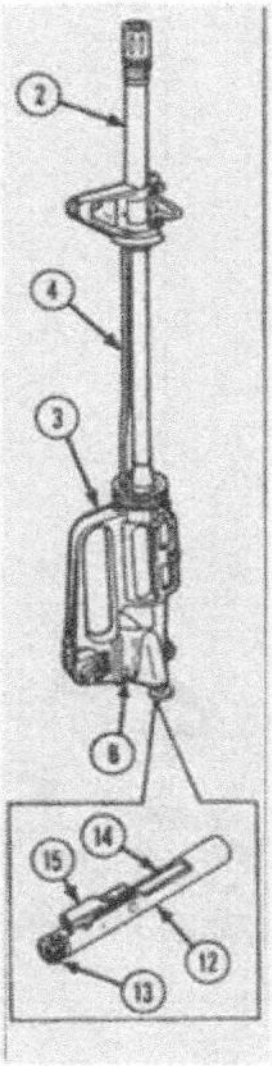	Inspect upper receiver (3) finish for scratches or worn shiny spots. If scratched or worn shiny in spots, disassemble and remove all lubricant from surface with dry cleaning solvent (item 16, app D). Wear rubber gloves (item 18, app D) and use a wash pan (item 24, app D) to apply solvent. Let parts dry thoroughly. Roughen the surface using abrasive cloth (item 1 3, app D) and apply solid film lubricant (item 16, app D). Allow 16 to 24 hours to dry before handling. Hold barrel (2) at 40-degree angle (muzzle down). Pull charging handle (8) to rear. Hold bolt carrier assembly (12) to rear and push charging handle forward. Release bolt carrier assembly (12). The bolt carrier assembly should close and lock under its own weight. If It does not, remove the bolt assembly (13) from the key and bolt carrier assembly (14) and slide the key and bolt carrier assembly (without bolt) back and forth In the upper receiver and barrel assembly. If the gas tube (4) hits the carrier key (1 5), or If the gas tube binds in the carrier key, try to correct the malfunction by adjusting (slightly bending) the gas tube In the area of the handguard assemblies. If unable to adjust, evacuate to support maintenance.	Adjustment does not correct the malfunction.

2-7. PREVENTIVE MAINTENANCE CHECKS AND SERVICES (CONT).

PREVENTIVE MAINTENANCE CHECKS AND SERVICES FOR M16A2 RIFLE (CONT)

Item No.	Interval	Item to be Checked or Serviced	Procedure	Not Fully Mission Capable if:
4	Quarterly (cont)	Upper receiver and barrel as sembly (service ability check) (cont)	**NOTE** If front or rear sight is moved, return to original position Check front sight post, de tent, and spring (7) for dam age and corrosion Clean and lubricate Check charg ing handle (8) and election port cover (9) for defects and proper function Check sling swivel (10) and rivet (11) for damage and proper function SH Other corn ponents are defective, re place as necessary.	Charging handle (8) is defective.

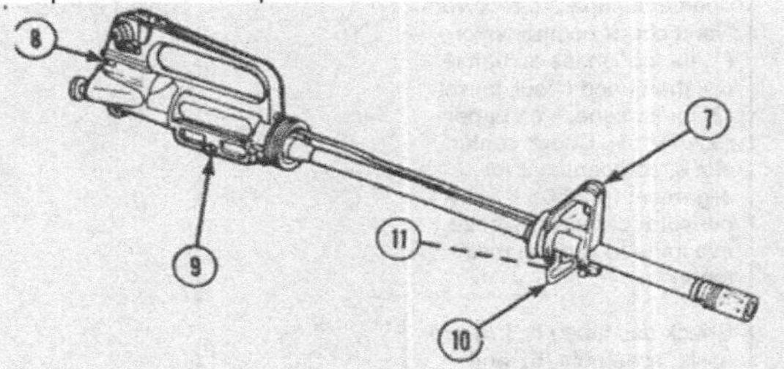

CAUTION

Do not use a wire brush to roughen surfaces Use a well ventilated area during cleaning and application of solid film lubricant. If solid film lubricant comes in contact with moving parts or functioning surfaces of the rifle, remove lubricant immediately by washing with dry cleaning solvent.

NOTE

Shiny metal exterior surfaces of the rifle should be recoated with solid film lubricant (Item 21, app D). Clean surface with dry cleaning solvent (item 16, app D); dry, roughen with abrasive cloth (Item 13, app D) and apply solid film lubricant.

2-7. PREVENTIVE MAINTENANCE CHECKS AND SERVICES (CONT).

PREVENTIVE MAINTENANCE CHECKS AND SERVICES FOR M16A2 RIFLE (CONT)

Item No.	Interval	Item to be Checked or Serviced	Procedure	Not Fully Mission Capable if:
4	Quarterly (cont)	Upper receiver and barrel assembly (serviceability) check) (cont)	**M4/M4A1** Inspect carrying handle assembly (16) and mounting surface (17) of upper receiver for damage. If the carrying handle is missing or can not be correctly mounted, repair as authorized or evacuate to support maintenance. Inspect carrying handle assembly (16) to insure unit applied identification (ID) code matches unit applied ID code on carbine. If it doesn't match locate correct carrying handle assembly and match up to correct M4/ M4A1 carbine. If a match can not be found, the weapon should be re-zeroed by the operator.	

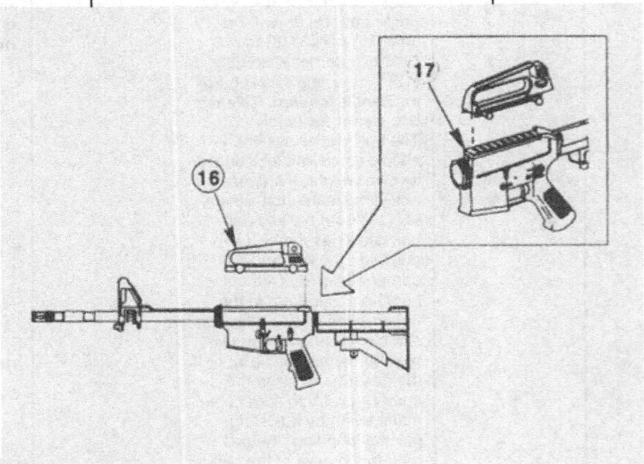

2-7. PREVENTIVE MAINTENANCE CHECKS AND SERVICES (CONT).

PREVENTIVE MAINTENANCE CHECKS AND SERVICES FOR M16A2 RIFLE (CONT)

WARNING

Below direct support maintenance, do not interchange bolt assemblies from one rifle to another. Doing so may result in injury to, or death of, personnel.

Item No.	Interval	Item to be Checked or Serviced	Procedure	Not Fully Mission Capable if:
5	Quarterly	Key and bolt carrier assembly and bolt assembly (serviceability check)	Remove and disassemble. Visually inspect bolt assembly (1) for cracks, especially in the area of the bolt cam pin hole (2). Check for cracks on locking lugs (3), for a cluster of pits or chipped bolt face (4), and for an elongated firing pin hole (5). If cracked or broken, evacuate to support maintenance for repair.	Defects are found.
			Check for worn or missing bolt rings (6) Check for proper staggering of bolt rings Insert the bolt assembly (1) into the key and bolt carrier assembly (7). Turn key and bolt carrier assembly (7) so the bolt assembly (1) points down. The bolt assembly must not drop out. Remove bolt assembly (p 2-35). Check for broken or missing firing pin retaining pin (8) and bolt cam pin (9); replace as necessary.	The bolt assembly drops out of the key and bolt carrier assembly due to its own weight. Missing or broken firing pin retaining pin or bolt cam pin.

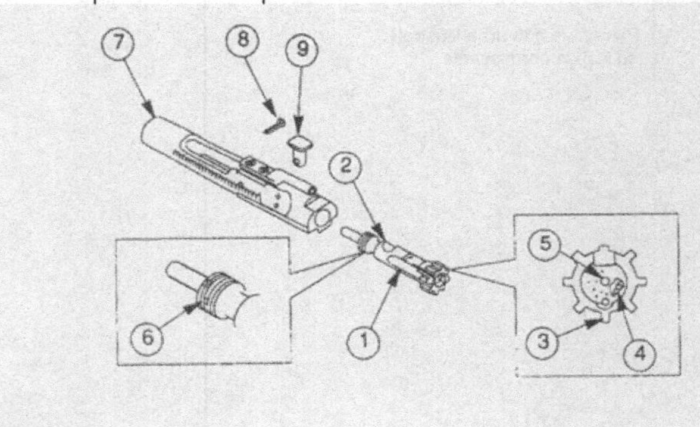

PREVENTIVE MAINTENANCE CHECKS AND SERVICES FOR M16A2 RIFLE (CONT)

Item No.	Interval	Item to be Checked or Serviced	Procedure	Not Fully Mission Capable if:
5	Quarterly (cont)	Key and bolt carrier assembly and bolt assembly (serviceability check) (cont)	Check cartridge extractor (10), extractor spring assembly (11), cartridge ejector (12), and ejector spring (13) for dirt and serviceability. If dirty, clean, lubricate and assemble. If unserviceable, replace as necessary.	Parts are missing or unserviceable
			Check key and bolt carrier assembly (7) and carrier key (14) for damage and looseness. If damaged or loose, evacuate to support maintenance **NOTE** If carrier key Is dented, evacuate to support maintenance.	Key and bolt carrier assembly or carrier key is damaged, or carrier key is loose.
			Check firing pin 15) for chips or breaks If damaged, evacuate to support maintenance. Pits or wear in area Illustrated (16) is permissable	Firing pin is damaged.

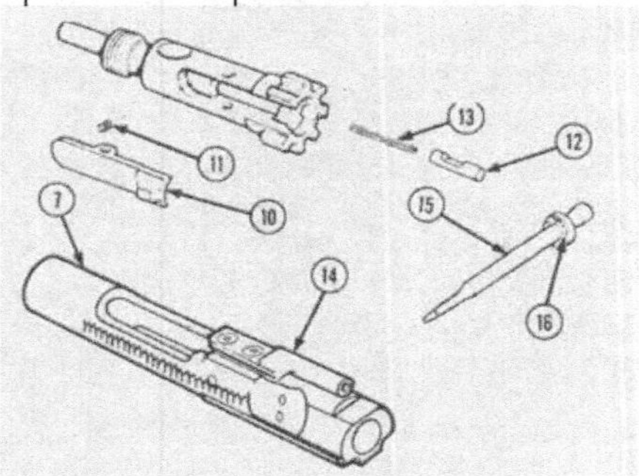

2-7. PREVENTIVE MAINTENANCE CHECKS AND SERVICES (CONT).

PREVENTIVE MAINTENANCE CHECKS AND SERVICES FOR M16A2 RIFLE (CONT)

Item No.	Interval	Item to be Checked or Serviced	Procedure	Not Fully Mission Capable if:
6	Quarterly	Lower receiver and buttstock assembly (serviceability check)	Remove buffer assembly (1) and action spring (2) Check buffer assembly for cracks SH - Buffer assembly is cracked Check action spring (2) for kinks and free length Free length should be **RIFLE**: 11 3/4 Inches (29 85 cm) minimum to 13 1./2 inches (34 29 cm) maximum SH - If action spring is kinked or does not meet free length require-ments **CARBINE**: 10 1/16 inches (25.56 cm) minimum to 11 1/4 inches (28.58 cm) maximum. Do not attempt to adjust spring length SH - If action spring is kinked or does not meet free length require-ments	

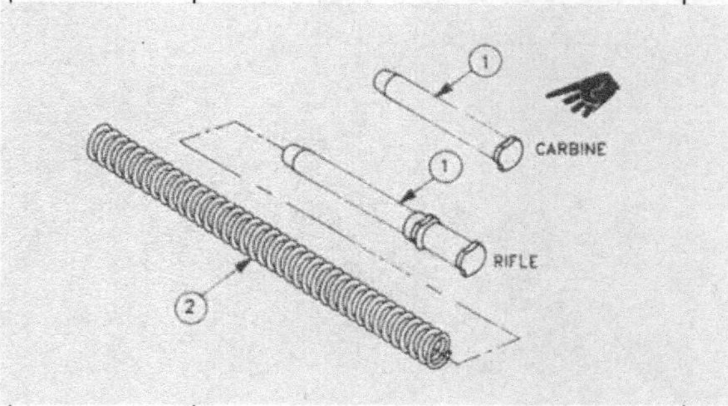

PREVENTIVE MAINTENANCE CHECKS AND SERVICES FOR M16A2 RIFLE (CONT)

Item No.	Interval	Item to be Checked or Serviced	Procedure	Not Fully Mission Capable if:
6	Quarterly (cont)	Lower receiver and buttstock assembly (serviceability check) (cont)	Remove pistol grip screw (3), lockwasher (4), pistol grip (5), helical spring (6), safety detent (7), pivot pin (8), pivot pin detent (9), and helical spring (10). Clean and lubricate metal components. Also clean and generously lubricate pivot pin holes and spring/detent holes. Replace defective/damaged components as necessary.	Components are defective/damaged.
			Disengage takedown pin (11) and pull out, push back in to re-engage takedown pin (an audible click should be heard). If an audible click is not heard, see page 2-57 for repair.	Components are defective/damaged.

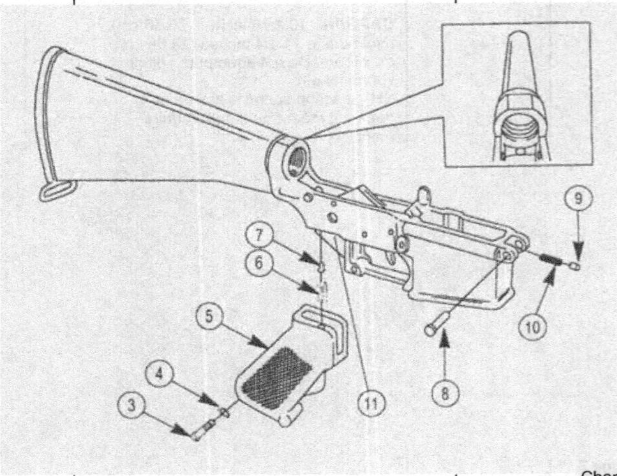

2-7. PREVENTIVE MAINTENANCE CHECKS AND SERVICES (CONT).

PREVENTIVE MAINTENANCE CHECKS AND SERVICES FOR M16A2 RIFLE (CONT)

Item No.	Interval	Item to be Checked or Serviced	Procedure	Not Fully Mission Capable if:
6	Quarterly (cont)	Lower receiver and buttstock assembly (serviceability check) (cont)	Lubricate helical compression spring and takedown pin detent (11) by placing one drop of lubricant on takedown pin detent and lowering the buttstock assembly (12) to vertical position. Allow the lubricant to work Its way around the helical compression spring and takedown pin detent (11). Check buttstock assembly(12) components for damage.	

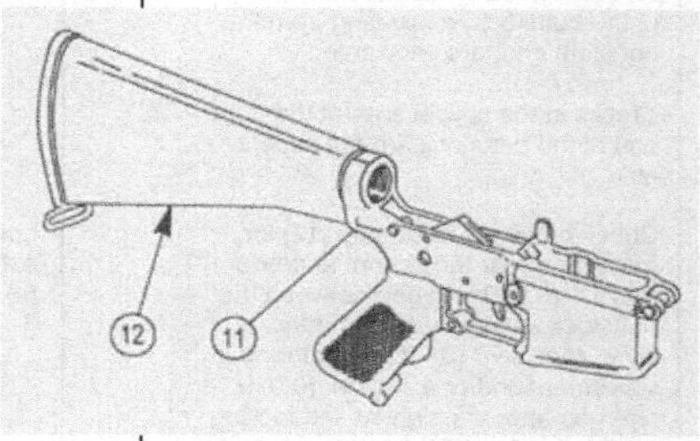

			RIFLE ONLY Under the following conditions, hairline cracks (no chipped away material allowed) originating from the buttplate end of the buttstock are acceptable.	Components are damaged.

PREVENTIVE MAINTENANCE CHECKS AND SERVICES FOR M16A2 RIFLE (CONT)

Item No.	Interval	Item to be Checked or Serviced	Procedure	Not Fully Mission Capable if:
			RIFLE ONLY	
6	Quarterly (cont)	Lower receiver and buttstock assembly (serviceability check) (cont)	a. One hairline crack, not to exceed 1 in. (2.54 cm) in length, per side of buttstock	The buttstock is cracked in the critical area or does not meet the crack criteria.
			b. Two additional hairline cracks up to 0.25 in. (0.64 cm) in length, per side of buttstock.	
			c. A total of three cracks per side of the buttstock, originating from the buttplate end, are allowable.	
			Cracks in the critical area at the front end of the buttstock are not acceptable.	
			Check buttstock assembly (12) for forward to rear movement and/or a 1/32 in. (0.079 cm) gap between the buttstock assembly (12) and the lower receiver (13). If forward to rear movement and/or a 1/32 in. (0.079 cm) gap appears, tighten self-locking screw. If still not tight, remove buttstock assembly and check for loose lower receiver extension. If loose evacuate to support maintenance. If not loose, replace buttplate (14).	Lower receiver extension cannot be tightened.

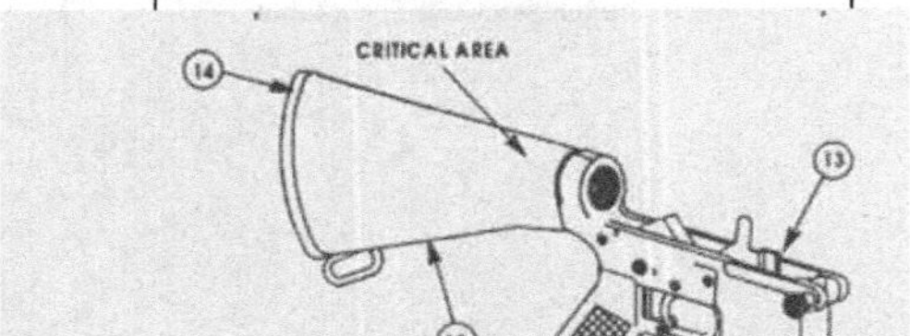

2-7. PREVENTIVE MAINTENANCE CHECKS AND SERVICES (CONS).

PREVENTIVE MAINTENANCE CHECKS AND SERVICES FOR M16A2 RIFLE (CONT)

Item No.	Interval	Item to be Checked or Serviced	Procedure	Not Fully Mission Capable if:
6	Quarterly (cont)	Lower receiver and buttstock assembly (serviceability check) (cont)	**RIFLE ONLY** Small amounts of side-to-side, up-and-down or rotational movement of the buttstock assembly is acceptable. (1) Cracks visible around the butt-plate mounting holes while screws are mounted SH - Cracks are visible around mounting holes when installed on rifle. (2) Cracks or separations around the door assembly are visible when the door assembly is closed SH - Cracks are visible when door assembly is closed. (3) If buttplate is cracked in excess of 0.25 in (O 64 cm) in length and extends through the buttplate (14). see pg. 2-64 for repair SH - If cracked in excess of 0 25 in. extends thru buttplate (4) The buttplate (14) should not be removed other than for repair or replacement of parts at which time a new self-locking screw, NSN 5305-01-147-8585, must be used.	

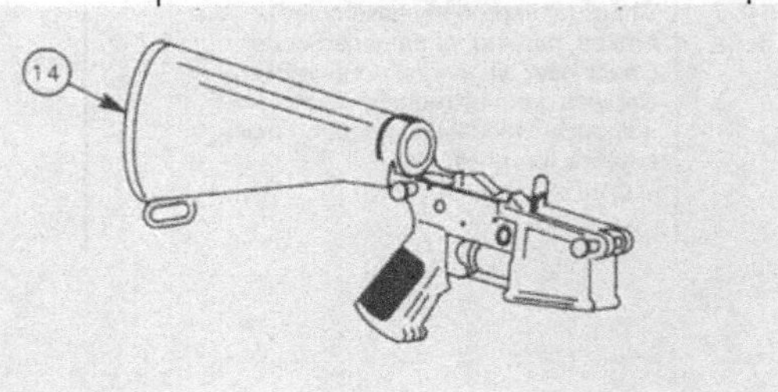

PREVENTIVE MAINTENANCE CHECKS AND SERVICES FOR M16A2 RIFLE (CONT)

Item No.	Interval	Item to be Checked or Serviced	Procedure	Not Fully Mission Capable if:
6	Quarterly (cont)	Lower receiver and buttstock assembly (serviceability check) (cont)	**NOTE** If a weapons lower receiver is missing one third or more of its exterior protective finish, resulting in an unprotected/light reflecting surface, it is candidate for overhaul. This missing finish will be considered a shortcoming. This shortcoming requires action to obtain a replacement weapon. Once a replacement has been received, evacuate the original weapon to depot for overhaul. If scratched or worn shiny in spots, repair in the same manner as outlined for upper receiver (see item 4 above).	
7	Quarterly	M16A2 Rifle	Assemble as in TM 9-1005-319-10 (operator's manual). Check sling for damage. If damaged, replace. Check for improperly assembled, broken, missing, or damaged parts. Check over all general appearance. Replace parts as required and authorize evacuation to support maintenance for repair.	

2-7. PREVENTIVE MAINTENANCE CHECKS AND SERVICES (CONT).

PREVENTIVE MAINTENANCE CHECKS AND SERVICES FOR M16A2 RIFLE (CONT)

Item No.	Interval	Item to be Checked or Serviced	Procedure	Not Fully Mission Capable if:
8	Quarterly	Annual DS safety and serviceability inspection and gaging	Check to ensure annual DS safety and serviceability inspection and gaging has been done and that the next gaging and inspection is scheduled. If annual gaging has not been performed within the last year, notify support maintenance.	Annual gaging has not been performed.

Section IV. TROUBLESHOOTING

2-8. GENERAL.

a. This section contains unit level troubleshooting information for locating and correcting most of the operating troubles which may develop in the M16A2 rifle, M4/M4A1 carbine. Each malfunction for the individual part or assembly is followed by a list of tests or inspections which will help you to determine the corrective actions in the order listed.

b. This manual cannot list all malfunctions that may occur, nor all tests or inspections and corrective actions. If a malfunction is not listed or is not corrected by listed corrective actions, see individual repair sections in the maintenance procedures on each major assembly.

2-9. TROUBLESHOOTING PROCEDURES. Refer to troubleshooting table for malfunctions, tests, and corrective actions. The symptom index is provided for a quick reference of the malfunctions covered in the table.

SYMPTOM INDEX

TROUBLESHOOTING

MALFUNCTION
TEST OR INSPECTION
CORRECTIVE ACTION

1. FAILURE OF MAGAZINE TO LOCK IN RIFLE.

Step 1. Dirty or corroded magazine catch (1).

Disassemble and clean.

Step 2 Defective magazine catch spring (2)

Evacuate to support maintenance.

Step 3 Worn or broken magazine catch (1).

Evacuate to support maintenance

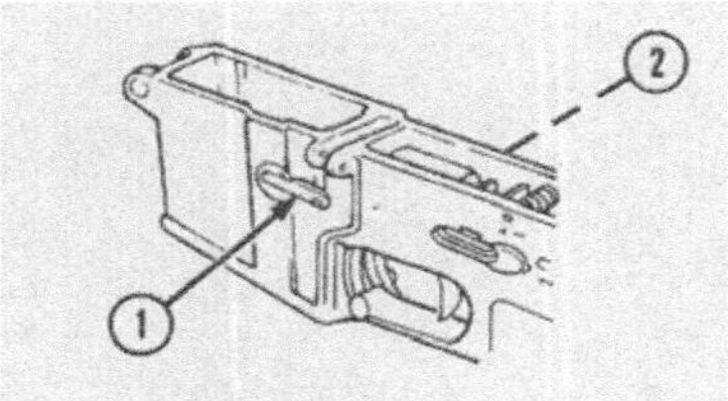

2-9. TROUBLESHOOTING PROCEDURES (CONT).

TROUBLESHOOTING (CONT)

MALFUNCTION
TEST OR INSPECTION
CORRECTIVE ACTION

2. FAILURE TO FEED

Step 1. Magazine catch spring weak or broken

Evacuate to support maintenance

Step 2. Magazine catch (1) defective.

Evacuate to support maintenance

Step 3. Magazine catch (1) out of adjustment (will not retain magazine)

Refer to operator's manual.

Step 4. Short recoil.

Refer to page 2-28

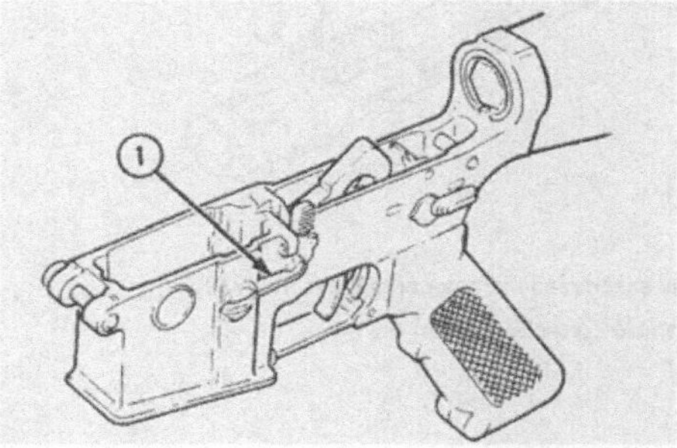

3. FAILURE TO CHAMBER.

Step 1. Weak or broken action spring (1), RIFLE ONLY (free length 11 3/4 inches (29.85 cm) minimum to 13 1.'2 Inches (34 29 cm) maximum) CARBINE ONLY (10 1 16 inches (2S 56 cm) minimum to 11 1,4 inches (28 58 cm) maximum)

Replace action spring (p 2-57)

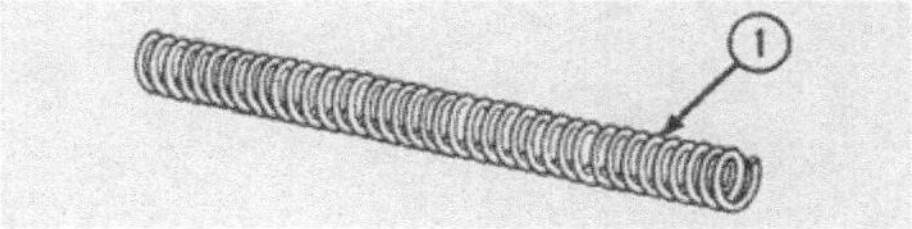

TROUBLESHOOTING (CONT)

MALFUNCTION
TEST OR INSPECTION
CORRECTIVE ACTION

Step 2. Short recoil

Refer to page 2-28

4. FAILURE TO LOCK

Step 1. Bolt cam pin (1) missing

Replace (p 2-35)

Step 2. Loose or damaged bolt carrier key (2).

a. Evacuate to support maintenance.

b. Dented bolt carrier key may be repaired (p 2 35).

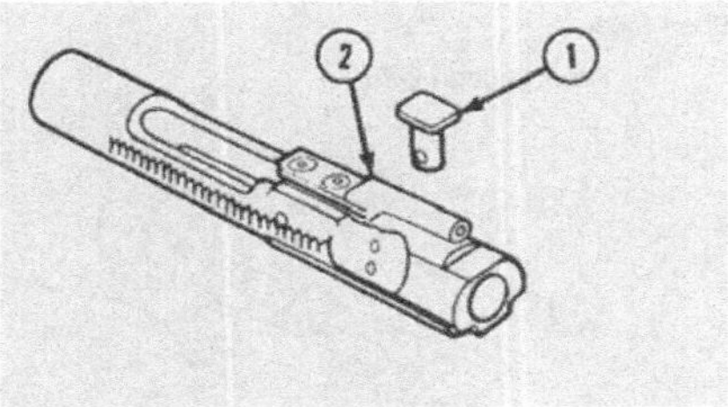

Step 3. Improperly assembled extractor spring assembly (3)

Assemble correctly (p 2-38).

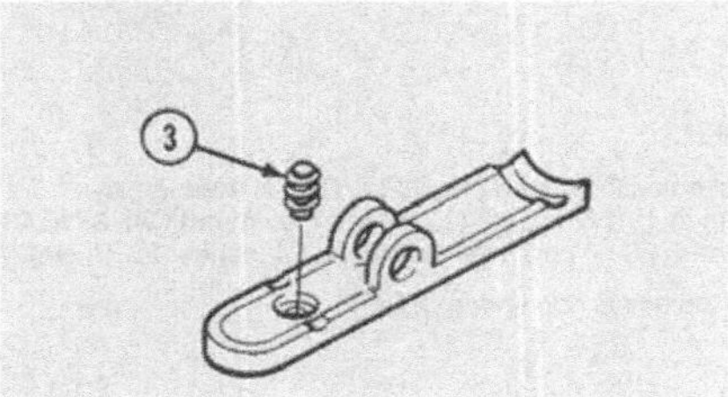

2-9. TROUBLESHOOTING PROCEDURES (CONT).

TROUBLESHOOTING (CONT)

MALFUNCTION
TEST OR INSPECTION
CORRECTIVE ACTION

4. FAILURE TO LOCK (CONT)

Step 4. Bent gas tube 14J

a Adjust to Its original configuration by bending in area of hand-guard assembly.

b If the gas tube cannot be returned to Its original configuration, evacuate the rifle to support maintenance

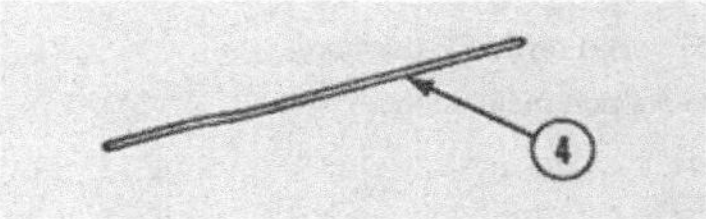

Step 5 Weak or broken action spring (5); RIFLE ONLY: (free length 11 3/4 inches (29,85 cm) minimum to 13 1,/2 Inches (34 29 cm) maximum). CARBINE ONLY (10 1/16 Inches (25 56 cm) minimum to 11 1,4 inches (28 58 cm) maximum

Replace action spring (p 2-571

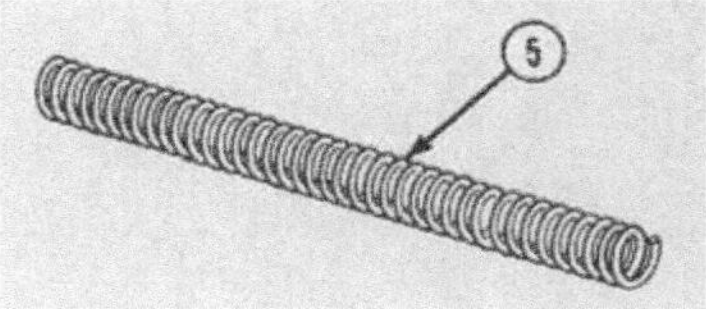

5. FAILURE TO FIRE.

Step 1 Broken or chipped firing pin (1).

Evacuate to support maintenance.

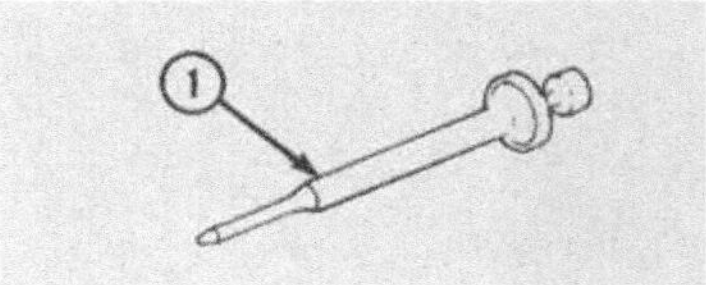

TROUBLESHOOTING (CONT)

MALFUNCTION
TEST OR INSPECTION
CORRECTIVE ACTION

Step 2 Carbon buildup in firing pin recess inside bolt assembly

Remove cartridge extractor and clean recess with pipe cleaner (Item 11, app D), refer to operator's manual.

Step 3. Firing mechanism (2) and or lower receiver assembly (31 improperly assembled or has worn, broken, or missing parts.

Evacuate to support maintenance

Step 4. Broken, defective, or missing firing pin retaining pin (4).

Replace (p 2-35)

Step 5. Selector lever (5) frozen on SAFE position.

Evacuate to support maintenance

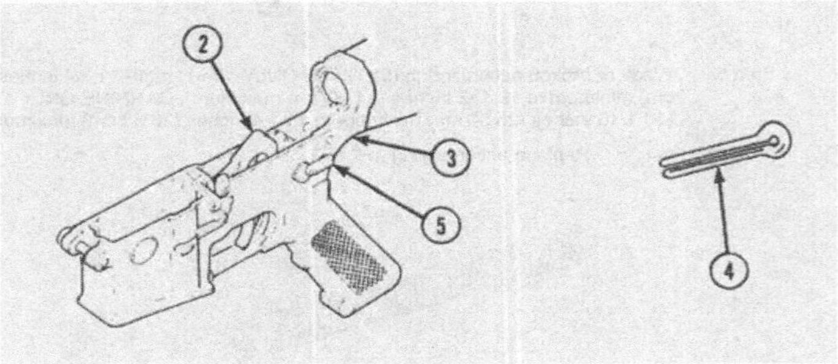

2-9. TROUBLESHOOTING PROCEDURES (CONT).

TROUBLESHOOTING (CONT)

MALFUNCTION
TEST OR INSPECTION
CORRECTIVE ACTION

6. FAILURE TO UNLOCK.

Step 1. Burred locking lugs (1) on bolt assembly.

Remove burrs.

Step 2. Burred lugs (2) on barrel extension

Remove burrs.

Step 3 Short recoil.

Refer to page 2-28.

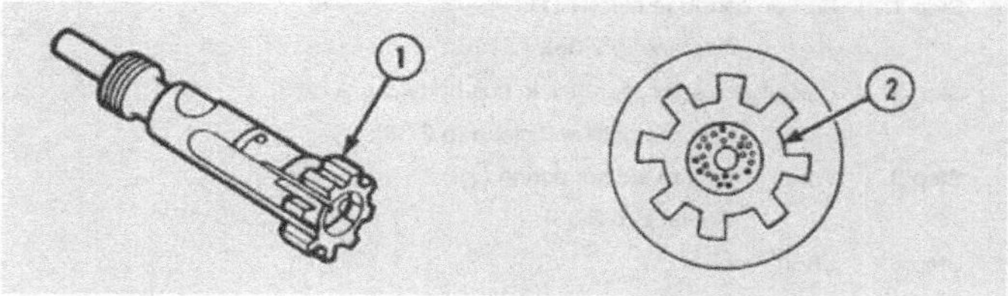

7. FAILURE TO EXTRACT.

Step 1. Defective extractor pin (1), cartridge extractor (2), and or extractor spring assembly (3).

Replace extractor pin (1), cartridge extractor (2), and or extra(:tor spring assembly (3) (p 2-381

Step 2. Short recoil

Refer to page 2-28

NOTE

Rubber insert and spring are an assembly Illustration shows Insert out of assembly for clarification only Do not remove the rubber Insert from the extractor spring assembly.

TROUBLESHOOTING (CONT)

MALFUNCTION
TEST OR INSPECTION
CORRECTIVE ACTION

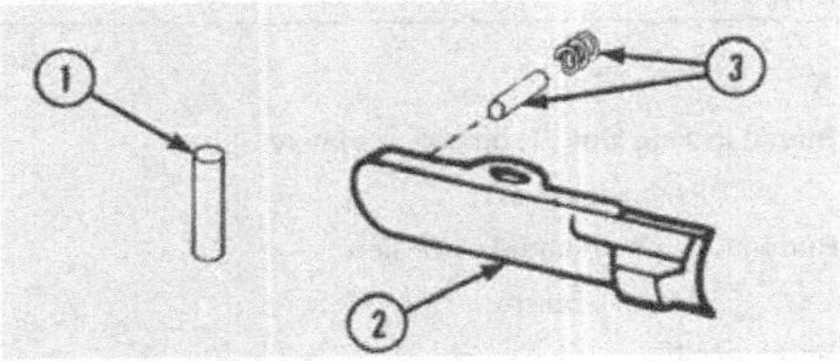

8. FAILURE TO EJECT.

Step 1. Broken cartridge ejector (1).

Replace (p 2-38).

Step 2. Cartridge ejector (1) stuck in bolt body (2).

Disassemble and clean (p 2-38).

Step 3. Weak or broken ejector spring (3).

Replace (p 2-38)

Step 4. Short recoil

Refer to page 2-28.

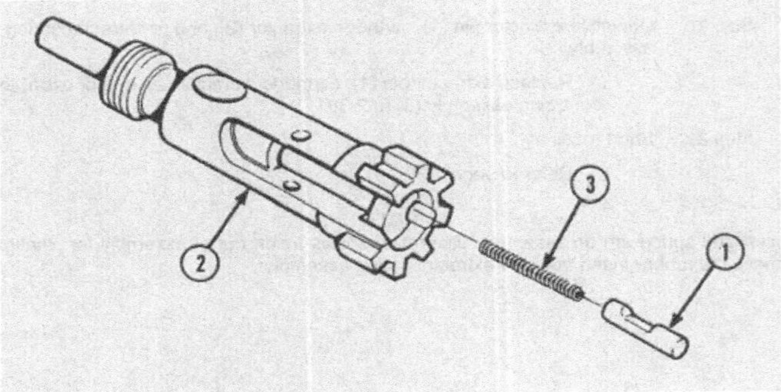

2-9. TROUBLESHOOTING PROCEDURES (CONT).

TROUBLESHOOTING (CONT)

MALFUNCTION
TEST OR INSPECTION
CORRECTIVE ACTION

9. FAILURE TO COCK

Step 1 Worn, broken, or missing parts of firing mechanism

Evacuate to support maintenance.

Step 2 Short recoil

Refer to below

10 SHORT RECOIL

Step 1 Broken or damaged action spring (1).

Replace action spring (p 2-57).

Step 2 Unlubricated or dirty action spring and receiver extension

Clean and lubricate

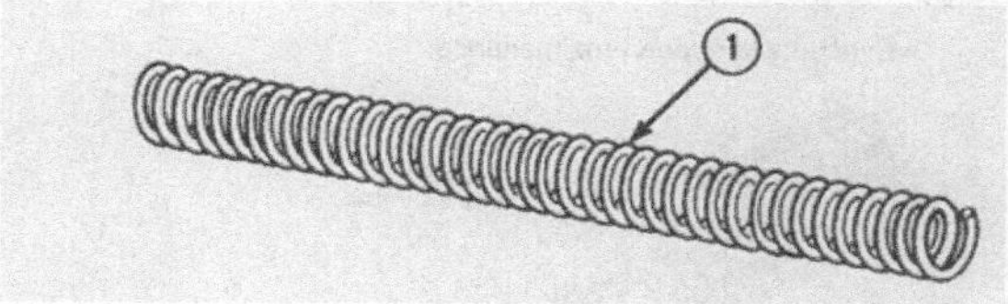

Step 3 Improper gap space or worn, missing, or broken bolt rings (2)

a. Stagger bolt ring gaps (approximately 13 turn apart}

b. Evacuate to support maintenance if bolt rings are worn, broken, or missing.

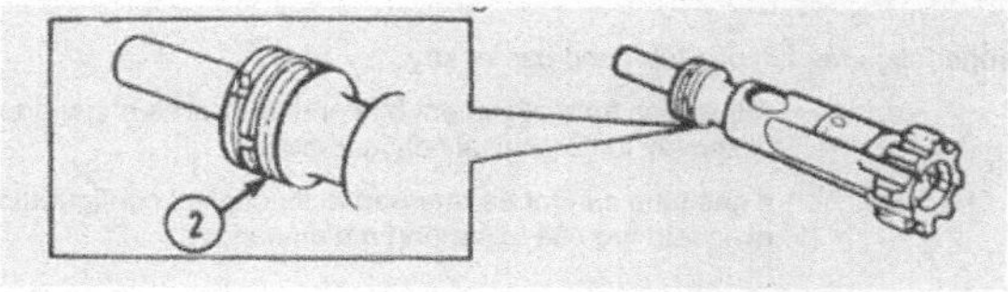

TROUBLESHOOTING (CONT)

MALFUNCTION
TEST OR INSPECTION
CORRECTIVE ACTION

Step 4. Carbon build-up or foreign matter In the narrow passage of the bolt carrier key (3).

Clean with CLP (Item 9, app D) and a pipe cleaner (Item 11, app D).

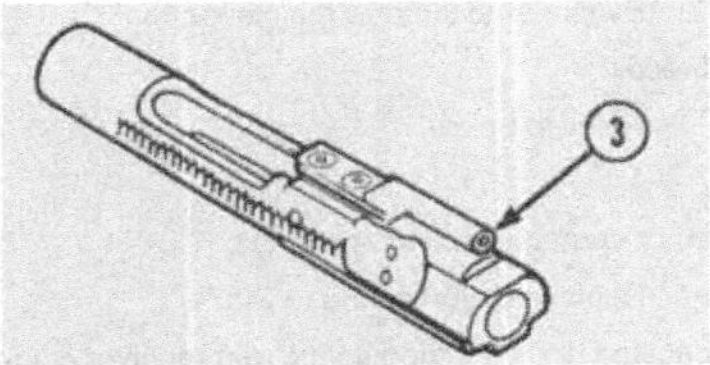

Step 5. Gas leakage caused by broken or loose gas tube (4) around front sight base.

Evacuate to support maintenance.

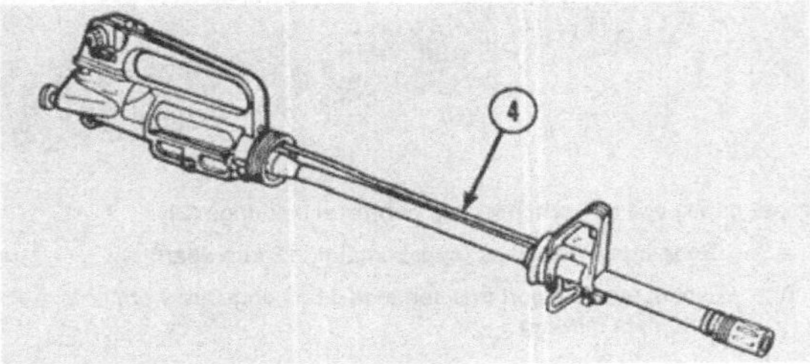

Step 6. Improper alignment of gas tube and carrier key.

a. Adjust gas tube alignment by bending in area of handguard assembly to Its original configuration

b. If gas tube cannot be returned to Its original configuration, evacuate the rifle to support maintenance.

2-9. TROUBLESHOOTING PROCEDURES (CONT).

TROUBLESHOOTING (CONT)

MALFUNCTION
TEST OR INSPECTION
CORRECTIVE ACTION

11. RIFLE CANNOT BE ZEROED.

Step 1. Defective barrel assembly (1).

Evacuate to support maintenance

Step 2. Barrel assembly out of alignment with rear sight assembly (2) on upper receiver.

Evacuate to support maintenance

Step 3. Defective front sight (3).

Remove front sight post (4), front sight detent (5), and helical spring (6). If damaged, replace.

Step 4. Defective -ear sight assembly (2).
Evacuate to support maintenance.

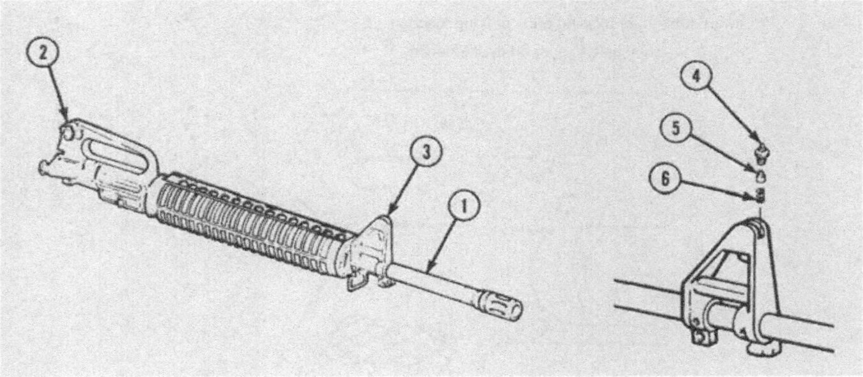

TROUBLESHOOTING (CONT)

MALFUNCTION
TEST OR INSPECTION
CORRECTIVE ACTION

12. FAILURE TO CYCLE WITH SELECTOR LEVER SET ON BURST (M16A2 AND M4).

Faulty selector lever or broken cam, cam clutch spring, or burst disconnector.

Evacuate to support maintenance.

12.1. FAILURE TO CYCLE WITH SELECTOR LEVER SET ON AUTO (M4A1).

Faulty selector lever.

Evacuate to support maintenance.

13. FIRES TWO ROUNDS WITH ONE PULL OF TRIGGER WITH SELECTOR LEVER SET ON SEMI (DOUBLE FIRING).

Perform function test.

If any part of function test (p 2-68) fails, evacuate to support maintenance.

14. FIRES WITH SELECTOR LEVER ON SAFE OR WHEN TRIGGER IS RELEASED WITH SELECTOR LEVER ON SEMI.

Worn, broken, or missing parts of firing mechanism.

Evacuate to support maintenance.

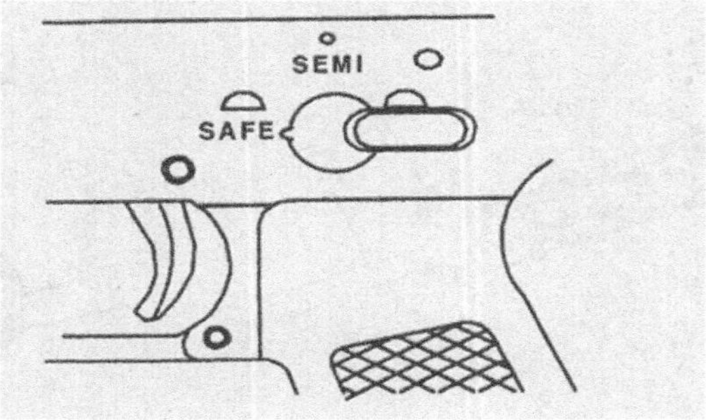

2-9. TROUBLESHOOTING PROCEDURES (CONT).

TROUBLESHOOTING (CONT)

MALFUNCTION
TEST OR INSPECTION
CORRECTIVE ACTION

15 BOLT ASSEMBLY FAILS TO LOCK TO REAR AFTER FIRING LAST ROUND.

Step 1. Magazine follower (1) worn or broken.

Replace magazine.

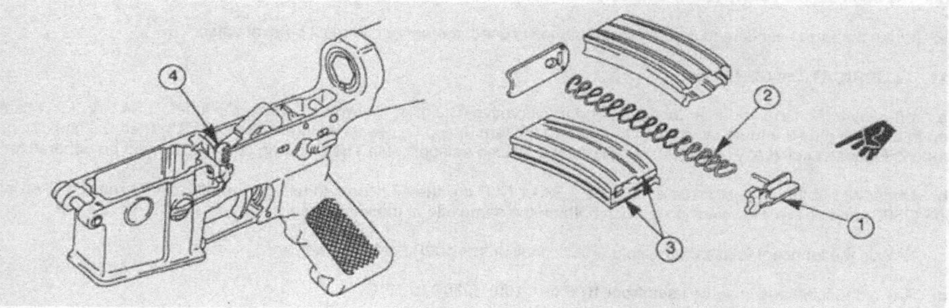

Step 2. Magazine catch spring (2) weak or broken.

Replace magazine.

Step 3. Magazine feeder lips (3) bent or broken.

Replace magazine.

Step 4. Magazine follower (1) binds during operation.

Replace magazine.

Step 5. Broken bolt catch (4) and/or spring.

Evacuate to support maintenance.

Section V. MAINTENANCE PROCEDURES

2-10. INITIAL SETUP. The following information will reduce the space required for the initial setup portion of the maintenance procedures.

a. Materials/Parts required are not listed unless they apply to the procedure

b. Personnel Required is listed only if the task requires more than one person If Personnel Required Is not listed, It means one person can do the job.

c. The normal standard equipment condition Is that the item is removed from the en(1 Item or next higher assembly and is in the assembled condition. Equipment Condition is not listed unless some other condition is required.

d. The approximate time required Is listed on the applicable Maintenance Allocation Chart (MAC).

e. When the term evacuate to support maintenance is used, the entire rifle must be evacuated.

2-11. LUBRICATION GENERAL.

a. Whenever the term "cleaner, lubricant, and preservative (CLP)" or the words "lubricant", "lube", "LSA", or "LAW" are cited In this TM, they are to be interpreted to mean CLP (Item 9, app D), weapons lubricating oil (LSA) (Item 23, app D), or weapons lubricating oil (LAW) (Item 22, app D) can be utilized as applicable The following constraints must be adhered to:

b. Under all but the coldest arctic conditions, LSA or CLP are the lubricants to use on the rifle. Either may be used at 100F (230C) and above. However, do not use both on the same rifle at the same time.

c. LAW is the lubricant to use during cold arctic conditions, + 100F (12C) and below

d. Any of the lubricants may be used from 10°F to + 100F (230C to 12°C)

e. **ARMY ONLY**: Do not mix lubricants on the same rifle The rifle must be thoroughly cleaned during change from one lubricant to another Dry Cleaning Solvent (SD) (Item 16, app D) is recommended for cleaning during change from one lubricant to another.

f. Rifle Bore Cleaner (RBC) (item 12, app D) may be used to remove carbon buildup In the bore and other portions of the rifle.

2-12. MAJOR COMPONENTS OF M16A2 RIFLE AND CARBINES.

This task covers disassembly

INITIAL SETUP:

References
TM 9-1005-319-10 (operator's manual)

Equipment Conditions
Rifle and Carbines assembled

General Safety Instructions
Before starting an inspection, be sure to clear the weapon. Do not keep live ammunition near the work area.

To avoid injury to your eyes, use care when removing and installing spring-loaded parts.

Below direct support maintenance, do not interchange bolt assemblies from one weapon to another. Doing so may result in injury to, or death of, personnel.

DISASSEMBLY

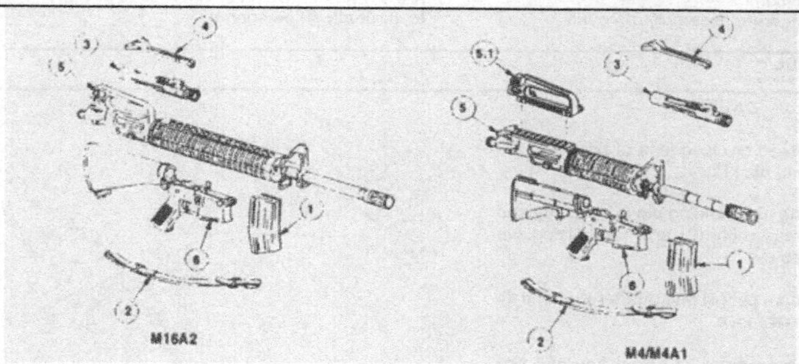

a. Refer to operator's manual.

b. Remove magazine (1), sling (2), bolt carrier assembly (3), charging handle assembly (4), and upper receiver and barrel assembly (5), carrying handle assembly (5.1), from lower receiver and buttstock assembly (6).

NOTE

Solid Film Lubricant (SFL) is the only authorized touchup for the MI 16A2 rifle and M4/M4A1 carbines and may be used on up to one third of the exterior finish of the weapon. **FOR ARMY CONUS USE ONLY AND AIR FORCE TRAINING WEAPONS ONLY**: SFL may be used as touchup without limitation on the upper receiver and barrel assembly. This is to say that units which DO NOT fall under the category of Divisional Combat Units or rapid deployment type units may have up to 100 percent of the exterior surface of the upper receiver and barrel assembly protected with SFL if necessary.

12-13. BOLT CARRIER ASSEMBLY.

This task covers:

a. **Disassembly**
b. **Cleaning**
c. **Inspection/Repair**
d. **Lubrication**
e. **Reassembly**

INITIAL SETUP:

Tools
(ARMY) Small Arms Repairman Tool Kit (item 3, app B)
Bolt Carrier Key Tool (item 11, figure C-17)

References
TM 9-1005-319-10 (operator's manual)

Equipment Conditions
2-34 Bolt carrier assembly removed

General Safety Instructions
Bolt cam pin must be installed or rifle will blow up while firing the first round. If the bolt cam pin is not installed, injury to, or death of, personnel may result.

Do not interchange bolt assemblies from one rifle to another. Doing so may result in injury to, or death of, personnel.

a. DISASSEMBLY

CAUTION

Do not spread or close legs of firing pin retaining pin (1).

1. Remove firing pin retaining pin (1). Tilt key and bolt carrier assembly (2) and catch firing pin (3) as it drops out.

2. Rotate bolt cam pin (4) one quarter turn and lift straight up to remove.

3. Remove bolt assembly (5) from key and bolt carrier assembly (2).

NOTE
For disassembly of bolt assembly (5), see page 2-38.

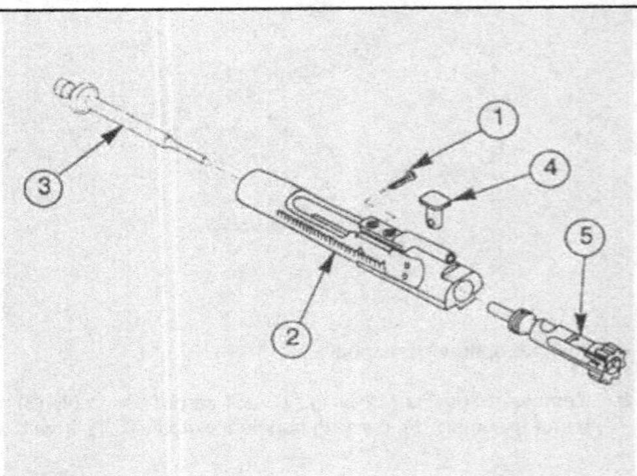

12-13. BOLT CARRIER ASSEMBLY (CONT).

b. CLEANING

Clean all items (operator's manual). Remove carbon deposits.

c. INSPECTION/REPAIR

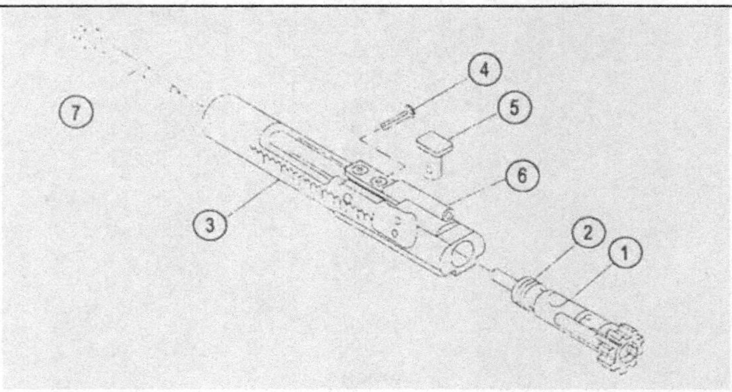

1. Inspect bolt assembly (1) as follows:

   a. Check to see that the bolt ring (2) gaps are staggered.

   b. Install bolt assembly (1) into key and bolt carrier assembly (3). Check for worn rings by holding the key and bolt carrier assembly (3) with the bolt assembly facing down. If bolt assembly falls out after firing pin retaining pin (4) and bolt cam pin (5) are removed, the rings are worn. Evacuate to support maintenance.

2.Inspect carrier key (6) for dents, distortion, or looseness. If dented or loose, evacuate to support maintenance.

3. Inspect firing pin retaining pin (4) and bolt cam pin (5) for cracks, damage, or excessive wear. Replace if unserviceable.

4. Inspect firing pin (7) for damage or If tip is chipped. Evacuate rifle to support maintenance if unserviceable

5. Inspect key and bolt carrier assembly (3) for damage or wear. If unserviceable, evacuate to support maintenance.

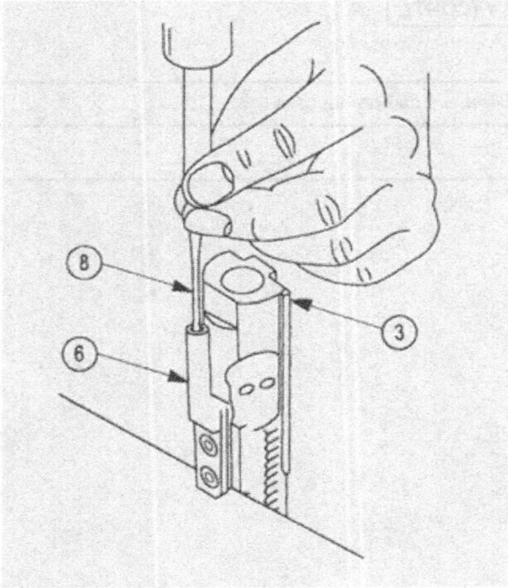

CAUTION

Extreme care must be exercised during the following procedure to assure that the striking force is not directed to the attaching screws and that the tube portion is not enlarged or flared beyond original requirement. Such enlargement would permit loss of gas pressure when the key and gas tube come together during functioning.

6. Repair small dents and/or distortions in carrier key (6) using bolt carrier key tool as follows:

a. Place the key and bolt carrier assembly (3) in a vertical position, supported In a manner that contact is made with the rear surface of the carrier key (6).

b. Insert the small end of the key tool (8) into the tube portion of the carrier key (6).

c. Strike the large end of the key tool (8) lightly with a 3 ounce, soft-brass hammer.

d. Repeat striking (gently) until carrier key (6) is reformed to original configuration.

e. If carrier key (6) cannot be reformed to original configuration, evacuate the weapon to support maintenance.

d. LUBRICATION

Lubricate all Items (p 2 33) (operator's manual)

e. REASSEMBLY

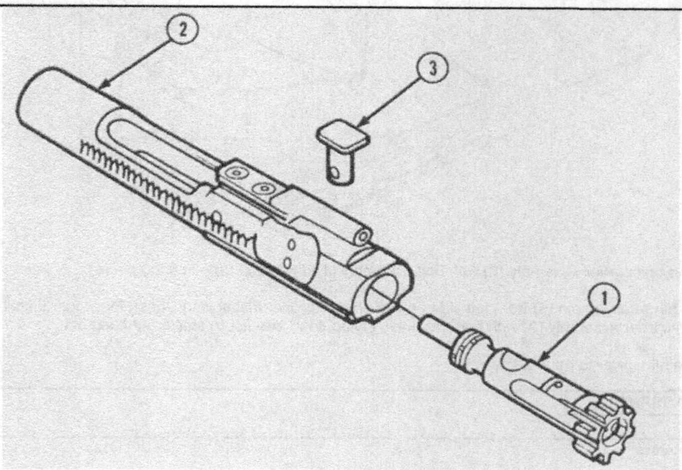

WARNING

Do not Interchange bolt assemblies from one rifle to another Doing so may result In Injury, or death of, personnel.

Bolt cam pin must be Installed or rifle will blow up while firing the first round. If the bolt cam pin is not Installed, Injury to, or death of, personnel may result

NOTE

Before installing bolt assembly, check to see that the bolt ring gaps are staggered to prevent loss of gas pressure

1. Install bolt assembly (1) In key and bolt carrier assembly (2)

2.Install bolt cam pin (3) and rotate one quarter turn

2-13. BOLT CARRIER ASSEMBLY (CONT).

e. REASSEMBLY (CONT)

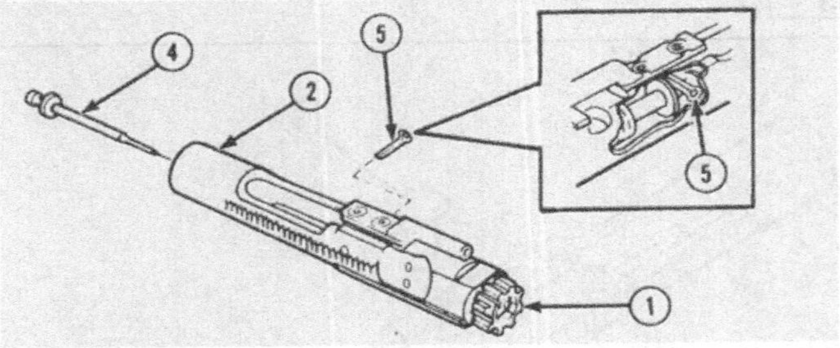

3. Hold key and bolt carrier assembly (2) with bolt assembly (1) down and drop in firing pin (4).

4. Install firing pin retaining pin (5) from left side only to ensure proper installation Check for proper installation by holding key and bolt carrier assembly (2) with bolt assembly (1) up, then attempt to shake out firing pin.

5. Reassemble rifle, refer to page 2-68.

2-14. BOLT ASSEMBLY.

This task covers:

a. **Removal**
b. **Cleaning**
c. **Inspection**
d. **Lubrication**
e. **Reassembly**

INITIAL SETUP:

Tools
(ARMY) Small Arms Repairman Tool Kit (item 3, app B)

Equipment Conditions
2-35 Bolt assembly removed

General Safety Instructions
Do not Interchange bolt assemblies from one rifle to another. Doing so may result in injury to, or death of, personnel.

To avoid injury to your eyes, use care when removing and installing spring-loaded parts.

a. DISASSEMBLY

NOTE

Do not separate cartridge extractor and extractor spring assembly unless replacement of either or both is required

Do not remove the rubber Insert from the extractor spring assembly

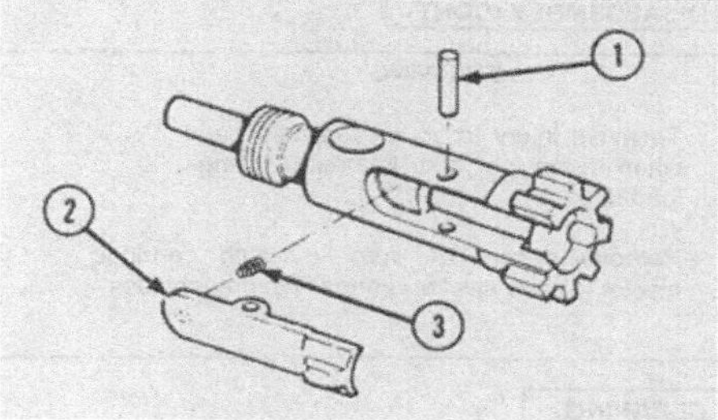

1. Push out extractor pin (1) and remove cartridge extractor (2) and extractor spring assembly (3) as a unit

2. If required, twist extractor spring assembly (3) counterclockwise to remove from cartridge extractor (2).

CAUTION

Be sure to use vise law protective caps

3. Hold bolt body In vise and remove spring pin (4) using 1 16 Inch punch and hammer

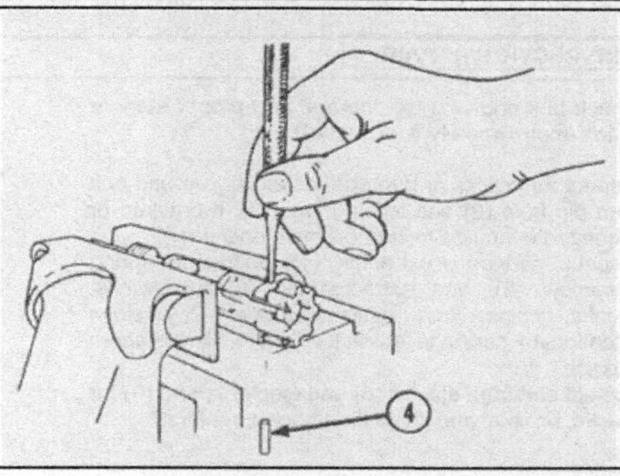

2-14. BOLT ASSEMBLY (CONT).

a. DISASSEMBLY (CONT)

WARNING

To avoid injury to your eyes, use care when removing and installing spring-loaded parts.

4. Remove punch, be sure to catch cartridge elector (5) and ejector spring (6) to prevent loss

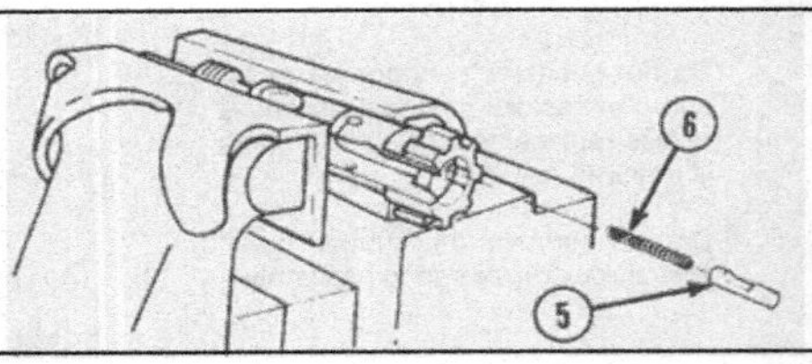

b. CLEANING

CAUTION

Do not distort extractor spring assembly during cleaning.

Clean all Items (operator's manual). Remove carbon deposits.

c. INSPECTION/REPAIR

1. Check bolt rings (1) for damage and proper stagger (gaps approximately 3 turn apart).

2. Inspect for cracks or damage, especially around bolt cam pin hole (2) and locking lugs (3). If cracked or damaged evacuate to support maintenance.
3. Inspect cartridge extractor (4), extractor spring assembly (5), and extractor pin (6) for cracks, breaks, chips, and other damage. Pay close attention to cartridge extractor lip (7). If damaged, replace.
4. Inspect cartridge ejector (8) and ejector spring (9) for cracks, breaks, and chips If damaged, replace.

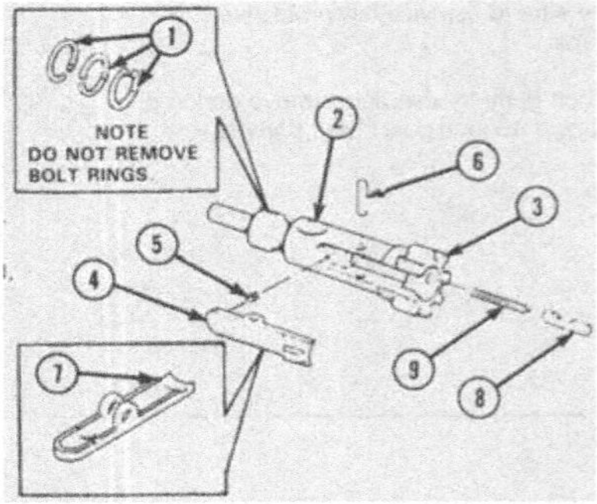

d. LUBRICATION

Lightly lubricate all items (p 2 33) (operator's manual)

e. REASSEMBLY

WARNING

To avoid Injury to your eyes, use care when removing and Installing spring loaded parts.

Do not Interchange bolt assemblies from one rifle to another Doing so may result In injury to, or death of, personnel.

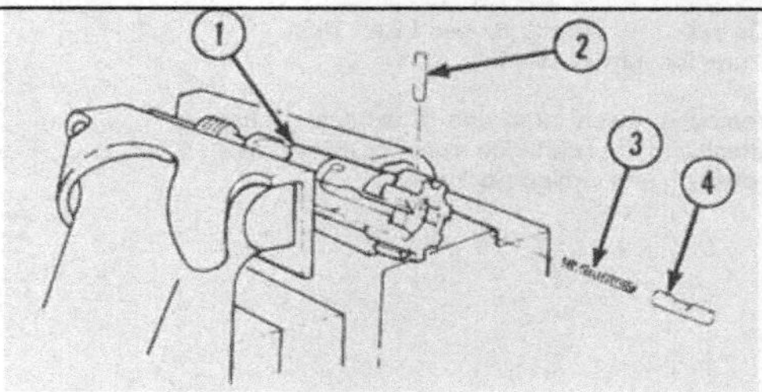

CAUTION

Be sure to use vise jaw protective caps

1. Place bolt body (1) in a vise and start spring pin (2) In hole.
2. Install ejector spring (3) and cartridge ejector (4). Align groove on cartridge ejector (4) so that spring pin 12) can be Installed
3. Compress and hold ejector spring and cartridge ejector In place with a 3 8 inch punch.
4. Using hammer and 1 16 Inch punch, complete Installation of spring pin (2) so that the ends are flush with the out side of bolt body (1).

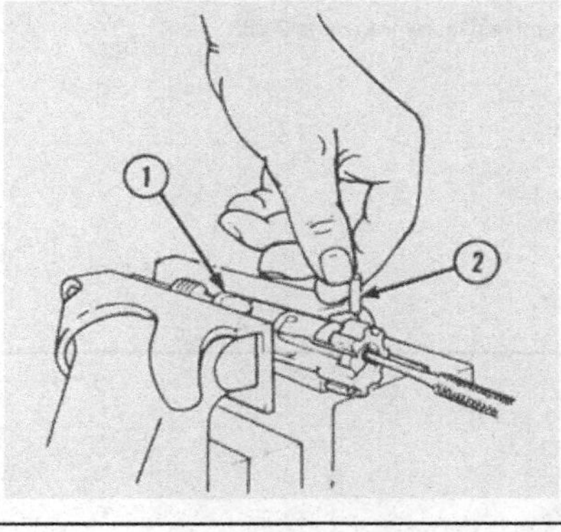

2-14. BOLT ASSEMBLY (CONT).

e. REASSEMBLY (CONT)

NOTE

Do not disassemble rubber Insert from extractor spring assembly.

5. If removed, Insert large end of extractor spring assembly (5) into cartridge extractor (6) and seat by pushing and turning clockwise.

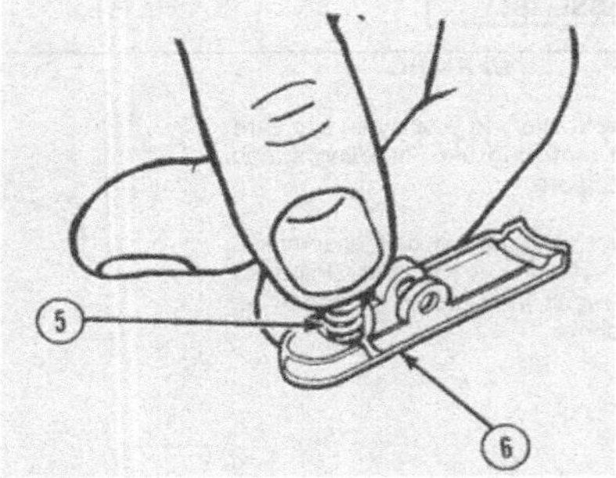

6 Position cartridge extractor (6) and ex tractor spring assembly (5) on bolt body (1).

7 Compress extractor spring assembly (5) and cartridge extractor (6) to align holes.

8. Install extractor pin (7) by hand.

9. Reassemble rifle, refer to page 2-68.

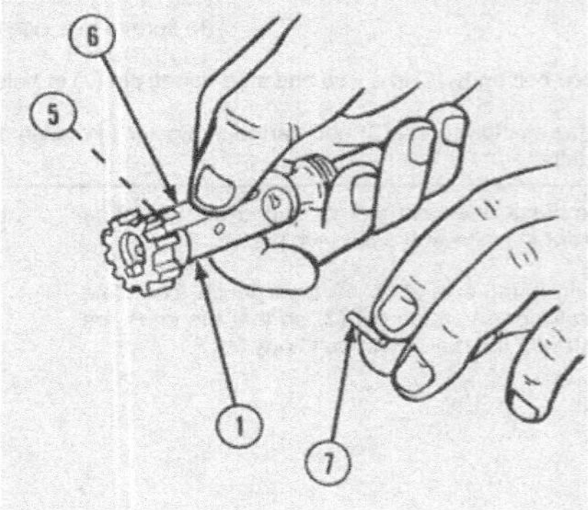

2-15. CHARGING HANDLE ASSEMBLY.

This task covers:
a. Disassembly **c. Inspection/Repair** **e. Reassembly**
b. Cleaning **d. Lubrication**

INITIAL SETUP:

Tools
(ARMY) Small Arms Repairman Tool Kit (item 3, app B)

Equipment Conditions
2-34 Charging handle assembly removed

References
TM 9-1005-319-10 (operator's manual)

a. DISASSEMBLY

WARNING

To avoid injury to eyes, use care when removing and installing spring-loaded parts.

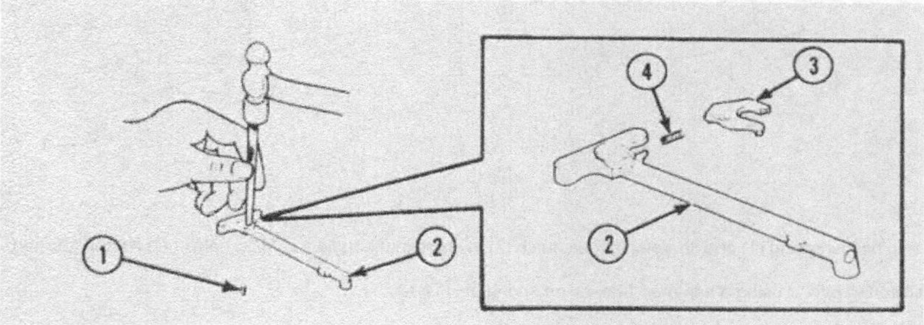

1. Remove spring pin (1) from charging handle (2) using a hammer and 1/16 inch punch.
2. As punch is withdrawn, catch charging handle latch (3) and helical spring (4) to prevent loss.

2-15. CHARGING HANDLE ASSEMBLY (CONT).

b. CLEANING

Clean all Items (operator's manual) Remove carbon deposits.

c. INSPECTION/REPAIR

Inspect all Items for breaks, cracks, or damage Replace all unserviceable Items

d. LUBRICATION

Lightly lubricate all Items (p 2-33) (operator's manual)

e. REASSEMBLY

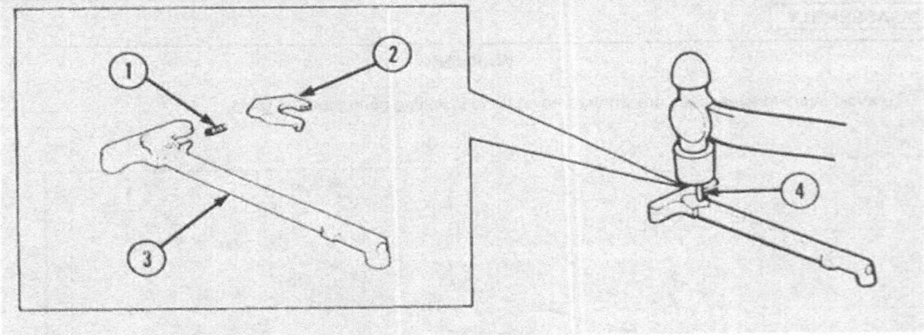

1. Position helical spring I1) and charging handle latch (2) in chdrgiin(J handle 13) Align holes and hold In position
2. Install spring pin (4) using a hammer Make sure spring pin is flush
3. Reassemble rifle, refer to page 2 68

2-16. UPPER RECEIVER AND BARREL ASSEMBLY, RIFLE BARREL ASSEMBLY, AND UPPER RECEIVER ASSEMBLY.

This task covers:

- **a.** Disassembly
- **b.** Cleaning
- **c.** Inspection
- **d.** Repair
- **e.** Lubrication
- **f.** Reassembly

INITIAL SETUP

Tools
(ARMY) Small Arms Repairman Tool Kit (item 3, app B)
Front sight post removal and installation tool (fig. E-2, app E)
Front sight detent depressor (fig. E-1, app E)

Materials/Parts
Lubricants (app D)

References
TM 9-1005-319-10 (operator's manual)

Equipment Conditions
2-34 Upper receiver and barrel assembly removed from lower receiver

General Safety Instructions
To avoid injury to your eyes, use care when removing and installing spring-loaded parts.

a. DISASSEMBLY

CAUTION

Do not use a screwdriver or any other tool when removing the handguard assemblies. Doing so may damage the handguard assemblies and/or slip ring. Do not remove heat shield for any reason. Doing so will damage the heat shield and the handguard assemblies will have to be replaced.

NOTE

Refer to operator's manual for "buddy system" procedure on removing handguard assemblies.

2-16. UPPER RECEIVER AND BARREL ASSEMBLY, RIFLE BARREL ASSEMBLY, AND UPPER RECEIVER ASSEMBLY (CONT).

a. DISASSEMBLY (CONT)

1. Push down on handguard slip ring (1) and lift upper handguard assembly (2) up and out.

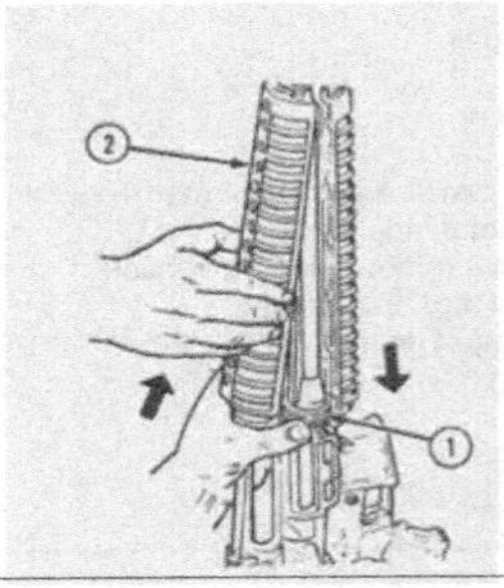

2. Push down on handguard slip ring (1) and lift the lower handguard assembly (3) up and out.

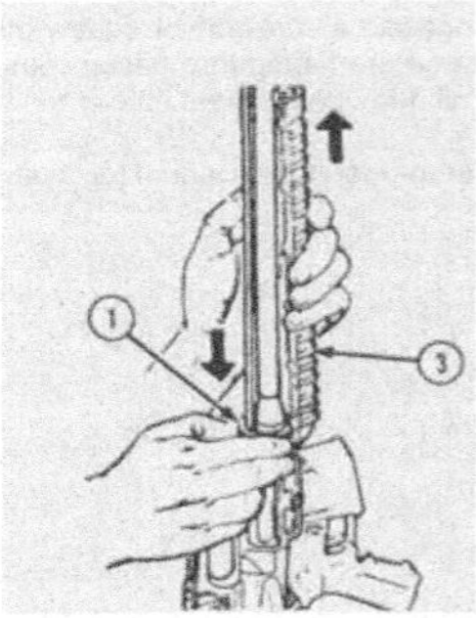

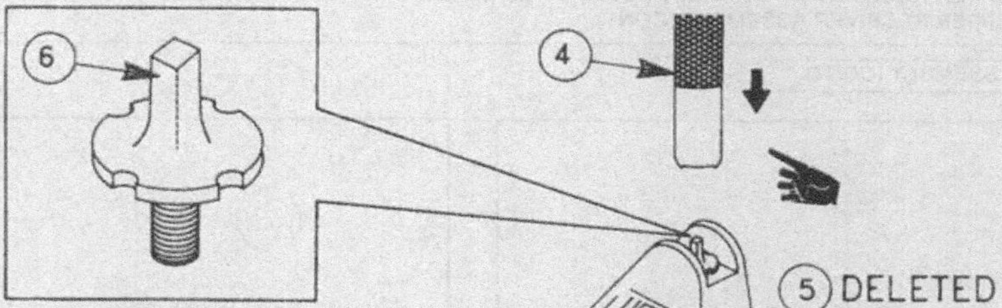

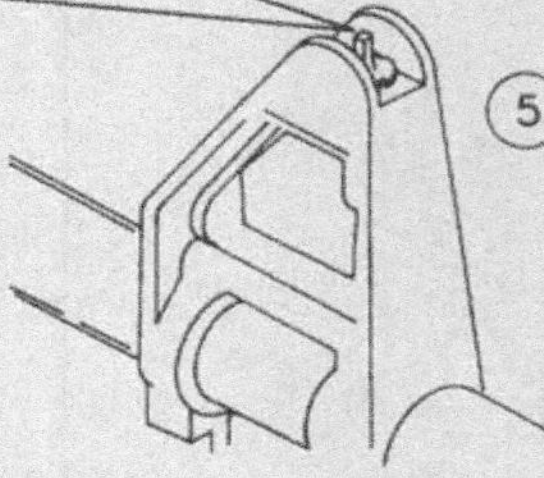

WARNING

To avoid injury to your eyes, use care when removing and installing spring-loaded parts.

NOTE

Remove front sight posts only if parts are damaged.

3. Using a dummy cartridge or a font sight post removal and installation tool (4).removed front sight post (6) by turning counter clockwise.

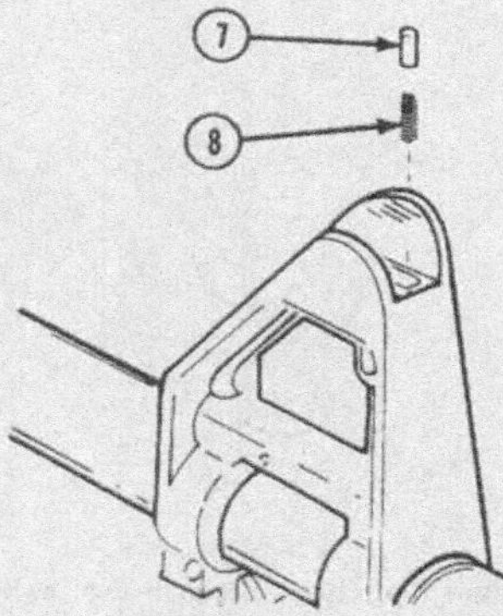

4. Catch front sight detent(7) and helical spring (8) to prevent loss.

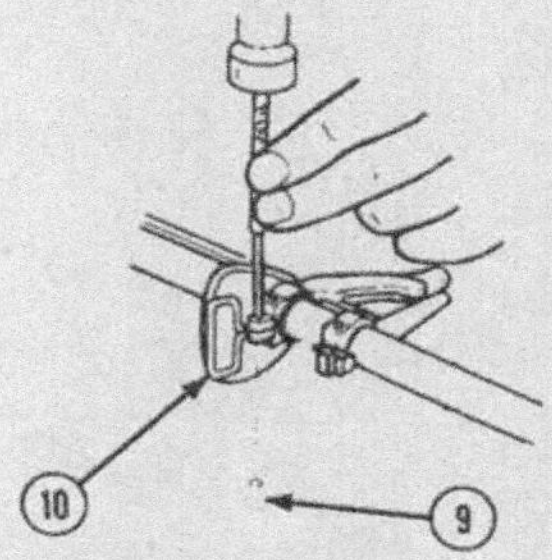

NOTE

RIFLE ONLY

Disassemble small sling swivel only if repair is necessary.

5. If necessary, knock out tubular rivet (9) with a hammer and punch and remove small sling swivel (10). Discard tubular rivet (9).

2-16. UPPER RECEIVE AND BARREL ASSEMBLY, RIFLE BARREL ASSEMBLY, AND UPPER RECEIVER ASSEMBLY (CONT).

a. DISASSEMBLY (CONT)

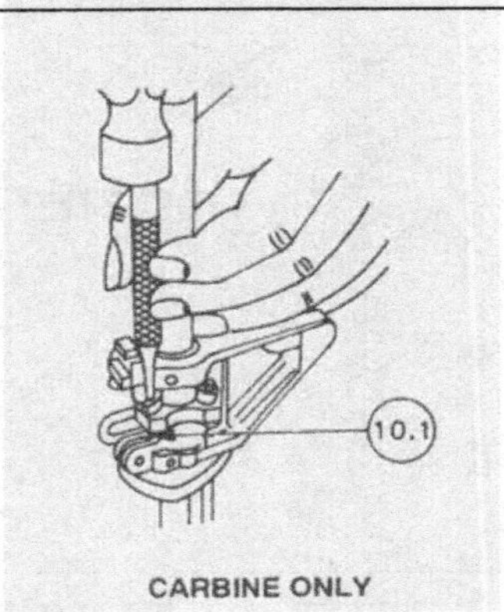

5A. Remove two spring pins (10 1)

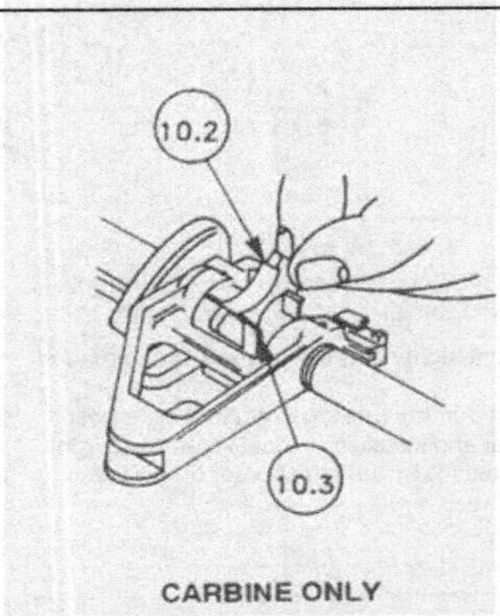

5B. Lift swivel locking bar (10 2) up and out of swivel mount (10 3)

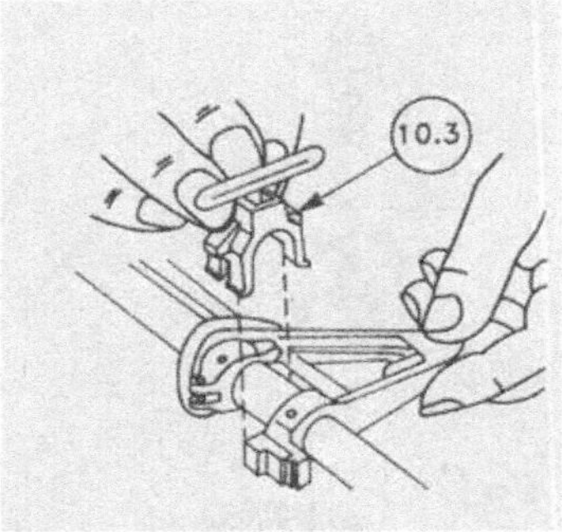

5C. Remove swivel mount (10.3).

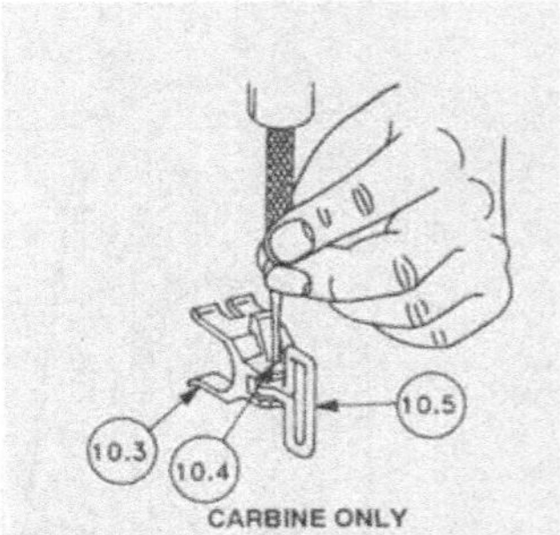

5D. Remove rivet (10 4) Separate front swivel (10 5) from swivel blank) mount (10 3)

2-16. UPPER RECEIVER AND BARREL ASSEMBLY, RIFLE BARREL ASSEMBLY, AND UPPER RECEIVER ASSEMBLY (CONT).

a. DISASSEMBLY (CONT)

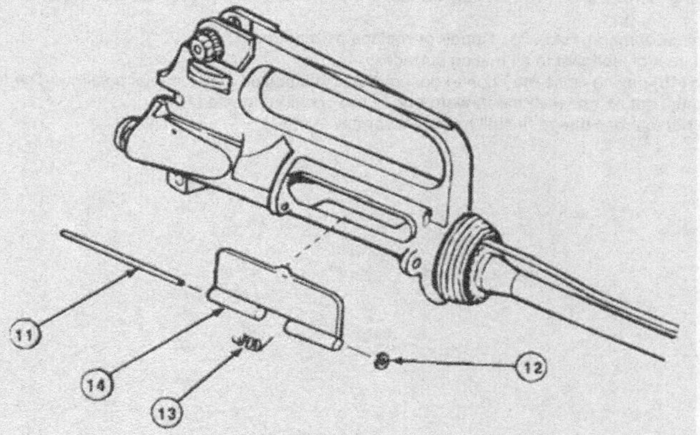

NOTE

Do not disassemble further unless repair is necessary. Headless grooved pin (11) may bind may bind against the forward assist housing and require some additional force to remove.

6. Remove retaining ring (12) and slide headles grooved pin (11) out to the rear.
7. Catch cover spring (13) and ejection port cover (14) to prevent loss as headless grooved pin is with-drawn.

M4/M4A1 ONLY

NOTE

Do not remove the carrying handle assembly nut (2 each) or clamp for normal maintenance. If carrying handle assembly nut or clamp are missing or the clamp is damaged to the point of being non-functional they may be replaced.

7.1. Disassemble as follows only if required to replace damaged or missing carrying handle assembly nut or clamp.

a. Observe the position of the clamp (15) in relation to the carrying handle (16).
b. Using care not to damage the surface finish, remove carrying handle assembly nut(s) (17) and clamp as required.
c. Inspect mating surfaces of the carrying handle (16) and the clamp (15) for damage. If damaged replace the damaged part.
d. Inspect for other damage, rust, etc. Repair or replace as required.
e. Apply a light coat of lubricant to all mating surfaces.
f. Install clamp (15) paying close attention to ensure that it is installed in its original position. The forward end of the clamp (15) must not extend past the forward end of the carrying handle (16).
g. Taking care not to cross thread, install handle assembly nut (17).

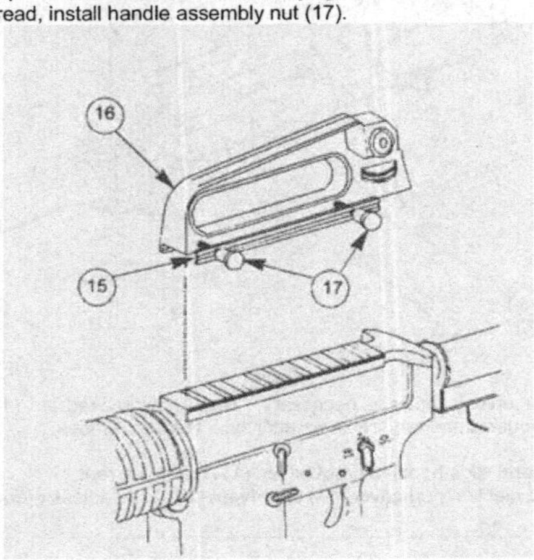

b. CLEANING

Clean all items (operator's manual).

c. INSPECTION

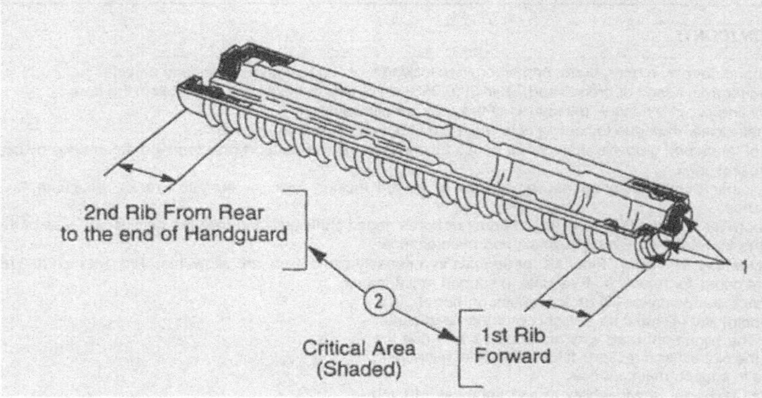

1. Inspect handguard assemblies for breaks separation, and cracks using the following guidelines:
   - **a.** Breaks and separations of material which prevent proper retention or interfere with functioning of the rifle will be cause for handguard assembly rejection and replacement.
   - **b.** M16A2 handguard assemblies may have up to two of the three front retaining tabs missing (1). M4/M4A1 may not have any front retaining tabs (1) missing. If all front retaining tabs for the M1 6A2 are missing, or any of the tabs for the M4/M4A1 are missing, handguard assemblies must be replaced.
   - **c.** Cracks up to one inch in length are acceptable provided they do not extend into the retaining flange (CRITICAL AREA) (2).
   - **d.** Replace severely cracked or damaged handguard assemblies. Handguard assemblies which have a heat shield which is loose enough to rattle when installed on the rifle must be replaced.

2. Inspect front sight assembly for chips, breaks, and cracks. Evacuate to support maintenance if broken, cracked, or bent.
3. Inspect front sight area for evidence of gas leakage around gas tube. Evacuate to support maintenance if short recoil results from gas leakage.
4. Inspect front sight post, front sight detent, and helical spring for damage. If damaged, replace.

2-16. UPPER RECEIVER AND BARREL ASSEMBLY, RIFLE BARREL ASSEMBLY, AND UPPER RECEIVER ASSEMBLY (CONT).

c. INSPECTION (CONT)

5. Inspect barrel for pits in bore, burrs, broken or worn locking lugs, and surface cracks and defects.
   a. Pits no wider than a land or groove and 3/8 inch (0.953 cm) or less in length are allowable in the bore.
   b. Uniformly fine pits in a densely pitted area of the bore are allowable.
   c. Lands that appear dark due to coating of gliding metal from projectiles are allowable.
   d. Striping of lands and grooves shall not be cause for rejection unless support maintenance determines by use of the barrel erosion gage.
   e. For pits other than mentioned above, broken or burred locking lugs, or surface cracks, evacuate to support maintenance.
6. Inspect bore for ringing. Definitely ringed bores or bores ringed sufficiently to bulge the outside surfaces of the barrel are causes for rejection. Evacuate to support maintenance.
7. Inspect chamber for pitting. Fine pits, or fine pits in a densely pitted area, are allowable. Pits 1/8 inch (0.318 cm) in length are cause for rejection. Evacuate to support maintenance.
8. Hand check compensator (3) for looseness on barrel. The third (middle) slot (4) must be straight up at top dead center (TDC). The alignment may vary as much as one half the width of the slot either direction. If loose or out of alignment, evacuate to support maintenance.
9. Inspect all items for serviceability in and tightness of latch assembly on ejection port cover. If items are damaged or nonfunctional, they are unserviceable.

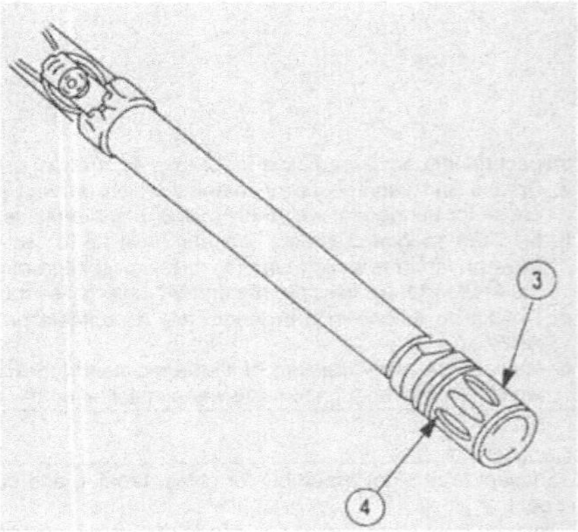

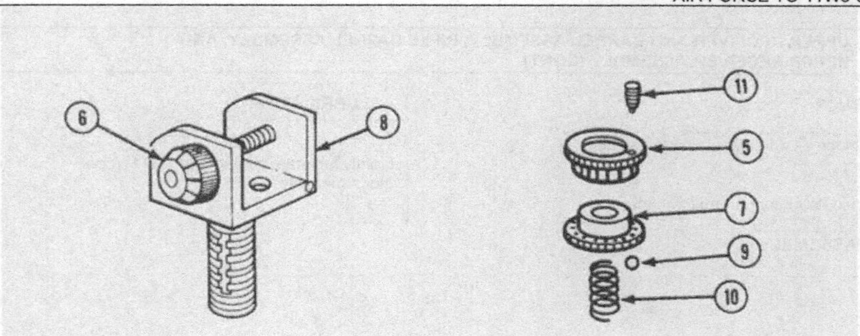

10. Rotate and test elevation index (5) and windage knob (6) for ease of functioning and legibility of markings.

11. Inspect elevation knob zero as follows:

a. Rotate elevation knob (7) counterclockwise until the rear sight assembly is all the way down. If a whole click is not felt as the rear sight assembly stops, the rear sight assembly has bottomed out and will not pivot freely.

b. Position elevation knob back slightly to Its last whole click so the rear sight assembly base (8) is under tension of the ball bearing (9) and helical spring (10). The 300 meter mark should align with the mark on the receiver.

c. If the 300 meter mark is not aligned with the mark on receiver, slip the range scale in the following manner:

(1) Position the 300 meter mark with the mark on the receiver.

(2) Insert a 1/16 inch allen wrench through the access hole of the rear sight assembly base and into the Index screw (11).

(3) Loosen the index screw three turns and leave the wrench in place.

(4) Rotate lower portion of elevation knob counterclockwise until it stops (range scale should not have moved). Elevation knob should be positioned on its last whole click.

(5) Tighten index screw and remove wrench.

(6) Check for proper setting.

2-16. UPPER RECEIVER AND BARREL ASSEMBLY, RIFLE BARREL ASSEMBLY, AND UPPER RECEIVER ASSEMBLY (CONT).

d. REPAIR

1. Replace all authorized unserviceable parts

2. Evacuate to support maintenance.

e. LUBRICATION

Lightly lubricate all Items (p 2 311 (oper ator's manual)

f. REASSEMBLY

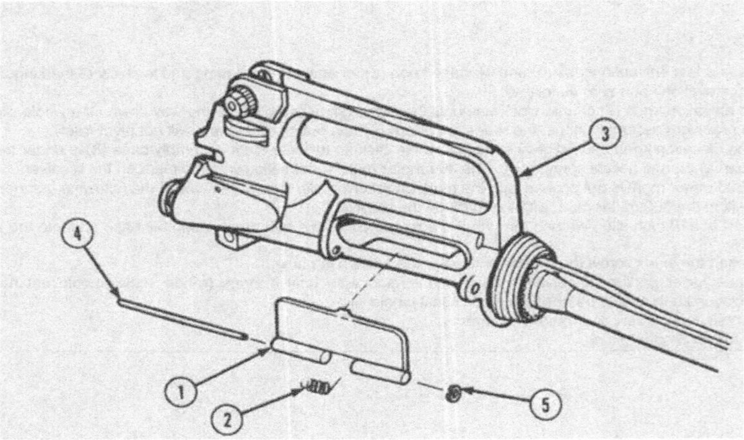

1. If previously disassembled, position election port cover 11) and helical spring (2) on Lip per receiver (3) with short leg of helical spring to the rear on Inside of ejection port cover

NOTE

Long legs of helical spring must be positioned and pretensioned before the headless grooved pin is installed

2. Hold helical spring (2) short leg In this position and turn long leg one half turn (180 degrees) with fingers of right hand

3. Position long leg of helical spring (21 against election port cover (1) Hold helical spring and election port cover in this position and Install headless grooved pin (4) Check for proper spring tension during installation of retaining ring (5)

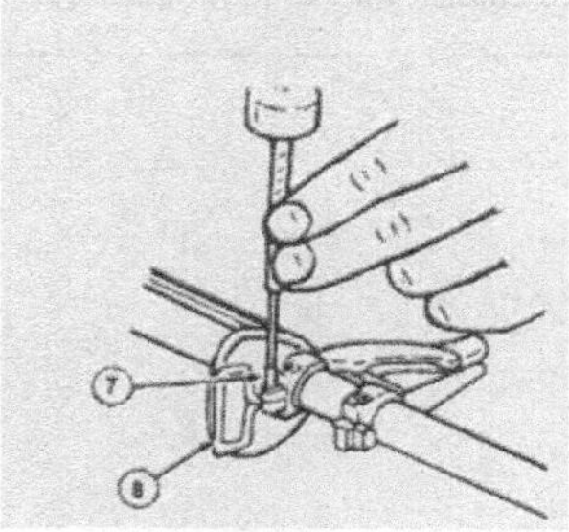

RIFLE ONLY

4. If previously disassembled, position small sling swivel (6) and install new tubular rivet (7) using center punch and hammer to spread and flare the hollow head of the tubular rivet.

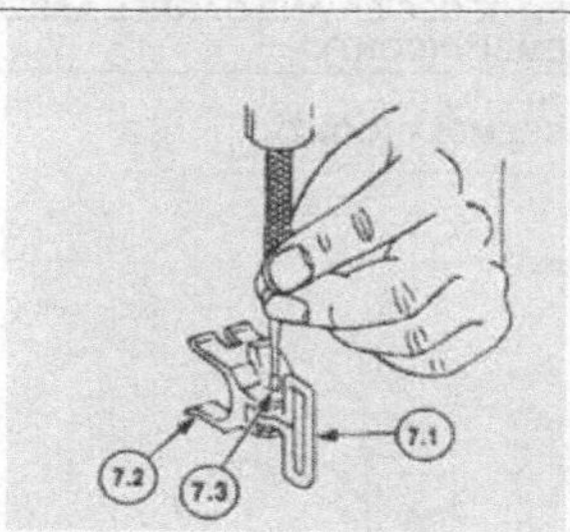

CARBINE ONLY

4A. Install front swivel (7.1) to swivel mount (7.2) with new rivet (7.3) using center punch and hammer to spread and flare the hollow head of the tubular rivet.

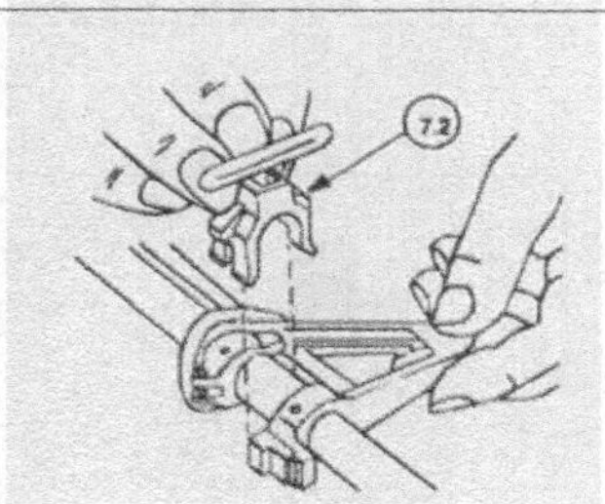

CARBINE ONLY

NOTE

It is recommended that the swivel mount be placed on the left side of the carbine for right handed operators and on the right side for left handed operators. This will keep the sling out of the way when the carbine is used.

4B. Install swivel mount (7.2).

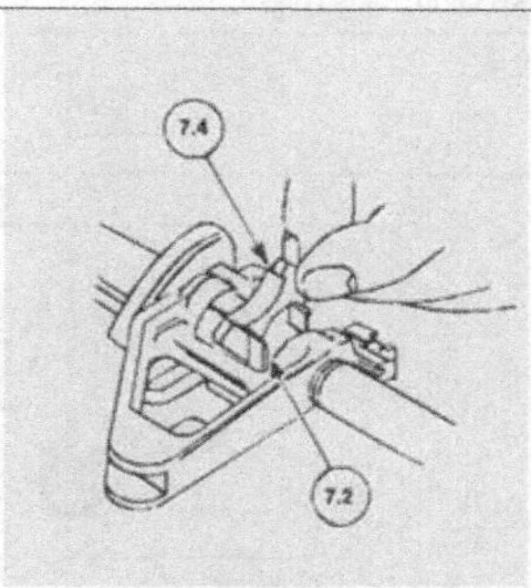

CARBINE ONLY

4C. Place swivel locking bar (7.4) in swivel mount (7.2).

2-16. UPPER RECEIVER AND BARREL ASSEMBLY, RIFLE BARREL ASSEMBLY, AND UPPER RECEIVER ASSEMBLY (CONT).

f. REASSEMBLY (CONT)

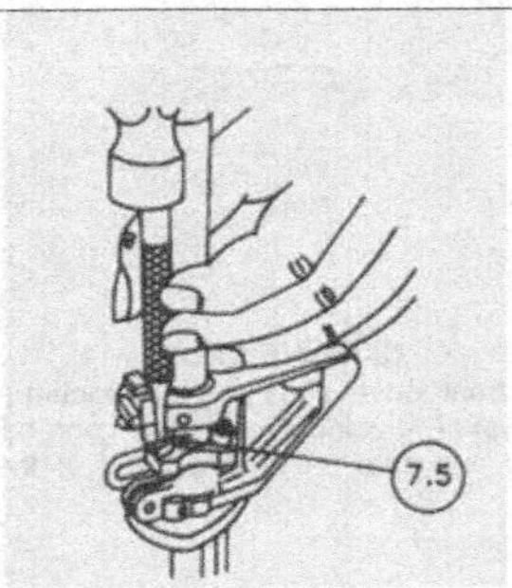

CARBINE ONLY

4D. Install two spring pins (7 5)

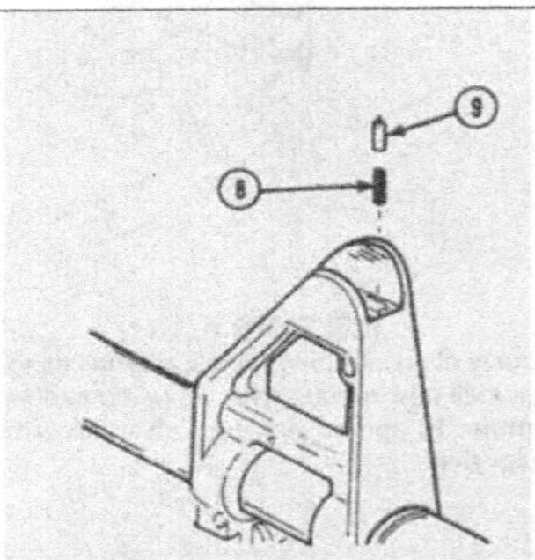

5. Position helical spring (8) and front sight detent (9).

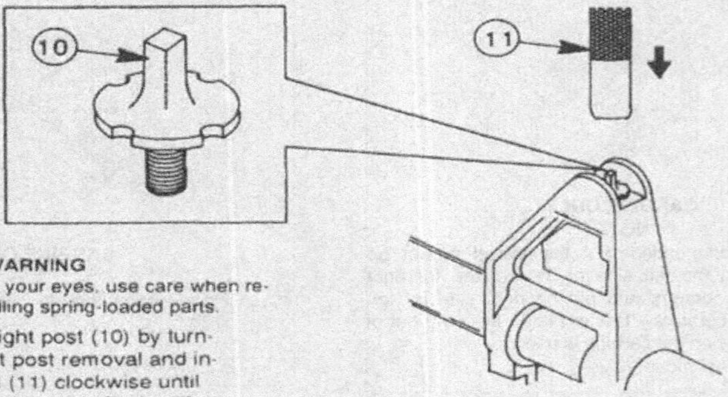

WARNING

To avoid injury to your eyes, use care when removing and installing spring-loaded parts.

6. Install front sight post (10) by turning front sight post removal and installation tool (11) clockwise until front sight post base is flush with or slightly below front sight frame.

2-16. UPPER RECEIVER AND BARREL ASSEMBLY, RIFLE BARREL ASSEMBLY, AND UPPER RECEIVER ASSEMBLY (CONT).

t. REASSEMBLY (CONT)

7. Mechanical Zero Procedures (A.F. Only).
   a. Mark a piece of plastic card stock or rigid paper with lines from 1 to 5mm in 1 mm increments Set the card on the front sight frame and check the height of the top of the front sight post.

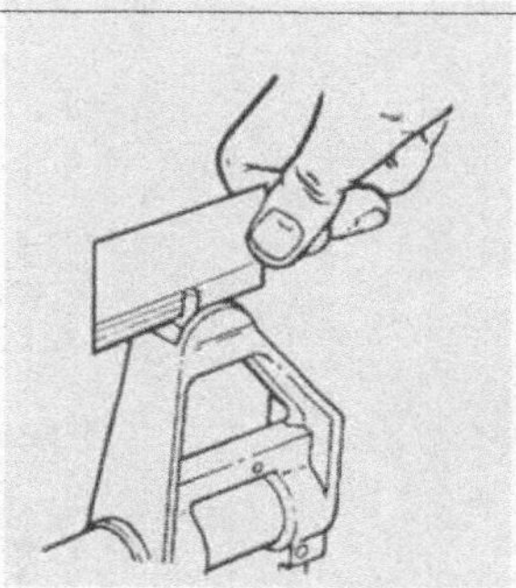

   b. Using a dummy round or front sight post tool adjust the front sight so the top of the front sight post is 5mm above the machined surfaces of the front sight frame.

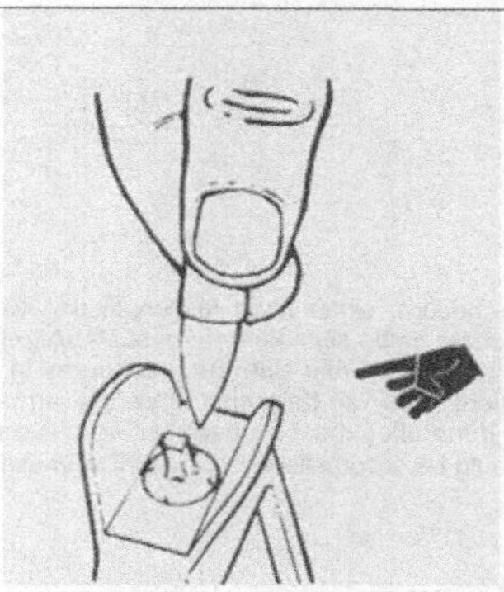

c. Visually check the front sight post top height by using the marked plastic or paper card. Card must set level on the machined surfaces of the front sight frame to obtain an accurate reading

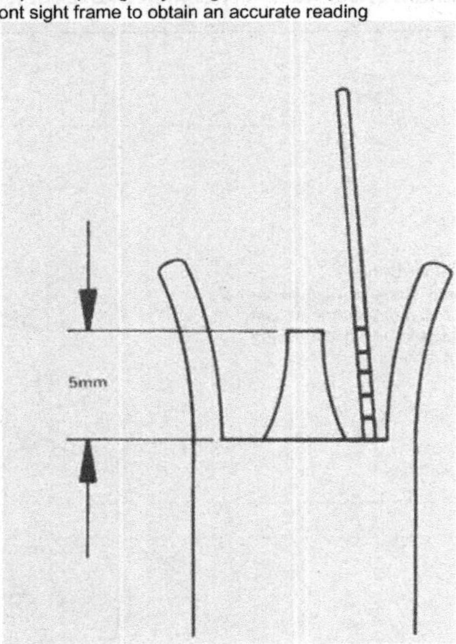

NOTE

This procedure, when used In conjunction with rear sight mechanical zero adjustment. will give an approximate battle sight zero to most M16A2 rifles The above steps can also be used before firing a new or newly assigned rifle Use the procedures to check rifles stored In preferred packaging during routine Inspections This will help ensure people armed with the rifles will stand a better chance of hitting an enemy If the rifles must be used before a live fire zero can be made Whenever possible, zeroing of the rifle should be accomplished using ball ammunition on a 25 meter zeroing target using the "L" aperture.

2-16. UPPER RECEIVER AND BARREL ASSEMBLY, RIFLE BARREL ASSEMBLY, AND UPPER RECEIVER ASSEMBLY (CONT).

f. REASSEMBLY (CONT)

NOTE
Refer to operator's manual for 'buddy system" procedure on installing hand guard assemblies

8. Install top of upper handguard assembly (12) In tube cap (13) while pushing down on handguard slip ring (14). Push bottom of upper handguard assembly (12) in place and release handguard slip ring (14) to lock handguard assembly In place

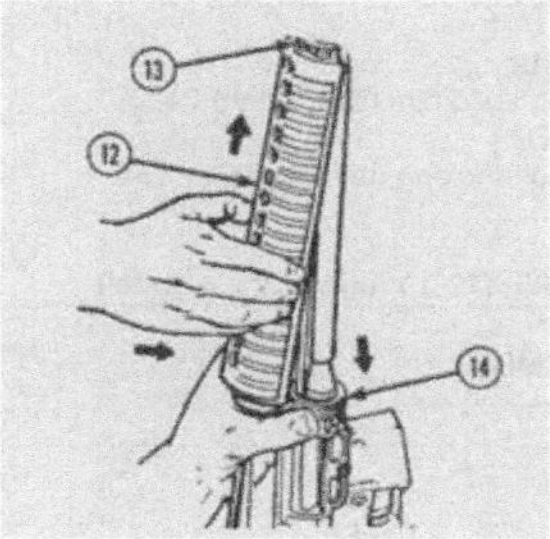

9. Install top of lower handguard assembly 1 5) In tube cap (13) while pushing down on handguard slip ring (14) Push bottom of lower handguard assembly (15) In place and release handguard slip ring {14) to lock both handguard assemblies In place

10. Reassemble rifle, refer to page 2-68

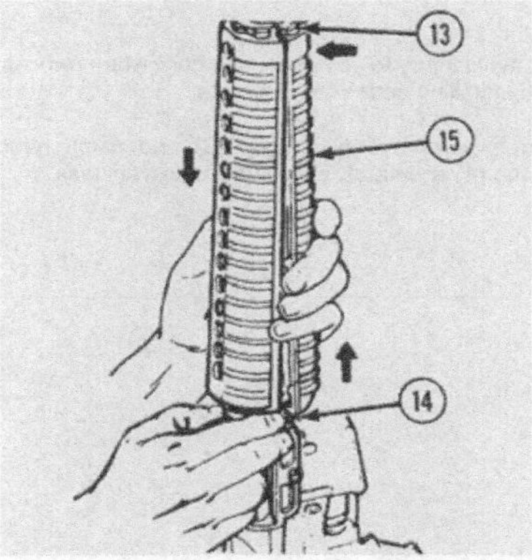

2-17. LOWER RECEIVER AND BUTTSTOCK ASSEMBLY.

This task covers:

a. Disassembly
b. Cleaning
c. Inspection
d. Repair
e. Lubrication
f. Reassembly

INITIAL SETUP

Tools
(ARMY) Small Arms Repairman Tol Kit (item 3, app B)
Pivot Pin Removal Tool (fig. E-3, app E)
Pivot Pin Installation Tool (fig. E-4, app E)

Materials/Parts
Lubricant, solid film (SFL) (item 21, app D)
Screw, self-locking (item 6, p C-1 1)

References
TM 9-1005-319-10 (operator's manual)

Equipment Conditions
2-34 Lower receiver and buttstock assembly removed

General Safety Instructions
To avoid injury to your eyes, use care when removing and installing spring-loaded parts.
When using solid film lubricant or dichloromethane, be sure the area is well ventilated.

a. DISASSEMBLY

1. Remove screw (1) and lockwasher (2).

WARNING
To avoid injury to you eyes, use care when removing and installing spring-loaded parts.

2. Carefully remove pistol grip (3) and catch helical spring (4) and safety detent (5) to prevent loss.

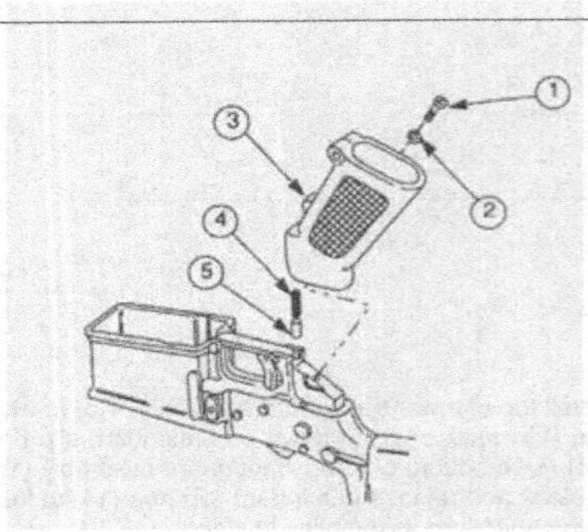

2-17. LOWER RECEIVER AND BUTTSTOCK ASSEMBLY (CONT).

a. DISASSEMBLY (CONT)

RIFLE ONLY
NOTE

If the self-locking screw is removed, it must be discarded and replaced with a new one

3. Remove self-locking screw (6).

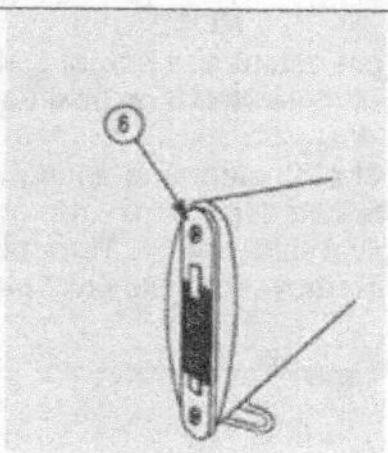

RIFLE ONLY
WARNING

To avoid injury to your eyes, use care when removing and installing spring-loaded parts.

4. Remove buttstock assembly (7) carefully and catch helical spring (8), takedown pin detent (9), takedown pin (10), and stepped spacer (11) to prevent loss.

NOTE

If detent (9) will not come out, use a wire to push it out.

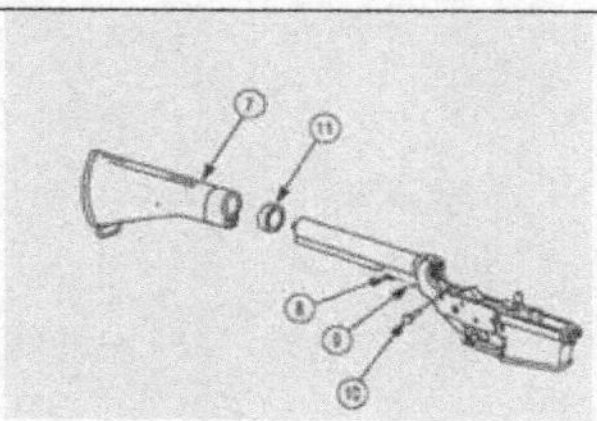

CARBINE ONLY

4A. Extend buttstock assembly (11.1).

4B Grasp the lock release lever (11.2) in the area of the retaining nut (11.3), pull down-ward, and slide buttstock to the rear to separate the buttstock assembly from the lower receiver extension

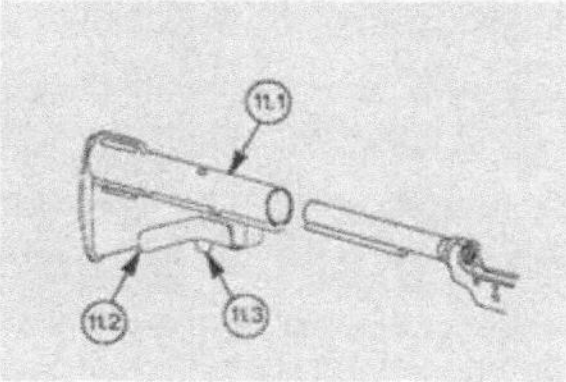

WARNING

To avoid injury to your eyes, use care when removing and installing spring-loaded parts

NOTE

Catch pivot pin detent and helical spring as pivot pin is removed (see step 6 on next page)

5. Depress pivot pin detent and spring and remove pin or Insert fabricated pivot pin removal tool (12) to compress pivot pin detent Turn pivot pin (13) a quarter turn Remove tool and pivot pin

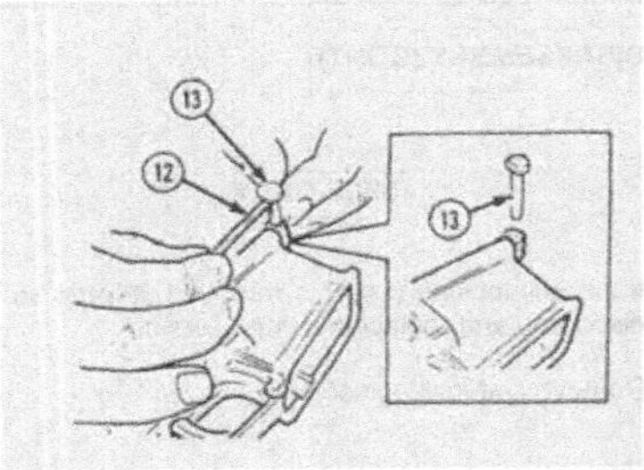

6. Be sure to hold cupped hand in front of pivot pin detent (14) and helical spring (15) to prevent loss of pivot pin detent and helical spring.

NOTE

If helical spring (15) will not come out, use a wire to pull it out.

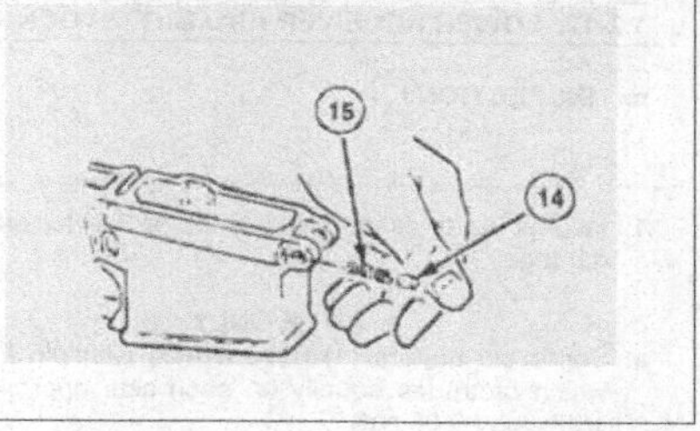

NOTE

Make sure the hammer is cocked and the selector lever is not set on BURST or AUTOMATIC before removing the buffer assembly.

7. Press buffer assembly (16) in about 1/4 inch (0.635 cm). Depress buffer retainer (17) and release buffer assembly (16) and action spring (18). Remove buffer assembly(16) and action spring (18) from receiver while depressing buffer retainer (17).

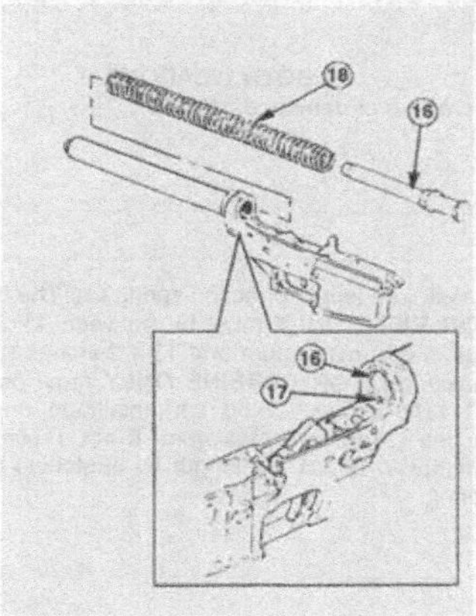

b. CLEANING

Clean all items (operator's manual). Remove carbon deposits

1 2-17. LOWER RECEIVER AND BUTTSTOCK ASSEMBLY (CONT).

c. INSPECTION I

1. Inspect buffer assembly (1, 2. 3, or 3 1) for cracks or damage

RIFLE ONLY

a. Some old buffers (1) have a hole with pin installed which protrudes equally on each side approximately 1/32 inch (O 08 cm)

b. Some buffers (2) have a hole In the housing but no pin

c. New buffers (3) do not have a hole In buffer body or a pin.

BOTH WEAPONS

2. If cracked or damaged, replace.

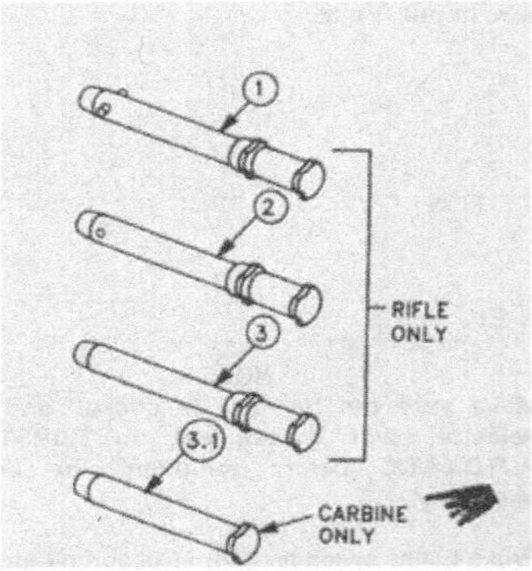

3. Check free length of action spring (4). The free length **FOR RIFLE ONLY** must be between 11 3/4 inches (29.85 cm) mini- mum and 13 1/2 inches (34 29 cm) maximum, **FOR CARBINE ONLY** must be between 10 1'/16 inches (25,56 cm) minimum and 11 114 Inches (28.58 cm) maximum, If not, replace Do not attempt to adjust the length by stretching the action spring

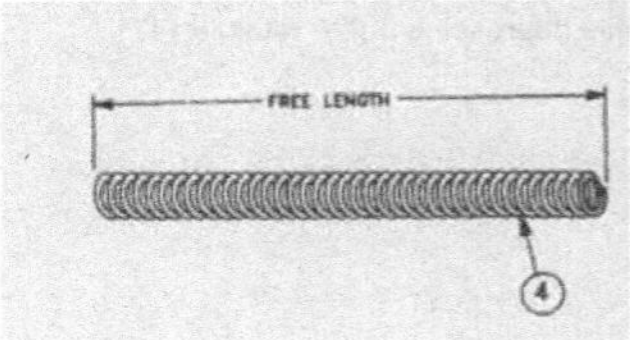

4. Inspect lower receiver (5) (without further disassembly) for legibility of serial number **ARMY ONLY**: If the serial number is hard to read, evacuate to direct support maintenance **AIR FORCE ONLY**: If the serial number is hard to read, evacuate to depot maintenance

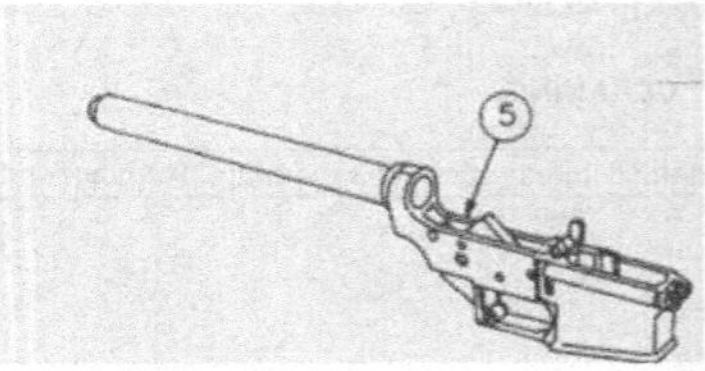

NOTE

AIR FORCE ONLY: Only depot maintenance is authorized to restamp the serial number.

5. Inspect for missing or damaged parts. Inspect finish of lower receiver for shiny spots. Touch up with solid film lubricant as required (p 2-34).

NOTE

If a M16A2 rifle or M4/M4A1 carbine lower receiver is missing one third or more of its exterior protective finish, resulting in an unprotected, light reflecting surface, it is candidate for overhaul. This missing finish will be considered a shortcoming. This shortcoming requires action to obtain a replacement weapon. Once a replacement has been received, evacuate the original weapon to depot maintenance for overhaul.

d. REPAIR

Replace all authorized unserviceable parts. If repair is not authorized at this level, evacuate to support maintenance.

e. LUBRICATION

Lightly lubricate all metal components (p 2-33) (operator's manual).

2-17. LOWER RECEIVER AND BUTTSTOCK ASSEMBLY (CONT).

f. REASSEMBLY

WARNING

To avoid Injury to your eyes, use care when removing and Installing spring loaded parts.

NOTE

Make sure the hammer Is cocked and the selector lever is not set on BURST before Installing the buffer assembly.

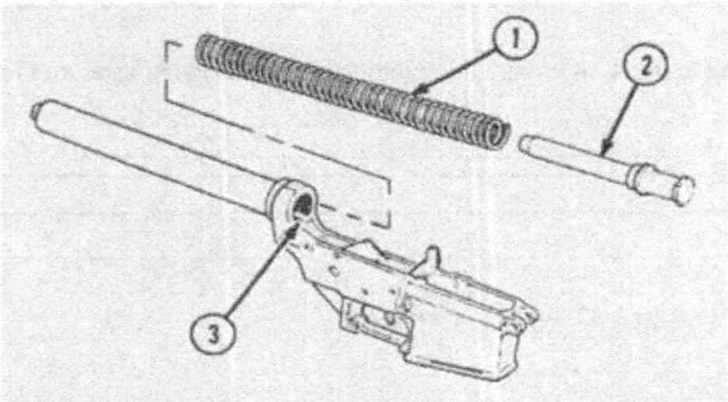

1. Press action spring (1) and buffer assembly (2) In until buffer retainer (3) snaps up and holds them In place.

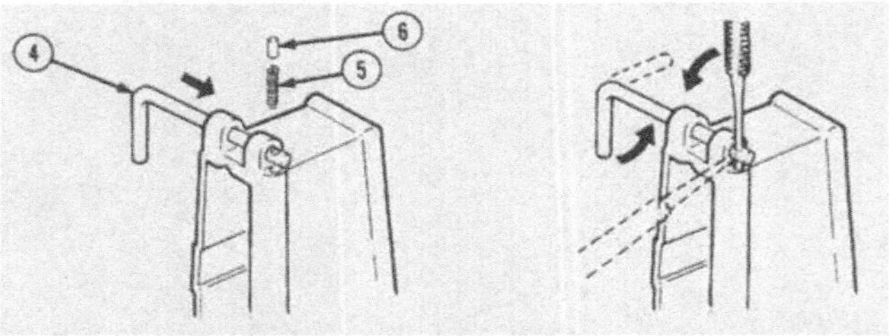

2. Install fabricated pivot pin installation tool (4) Insert helical spring 151 and pivot pin detent (6) Compress pivot pin detent in recess with 3 32 inch punch and rotate tool Remove 3'32 Inch punch

NOTE

Rounded end of pivot pin detent must be In the groove of the pivot pin (7) when assembly is complete.

3. Install pivot pin (7) while removing fabricated pivot pin Installation tool (4) Maintain pressure while sliding pivot pin (7) into hole Rotate pivot pin until pivot pin detent is inserted Into pivot pin groove

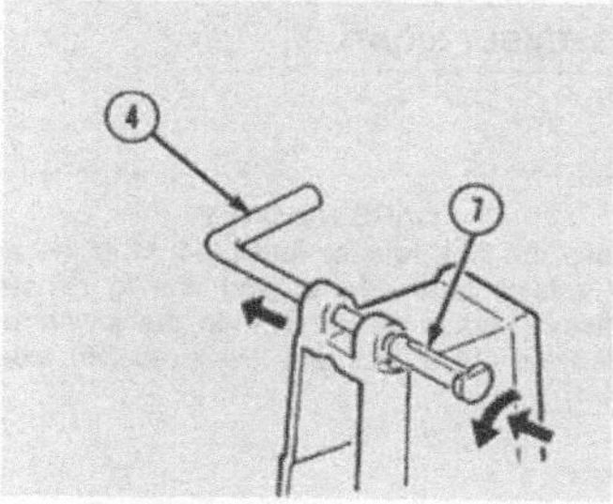

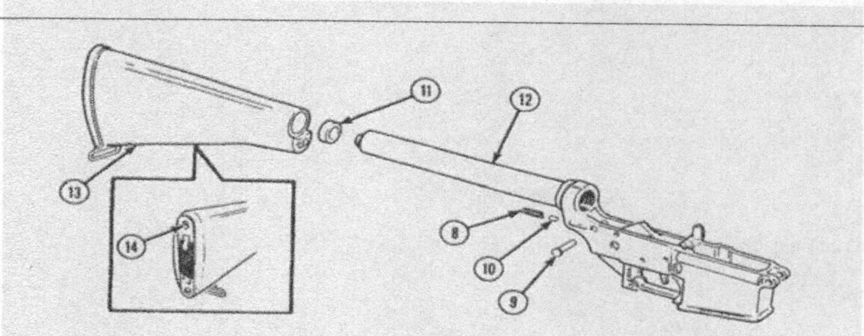

RIFLE ONLY

CAUTION

Do not kink helical spring (8) during assembly

4. Install takedown pin (9) with groove toward the rear Install takedown pin detent (10)and helical spring (8) from the rear
5. Install stepped spacer (11) on receiver extension (12) and carefully slide buttstock assembly (13) on to compress helical spring (8)

NOTE

Self-locking screw (14), If removed must be discarded and replaced with a new one 6. Install self-locking screw (14) to secure buttstock assembly (13)

6. Install self-locking screw (14) to secure buttstock assembly (13)

1 2-17. LOWER RECEIVER AND BUTTSTOCK ASSEMBLY (CONT).

f. REASSEMBLY (CONT)

CARBINE ONLY

6A. Grasp the lock release lever (14 1) In the area of the retaining nut (14 2) and pull to reinstall the buttstock assembly (14.3) onto the lower receiver extension (14 3) onto the lower receiver extension (14 4)

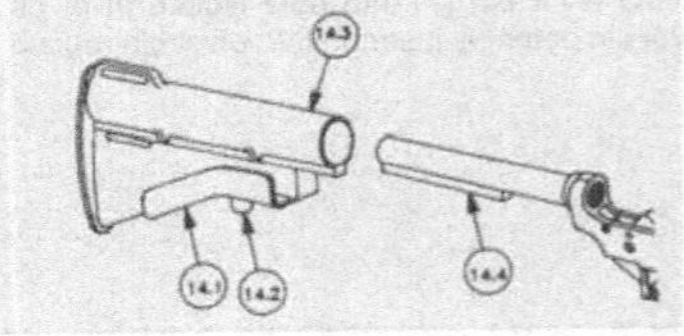

2-17. LOWER RECEIVER AND BUTTSTOCK ASSEMBLY (CONT).

f. REASSEMBLY (CONT)

WARNING

When utilizing the enhanced rifle grip (it has a bump between the second third finger for a better grip) PN 9349127, rifle grip screw. PN AN 501D416 18 (1-1/8 in.) or AN 501 D416 16 (1 in), is authorized to be(used with the enhanced grip Any screw longer than 1-1 8 In used with 9349127 could cause a hazardous situation Also, ensure the washer Is in 11place

CAUTION

Do not kink helical spring (15) during assembly

7. Install safety detent (16), pointed end first, and helical spring (1 5) into bottom of lower receiver (17)

NOTE

A portion of the helical spring will fit In a hole in the pistol grip

8. Carefully install pistol grip (18) to compress helical spring 115). Secure pistol grip 118) in place with lockwasher (19) and screw 120)
9. Reassemble rifle, refer to page 2 68

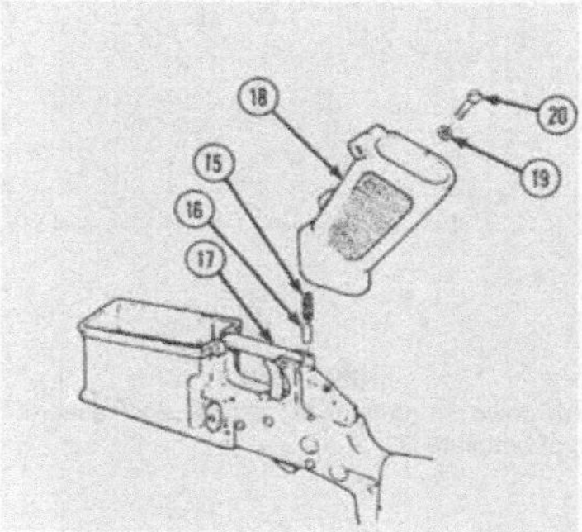

2-18. BUTTSTOCK ASSEMBLY.

This task covers:

a. Disassembly
b. Cleaning
c. Inspection
d. Repair
e. Lubrication
f. Reassembly

INITIAL SETUP

Tools

(ARMY) Small Arms Repairman Tool Kit (item 3, app B)

Equipment Conditions

2-57 Buttstock assembly removed from lower receiver and buttstock assembly

a. DISASSEMBLY

RIFLE ONLY

1. Remove self-locking screw (1), small sling swivel (2), and buttplate group (3) from buttstock (4)

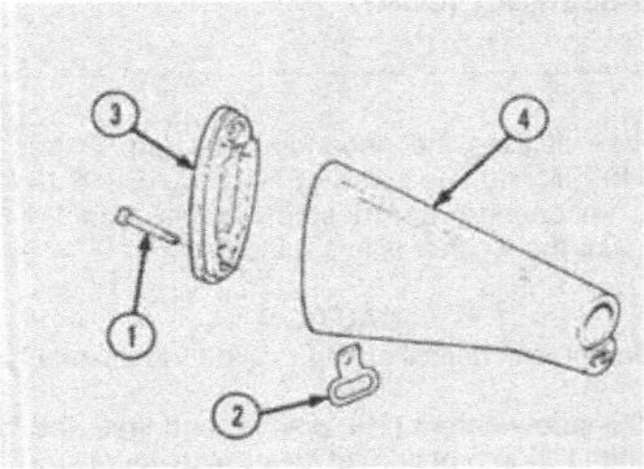

RIFLE ONLY

2. Push down on plunger (5) and lift door assembly (6) out of buttplate (7).

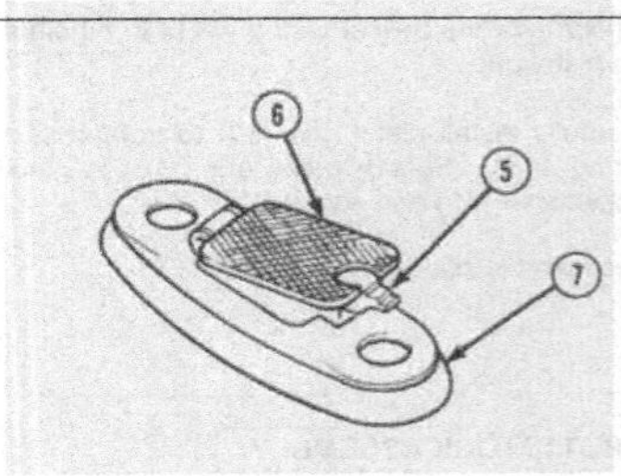

RIFLE ONLY

3. Remove straight pin 181 and separate hinge (9) and door assembly (61

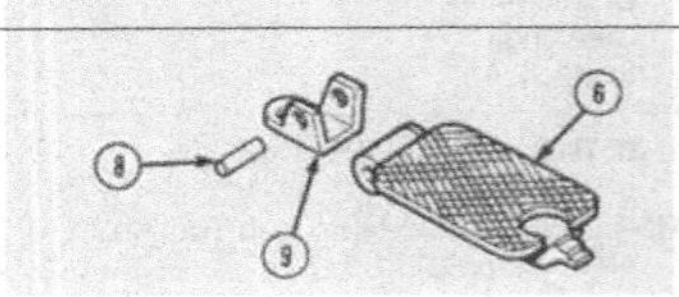

2-18. BUTTSTOCK ASSEMBLY (CONT).

a. DISASSEMBLY (CONT)

CARBINE ONLY

4. Disassemble the buttstock assembly by tapping out the spring pin (10) located in ,he oval slot of the retaining nut (1 1) This is done using a 1 16 Inch punch

5. Insert your Index finger into the forward end of the buttstock 112) and push down on the locking pin (13 Unscrew the retaining nut 11 and remove the release lever (14) locking pin (13) and spring (15)

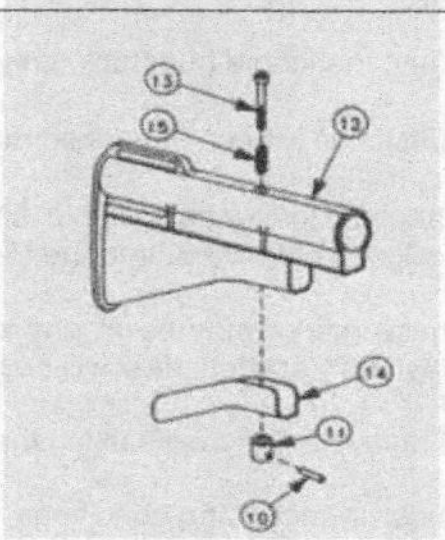

b. CLEANING (RIFLE ONLY)

Clean all parts with CLP (operator s manual) Use brush to clean knurled surface of door assembly

2-18. BUTTSTOCK ASSEMBLY (CONT).

c. INSPECTION

NOTE

M16A2 buttstocks, PN 9349121, with unauthorized markings may be used under the following conditions:

a. The only authorized markings are those which are temporary In nature, i e , paint, tape, etc

b. When marking a buttstock, only use temporary markings

c. Buttstocks with unauthorized markings that have been stamped into the surface of the buttstock will not be used

d. Unauthorized markings that have previously been scratched, etched, carved, etc may continue In use if the marks do not extend Into the fiber of the buttstock Cutting Into the fiber of the buttstock may weaken it

e. These marks may be at any location on the buttstock Unauthorized markings are not desirable. However, if previously applied, they will be allowed to continue In use due to the cost of the buttstock.

1. Inspect buttstock for cracks using the following guidelines:

a. Under the following conditions, hairline cracks (no chipped away material allowed) originating from buttplate end of buttstock are acceptable.

(1) One hairline crack, not to exceed 1 in. 12.54 cm) In length, per side of buttstock.

(2) Two additional hairline cracks up to 0 25 in (0 64 cm) In length, per side of buttstock

(3) A total of three cracks per side of buttstock, originating from buttplate end, are allowable.

b. Cracks In the critical area at the front end of the buttstock are not acceptable and these buttstocks must be replaced.

2. While buttplate Is Installed on rifle, Inspect for cracks around the mounting holes Check for cracks In excess of 0 25 In. (O 64 cm) In length which extend through the buttplate Replace If cracked.

3. Inspect door assembly for cracks, corrosion, stuck plunger, separations on outer face, or other damage Replace If defective.

d. REPAIR

Replace all authorized unserviceable Items Unserviceable Items are those Items which are damaged.

e. LUBRICATION

Lubricate all metal components (p 2-33) (operator's manual)

f. REASSEMBLY

RIFLE ONLY

1. Position hinge (1) on door assembly (2) and Install straight pin (3)

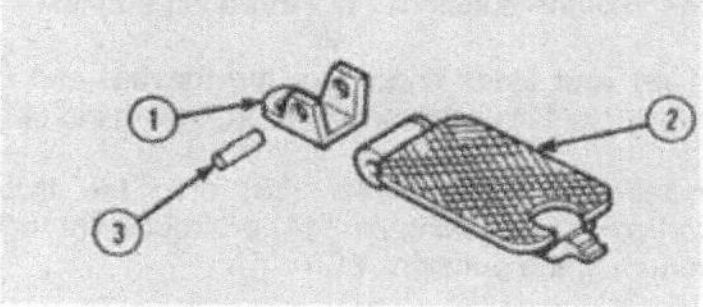

RIFLE ONLY

2. Install door assembly (2) into buttplate (4) and press plunger (5) to lock

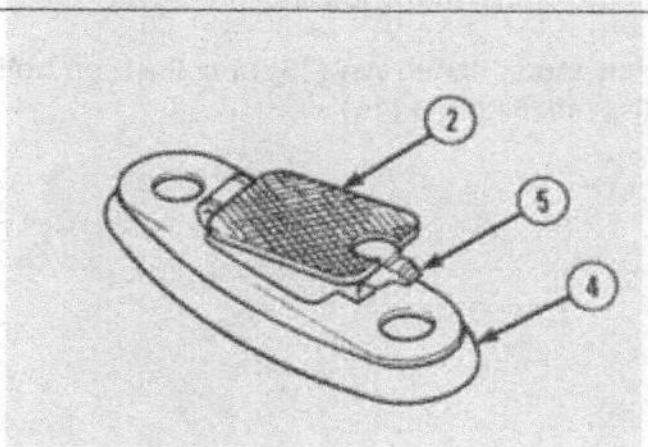

RIFLE ONLY

3. Position buttplate group (6) and small sling swivel (7) to the buttstock (8) and secure with self-locking screw (9).

NOTE

See page 2-62, reassembly, for reassembly of buttstock assembly to lower receiver

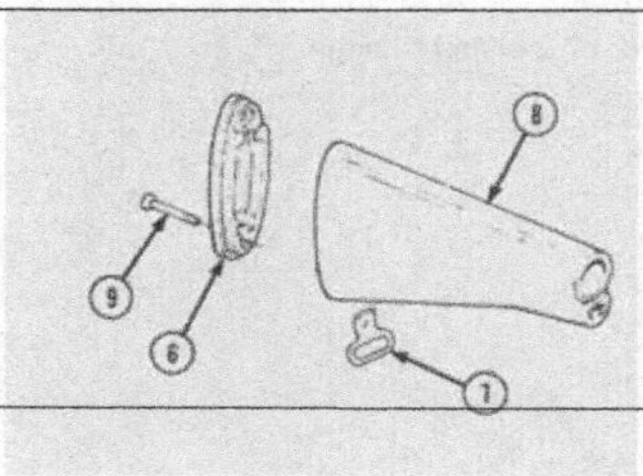

2-18. BUTTSTOCK ASSEMBLY (CONT).

f. REASSEMBLY (CONT)

CARBINE ONLY

4. Insert locking pin spring (10) onto locking pin (11)

5. Insert locking pin (11) and spring (10) Into the hole on top of the buttstock (12), threaded end first

6. Insert your index finger into the forward end of the buttstock (12) and push down on locking pin (11)

7. Install the release lever (13) onto the threaded portion of the locking pin (11) protruding through the bottom of the buttstock (12)

8. Screw on the locking nut (14) until flush with locking pin (11) Align the slot In locking nut (14) with the spring pin hole in the locking pin (11)

9. Lightly tap in spring pin (15) until flush on both sides of the retaining nut (14)

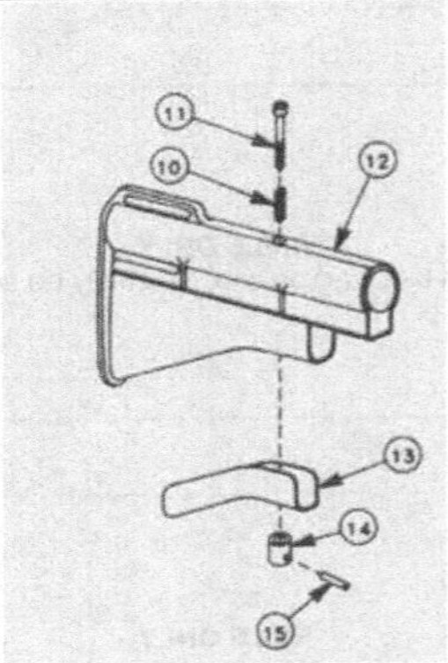

2-19. MAJOR COMPONENTS OF M16A2 RIFLE.

This task covers

a. Reassembly
b. Inspection
c. Stowage

INITIAL SETUP

References
TM 9 1005-319 10

Equipment Conditions
Rifle disassembled into major components

General Safety Instructions
To avoid injury to eyes, use care when removing and installing spring-loaded parts

Do not Interchange bolt assemblies from one rifle to another Doing so may result In Injury to, or death of. personnel

Do not keep live ammunition near the work area

a. REASSEMBLY

Refer to operator's manual and assemble lower receiver and buttstock assembly (1) upper receiver and barrel assembly (2). charging handle assembly (3), bolt carrier assembly (4), sling (5), and magazine(6).

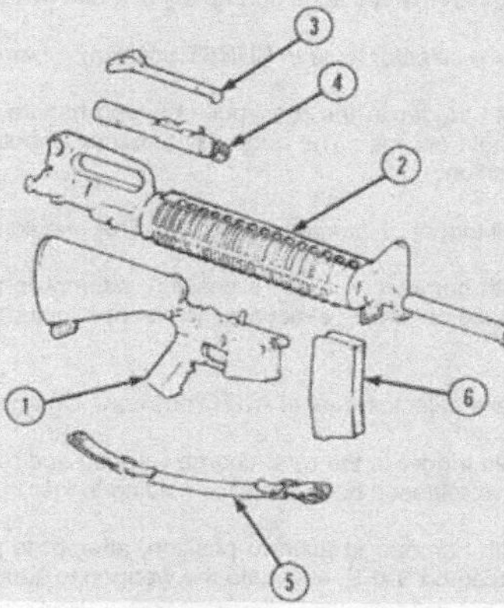

b. INSPECTION

Perform the following function checks on assembled weapon:

1. Remove magazine if installed. Pull charging handle assembly to rear. Check that chamber is clear. Let bolt and bolt carrier close. Do not pull trigger. Leave hammer in cocked position.

WARNING

If rifle fails any of the following tests, continued use of the rifle could result in injury to, or death of, personnel.

2. Place selector lever in SAFE position and pull trigger. Hammer should not fall.

3. Place selector lever in SEMI position. Pull trigger. Hammer should fall.

M16A2 and M4 ONLY
NOTE

For the purpose of the following check "SLOW" is defined as 1/4 to 1/2 the normal rate of trigger release.

4. Hold trigger to the rear, charge weapon, and release the trigger with a slow, smooth motion, without hesitations or stops, until the trigger is fully forward; an audible click should be heard. Hammer should not fall.

5. Repeat the SEMI position test five times. The weapon must not malfunction during any of these five tests. If the weapon malfunctions during any of these five tests, evacuate rifle to support maintenance for repair.

6. Place selector lever in BURST position. Charge weapon and pull trigger. Hammer should fall.

7. Hold trigger to the rear, pull charging handle assembly to rear and release three times. Release trigger. Hammer should not fall. The burst disconnector should have held the hammer to the rear while the trigger was in the pulled position.

8. Pull trigger. Hammer should fall. This should be the first round of a three round burst.

9. With hammer in forward position, attempt to place the selector lever in the SAFE position. If selector lever can be placed on SAFE, evacuate the weapon to support maintenance.

M4A1 ONLY

10. Place selector lever in AUTO position. Charge weapon and pull trigger. Hammer should fall.

11. Hold trigger to the rear, charge weapon and release trigger. Pull trigger. Hammer should not fall. AUTO sear should have released hammer while holding trigger in the squeezed position before releasing and resqueezing the trigger.

12. With hammer in forward position, attempt to place the selector lever in the SAFE position. If selector lever can be placed on SAFE, evacuate the weapon to support maintenance.

1 2-19. MAJOR COMPONENTS OF M16A2 RIFLE (CONT).

c. STOWAGE

Prior to stowing the weapon in arms room, perform the following procedures:

NOTE

Weapon passed into arms room issue window should be passed butt first with the bolt locked to the rear.

1. Clear. Refer to operator's manual.
2. Place selector lever in SEMI position.
3. Pull trigger. Hammer should fall.
4. Close ejection port (dust) cover.
5. Place weapon in rack.

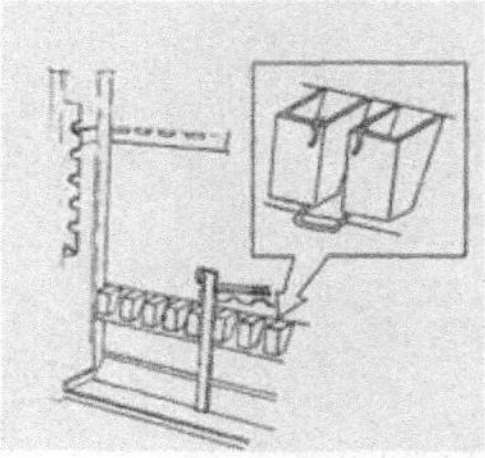

6. M4/M4A1 Carbine Only, use M12 Arms Rack.

a. The M12 arms rack is the correct arms rack in which to store the M4/M4A1 Carbine. The carbine must be stored with buttstock extended. When storing the M4/M4A1 Carbine in the M12 arms rack a mounting bracket, NSN 5340-01-230-3181, (app D, item 23.1) may be used for each M4/M4A1 Carbine being stored. This option is for the convenience of the person who opens and closes the arms rack to store the carbines.

b. To install the mounting bracket on the M12 arms rack, for use with the M4/M4A1 Carbine, install the bracket with the hooks of the bracket facing toward the carbine, so that the lower receiver extension will contact the bent end of the bracket. The bent end of the bracket will hold the carbine upright when the arms rack is opened. The bracket can be turned around when not in use for the carbine to allow storage of the M16A2 rifle.

c. When storing the M4/M4A1 Carbine in the M12 arms rack, an adapter bar (fig E-6, app E), MUST be used for security reasons. To install the adapter bar to the M12 arms rack:

(1) Remove all weapons from the rack and position the rack to gain access to the back.

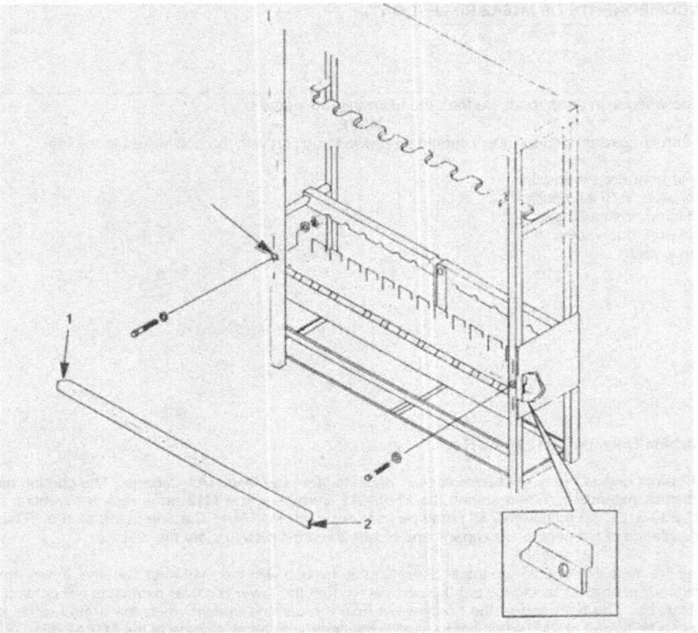

(2) The side of the adapter bar with the corners (1) cut off is the top and the side with square corners (2) is the back. The portion of the bar must be placed so the cut off corners (1) face the front of the rack.

(3) Holding the adapter bar at an angle, place one end into position inside the rear leg of the arms rack. Lower the other end of the bar into position. Allow the adapter bar to rest on the rack.

(4) Clamp both ends of the adapter bar into position. Mark the center line of the leg and adapter bar where they meet (see illustration) Using a center punch, mark the location of the holes to be drilled where the center lines cross. The holes must be centered on both the leg and the adapter bar. Drill a 1/8 inch pilot hole through both arms rack legs and the ends of the adapter bar. Drill a 3/8 inch hole through both arms rack legs and the ends of the adapter bar using the pilot hole as a guide. Remove the adapter bar. File the edges of all holes smooth Paint all bare metal surfaces with olive drab enamel paint, NSN 8010-01-350-5249, or equivalent.

2-19. MAJOR COMPONENTS OF M16A2 RIFLE (CONT).

c. STOWAGE (CONT)

(5) Reinsert the adapter bar into position on the arms rack. Using two 3/8 inch x 2 inch machine screws (MS35206-315, NSN 5305-00-984-5695) or equivalent, four washers (MS21306-1C, NSN 5310-01- 327-9713) or equivalent and two nuts (MS35649-2382, NSN 5310-00-056-3395) or equivalent assemble adapter bar to arms rack and tighten securely. The bolts can be inserted from either the back or the front to meet your requirements. If the rack is placed close to a wall or another rack It is recommended that the bolts be inserted from the back

(6) Tack weld, braze, or peen the threaded end of the bolt to the nut to prevent easy removal

(7) Place rack back into position and replace the weapons

NOTE

Adapter bar must be removed from arms rack prior to turning in the arms rack to the supply system.

d. To remove the adapter bar from the arms rack, remove the head of the bolt (or the nut), using a hand held grinder or file to avoid damage to the aluminum arms rack legs. A hammer and cold chisel may be used If no other way to remove the adapter exist

e. It is recommended that the rail protector NSN 1005-01-394-7677, (app D, item 26.1) be used during storage of the carbines when the carrying handle assembly or some other accessory is not installed on the upper receiver to prevent damage to the mounting surface on the upper receiver.

Section VI. PREPARATION FOR STORAGE OR SHIPMENT

2-20. PREPARATION FOR STORAGE OR SHIPMENT.

a. Packaging of the M16A2 Rifle and the M4/M4A1 Carbine shall be in accordance with the following

ARMY ONLY: Army users shall package the rifle and the carbine in accordance with each respective Packaging Data Sheet (PDS) for shipment or storage which may exceed 90 days The PDS is part of the Army Master Data File Retrieval Microform System (ARMS) Packaging
File

AIR FORCE ONLY: Air Force users shall package the rifle in accordance with each respective Special Packaging Instruction (SPI) 00-856-6885 for shipment or storage which may exceed 90 days The SPIs are part of the Army Master Data File Retrieval Microform System (ARMS) Packaging File

b. Packaging, if required, for shipping/storage which will not exceed 90 days shall be as follows

(1) Clean in accordance with operator's manual

(2) Wrap with MIL-B-121 waterproof material.

(3) Place in barrier bag MIL-B-117, Type I, Class C, or wrap with MIL-B-121, Type I, Grade A, and seal with tape, PPP-T-76.

(4) Place one or more of item in minimum size container. Block and brace in accordance with MIL-STD-1186. Cushion the M16 and similar weight items with PPP-C-843, and use PPP-F-320 as filler, to create a tight pack.

(a) Fiber board containers shall be in accordance with PPP-B-636 and may be Class Domestic. Gross weight and size of material shall determine grade of fiberboard container. PPP-B-640 may also be used.

(b) Wood containers shall be in accordance with PPP-B-601 or PPP-B-621.

(5) Equivalent materials may be used.

c. NSNs are not assigned to all the specified material. If it is necessary to specify an NSN in the TMs, the packing materials will have to be spared and part numbers and NSNs assigned.

d. The specifications used are:

(1)	MIL-B-121	Barrier Material, Greaseproofed, Waterproofed, Flexible (NSN 8135-00-753-4661)
(2)	MIL-B-117	Bag, Sleeve and Tubing -- Interior Packaging (NSN 8135-00-543-6574)
(3)	PPP-B-636	Boxes, Shipping, Fiberboard
(4)	PPP-B-601	Boxes, Wood, Cleated Plywood
(5)	MIL-STD-129	Marking for Shipment and Storage
(6)	PPP-T-76	Tape, Packaging, Paper
(7)	MIL-STD-1186	Cushioning, Anchoring, Bracing, Blocking and Waterproofing; with Appropriated Test Methods
(8)	PPP-C-843	Cushioning Material, Cellulosic
(9)	PPP-F-320	Fiberboard, Corrugated and Solid Sheet Stock (Container Grade), and Cut Shapes
(10)	PPP-B-640	Boxes, Fiberboard, Corrugated, Triple-Wall
(11)	PPP-B-621	Boxes, Wood, Nailed and Locked -- Corner

CHAPTER 3

DIRECT SUPPORT MAINTENANCE INSTRUCTIONS

CHAPTER OVERVIEW

This chapter provides information and instructions to keep the weapon in good repair and contains the following sections:

a. Repair Parts and Special Tools
b. Direct Support Troubleshooting
c. Direct Support Maintenance Procedures for the M1 6A2 Rifle and M4/M4A1 Carbine
d. Preparation for Storage or Shipment
e. Preembarkation Inspection of Material in Units Slated for Overseas Movement

Section I. REPAIR PARTS, SPECIAL TOOLS, TMDE, AND SUPPORT EQUIPMENT

3-1. COMMON TOOLS AND EQUIPMENT. For authorized common tools and equipment, refer to the Modified Table of Organization and Equipment (MTOE) applicable to your unit. Air Force users must maintain the following common tools:

Flat tip screwdriver
Socket wrench handle and socket head screw socket wrench
Vise jaw caps
Machinist's vise
Solid center punch
Hammer
Combination wrench
Torque wrench
Retaining ring pliers
(Two) 8-inch adjustable wrenches
Flat file
Ball-peen hammer
Trigger pull test fixture rod and weights
5/64-inch drive pin punch
1/8-inch drive pin punch
1/16-inch drive pin punch
3/32-inch drive pin punch

3-2. SPECIAL TOOLS, TMDE, AD SUPPORT EQUIPMENT. Special tools required for direct support maintenance are listed in appendixes B and C. Fabricated tools are listed and illustrated in appendix E.

3-3. REPAIR PARTS. Repair parts are listed and illustrated in appendix C of this manual.

NOTE

Bolt assemblies, and/or barrel assemblies may be interchanged , at the Direct Support Maintenance level, from one rifle to another, under the provisions of the note on page C-3: If these parts are interchanged the rifle must be checked/inspected as depicted on pages 3-17, 3-21, and 3-33. While performing these checks and inspections, pay special attention to the headspace requirements on page 3-47.

Section II. TROUBLESHOOTING

3-4. GENERAL.

a. This section contains direct support level troubleshooting information for locating and correcting most of the operating troubles which may develop in the M16A2 rifle and M4/M4A1 carbine. Each malfunction for the individual part or assembly is followed by a list of tests or inspections which will help you to determine the corrective action to take. You should perform the tests/inspections and corrective actions in the order listed.

b. This manual cannot list all malfunctions that may occur, nor all tests or inspections and corrective actions. If a malfunction is not listed or is not corrected by listed corrective actions, see individual repair sections in the maintenance procedures on each major assembly.

3-5. TROUBLESHOOTING PROCEDURES. Refer to troubleshooting table for malfunctions, tests, and corrective actions. The symptom index is provided for a quick reference of the malfunctions covered in the table.

SYMPTOM INDEX

TROUBLESHOOTING

MALFUNCTION
TEST OR INSPECTION
CORRECTIVE ACTION

1. FAILURE OF MAGAZINE TO LOCK IN RIFLE.
   Step 1. Dirty or corroded magazine catch (1).
   Disassemble and clean.
   Step 2. Defective magazine catch spring (2).
   Replace magazine catch spring (2) (p 3-62).
   Step 3. Worn or broken magazine catch (1).
   Replace magazine catch (1) (p 3-62).

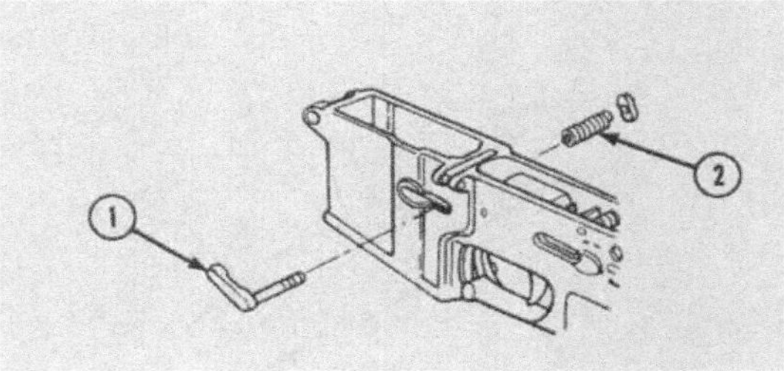

2 FAILURE TO FEED
   Step 1 Magazine catch spring (1) weak or broken.
   Replace magazine catch spring I1) (p 3-62)

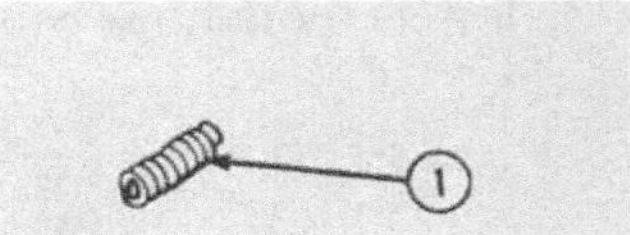

   Step 2 Short recoil
   Refer to page 3-9

3 FAILURE TO CHAMBER
   Short recoil.
   Refer to page 3-9.

3-5. TROUBLESHOOTING PROCEDURES (CONT).

TROUBLESHOOTING (CONT)

MALFUNCTION
TEST OR INSPECTION
CORRECTIVE ACTION

4. FAILURE TO LOCK.

Step 1. Damaged bolt carrier key (1).
Repair or replace bolt carrier key (1) (p 3-25)

Step 2 Loose screws (2) on bolt carrier key (1).
a. Disassemble and repair (p 3-25)
b. Reassemble using new screws.

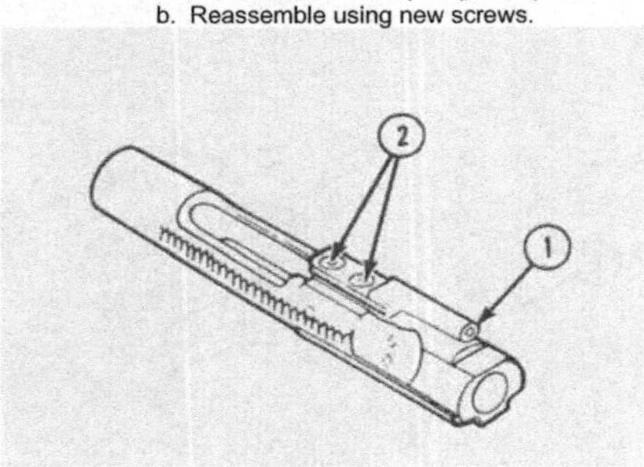

Step 3 Bent gas tube (3).
a Adjust by bending gas tube 13) In area of handguards
b Replace gas tube (3) and check alignment (p 3-29)

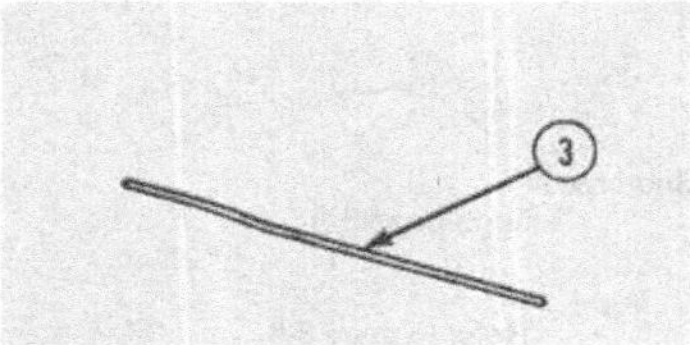

Step 4. Short recoil
Refer to page 3 9

TROUBLESHOOTING (CONT)

MALFUNCTION
TEST OR INSPECTION
CORRECTIVE ACTION

5. FAILURE TO FIRE.

Step 1. Broken hammer (1).
Replace hammer (1) (p 3-73).

Step 2. Weak or broken hammer spring (2).
Replace hammer spring (2) (p 3-73).

Step 3. Hammer spring (2) improperly assembled.
Assemble properly.

Step 4. **(M16A2 and M4 ONLY)** Burst cam (3) and/or cam spring (4) frozen or improperly assembled
Disassemble, clean, lubricate, and reassemble correctly (p 3-73).

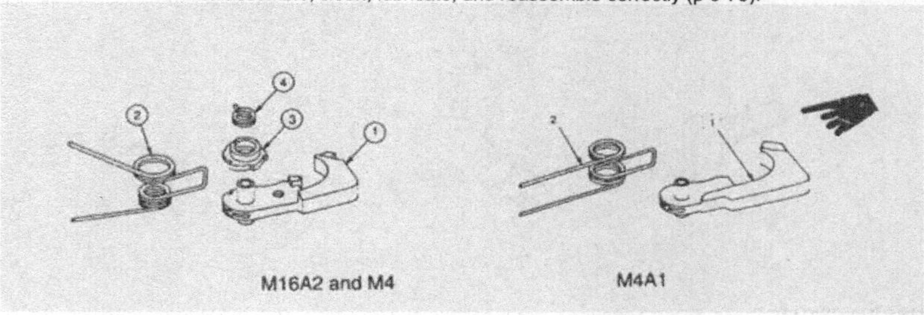

Step 5. Selector lever (5) frozen on SAFE position.
Disassemble and clean (p 3-62)

Step 6. Broken firing pin (6) or firing pin does not meet gage protrusion requirement.
Replace firing pin (6) (p 3-16).

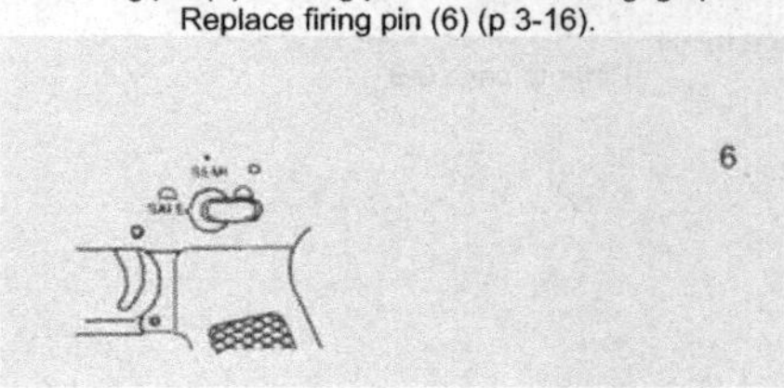

3-5. TROUBLESHOOTING PROCEDURES (CONT).

TROUBLESHOOTING (CONT)

MALFUNCTION
TEST OR INSPECTION
CORRECTIVE ACTION

6. FAILURE TO UNLOCK.

Step 1 Burred locking lugs (1) on bolt assembly.
Remove burrs.

Step 2 Burred lugs (2) on barrel extension.
Remove burrs.

Step 3. Short recoil.
Refer to page 3-9.

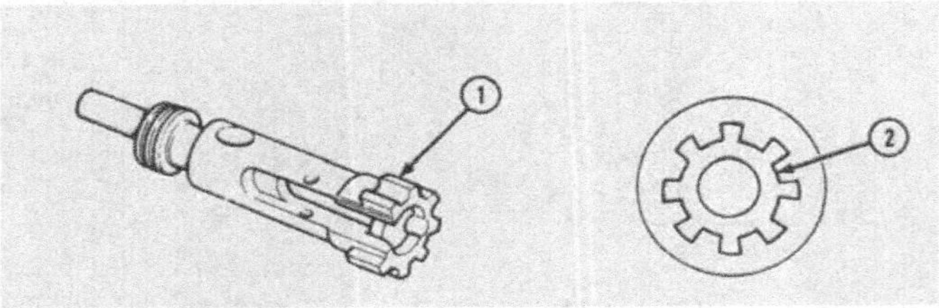

7. FAILURE TO EXTRACT

Step 1. Inspect cartridge extractor and extractor spring assembly.
Replace if cracked or broken (p 2-35).

Step 2. Inspect badly pitted chamber with reflector tool (Item 2, fig. C-1 7).
Replace barrel assembly If chamber Is badly pitted (p 3-29).

8 FAILURE TO EJECT.

Short recoil.
Refer to page 3-9.

TROUBLESHOOTING (CONT)

MALFUNCTION
TEST OR INSPECTION
CORRECTIVE ACTION

9. FAILURE TO COCK.

Step 1. Worn or broken trigger nose (1) or trigger spring (2).
Replace trigger (3) or defective trigger spring (2) (p 3-62).

Step 2. Worn or broken hammer trigger notch (4).
Replace hammer (5) (p 3-62).

Step 3. Worn or broken hammer disconnector hook (6).
Replace hammer (5) (p 3-62).

Step 4. Worn or broken hammer automatic sear hook (7).
Replace hammer (5) (p 3-62).

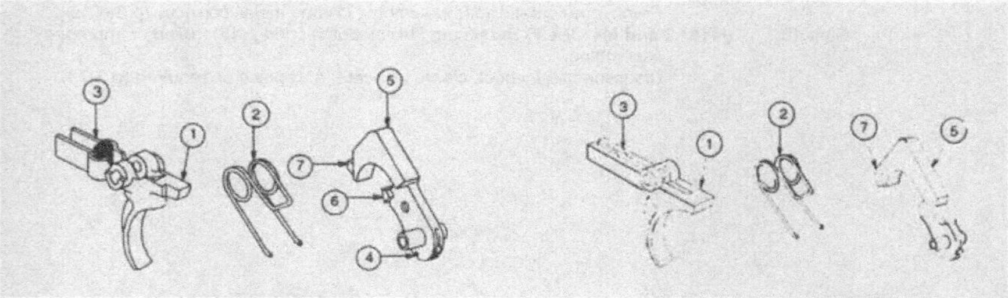

M16A2 and M4 ONLY M4A1 ONLY

3-5. TROUBLESHOOTING PROCEDURES (CONT).

TROUBLESHOOTING (CONT)

MALFUNCTION
TEST OR INSPECTION
CORRECTIVE ACTION

Step 5 **(M4A1 has only 1 disconnector)** Worn or broken disconnector hooks (8).
Replace defective disconnectors (9) (p3-62).

Step 6 **(M4A1 has only 1 spring)** Weak, broken, or missing disconnector springs (10).
Replace disconnector springs (10) (p 3-62).

Step 7. Worn, broken, or missing automatic sear (11).
Replace automatic sear (11) (p 3-62).

Step 8 Weak or broken automatic sear spring (12).
Replace automatic sear (11) (p 3-62).

Step 9. Long leg (13) of automatic sear spring incorrectly assembled in receiver.
Remove automatic sear assembly (11) and install correctly (p 3-62).

Step 10. **(M16A2 and M4 ONLY)** Burst cam (14) or clutch spring (15) frozen or improperly assembled.
Disassemble, inspect, clean, lubricate, or replace as required (p 3-73).

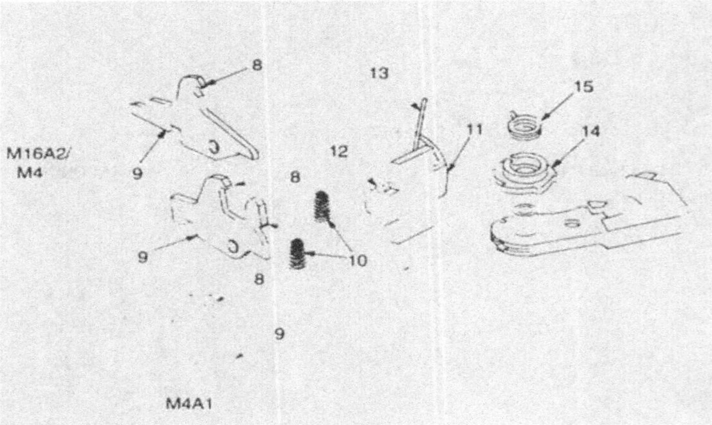

TROUBLESHOOTING (CONT)

MALFUNCTION
TEST OR INSPECTION
CORRECTIVE ACTION

10. SHORT RECOIL.

Step 1. Improper gap space or worn, missing, or broken bolt rings (1).

a. Stagger bolt ring gaps (p 3-21).

b. Replace bolt rings (1) and stagger gaps (p 3-21).

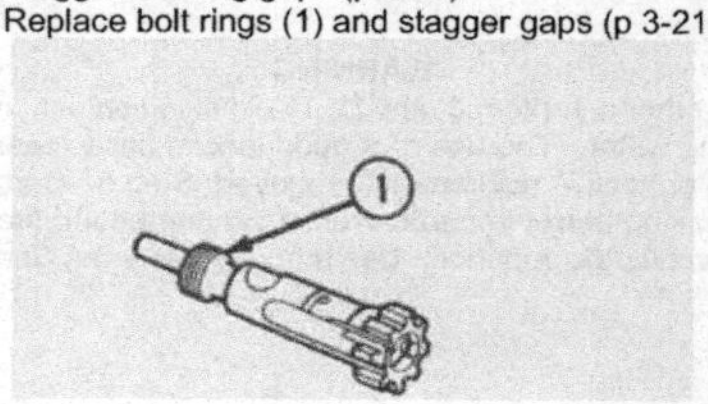

Step 2. Broken or bent gas tube (2).

Adjust by bending in area of handguards or replace gas tube (2) (p 3-29).

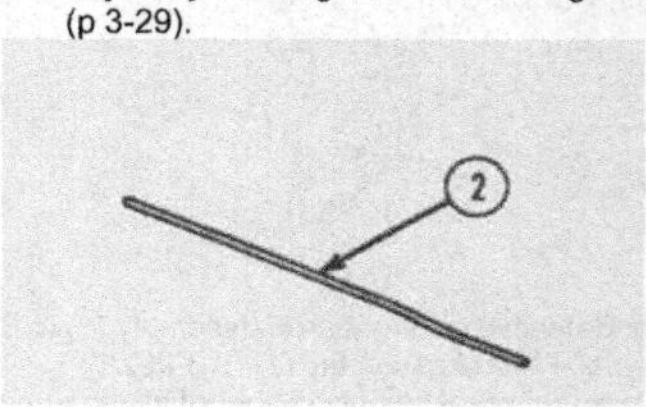

Step 3. Gas tube spring pin (3) missing from front sight (4).

Replace gas tube spring pin (3) (p 3-29).

Step 4. Partially plugged gas system because of carbon build-up in the gas tube (2).

Replace gas tube (2) (p 3-29).

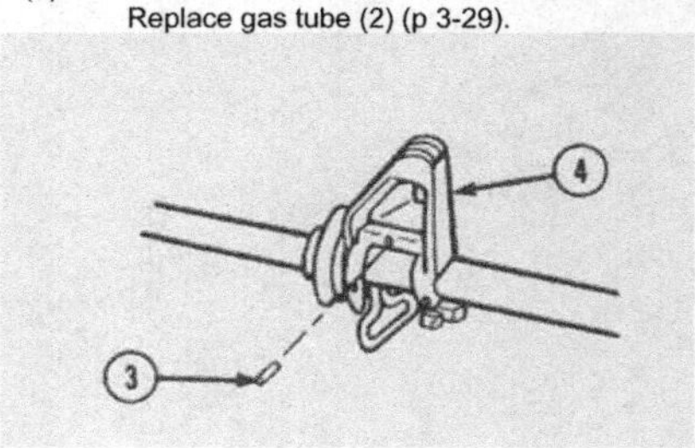

3-5. TROUBLESHOOTING PROCEDURES (CONT).

TROUBLESHOOTING (CONT)

MALFUNCTION
TEST OR INSPECTION
CORRECTIVE ACTION

10 SHORT RECOIL (CONT).

WARNING

When using carbon removing compound (Item 8, app D), avoid skin contact. If it comes in contact with the skin, wash off thoroughly with running water. The use of a good lanolin base cream after exposure to compound Is helpful. The use of gloves and protective equipment Is required. Step 5. Carbon build-up in barrel gas port (5). Remove carbon build-up by soaking barrel In carbon removing compound (item 8, app D). Use rubber gloves (Item 18, app D) with carbon removing compound. Use a bore small arms cleaning brush (Item 4, app D).

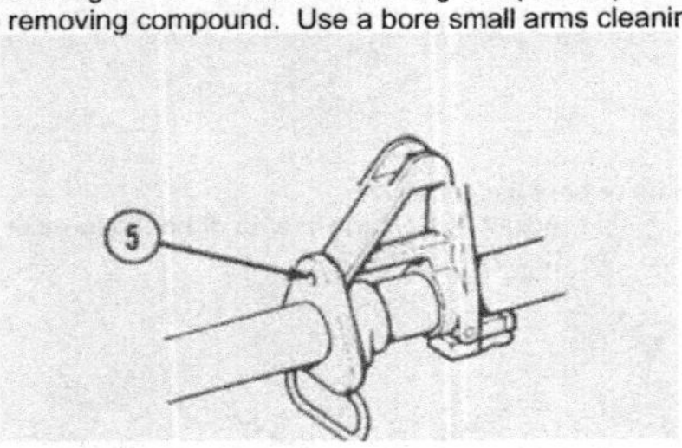

11. RIFLE CANNOT BE ZEROED.

Step 1 Inspect for defective or bent barrel assembly (1) (p 3-29).
Replace barrel assembly (1) (p 3-29).

Step 2 (For windage) barrel assembly (1) out of alignment with rear sight on up per receiver.
Align barrel assembly (1) and upper receiver (p 3-291

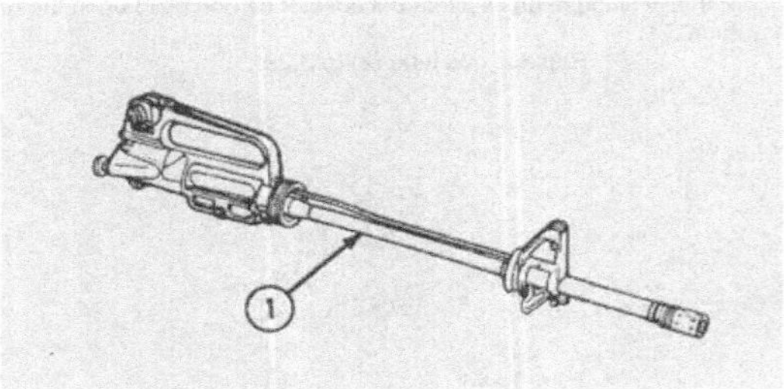

TROUBLESHOOTING (CONT)

MALFUNCTION
TEST OR INSPECTION
CORRECTIVE ACTION

Step 3. (For elevation) defective front sight (2) or rear sight (3).
Repair as required (p 3-48).

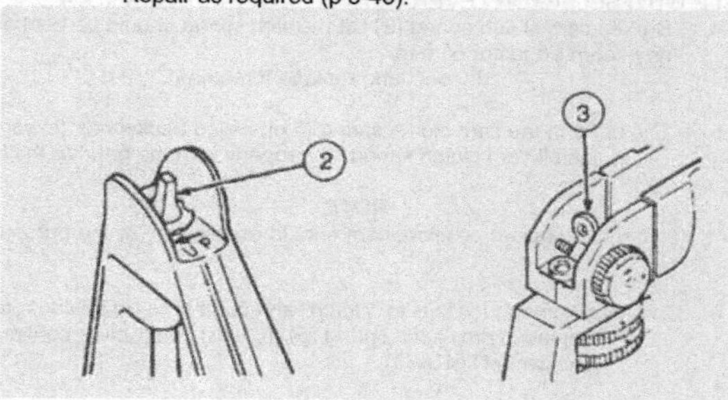

12. FAILURE TO CYCLE WITH SELECTOR LEVER SET ON BURST (M16A2 and M4 ONLY).

Step 1. Broken automatic sear (1) or spring (2).
Replace automatic sear assembly (1) (p 3-62).

Step 2. Faulty selector lever (3).
Replace selector lever (3) (p 3-62).

Step 3. Broken tooth on burst cam (4).
Replace burst cam (4) (p 3-73).

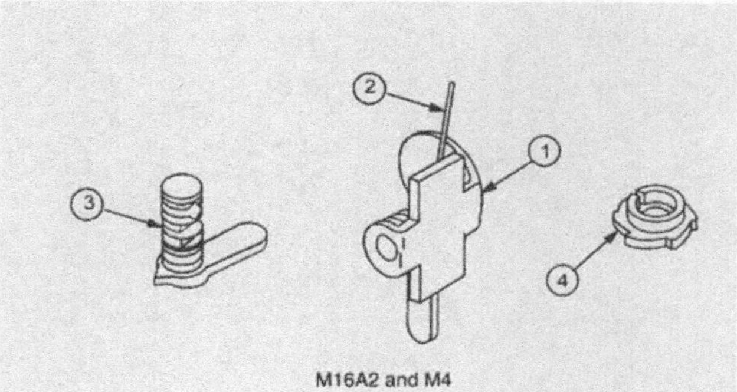

M16A2 and M4

3-5. TROUBLESHOOTING PROCEDURES (CONT).

TROUBLESHOOTING (CONT)

MALFUNCTION
TEST OR INSPECTION
CORRECTIVE ACTION

12 FAILURE TO CYCLE WITH SELECTOR LEVER SET ON BURST (M16A2 and M4 ONLY) (CONT).

Step 4 Broken cam clutch spring (5) Cam clutch spring should be bent and properly formed without any sharp edges or corners
Inspect and replace If required.

Step 5 The bend In the cam clutch spring (5) installed backwards (toward outside).
Install cam clutch spring (5) properly with the bend to the inside (p 3-73).

NOTE

When hammer Is rotated back to cocked position, cam should rotate to allow the burst disconnector to latch in the next notch.

Step 6. Cam clutch spring (5) fails to "clutch" and burst cam (4) fails to rotate back with hammer (6).
Replace cam clutch spring (5) (p 3-73). If problem continues, replace hammer (6) and cam (4) (p 3-73).

Step 7. Broken front hook (7) on burst disconnector (8).
Replace burst disconnector (8) (p 3-62).

Step 8. Short recoil.
Refer to page 3-9

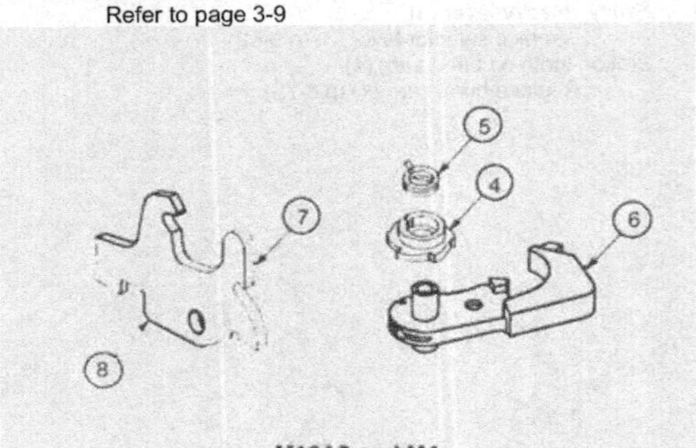

M16A2 and M4

TROUBLESHOOTING (CONT)

MALFUNCTION
TEST OR INSPECTION
CORRECTIVE ACTION

12.1. FAILURE TO CYCLE WITH SELECTOR LEVER SET ON AUTO (M4A1 ONLY)

Step 1. Broken automatic sear (1) or spring (2).
Replace automatic sear assembly (1).

Step 2. Faulty selector lever (3).
Replace selector lever (3).

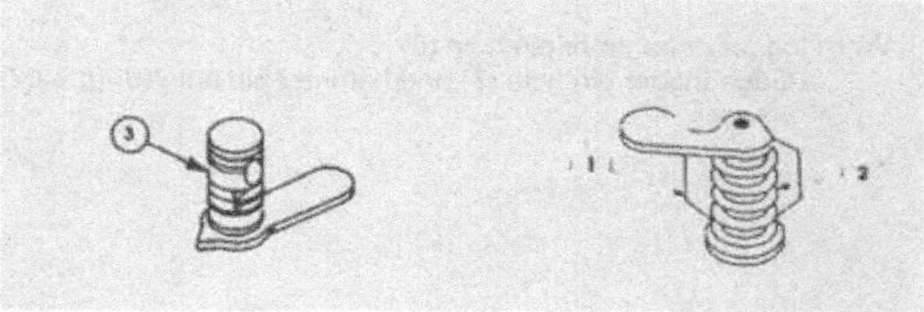

Step 3. Short recoil.
Refer to page 3-9.

13. FIRES TWO ROUNDS WITH ONE PULL OF TRIGGER WITH SELECTOR LEVER SET ON SEMI (DOUBLE FIRING).

Step 1. Defective semiautomatic disconnector (1). Replace semiautomatic disconnector (1) (p 3-62).

Step 2 Worn or broken trigger notch (2) of hammer (3) (searing portion). Replace hammer (3) (p 3-62).

Step 3. Worn or broken disconnector notch (4) of hammer (3). Replace hammer (3) (p 3-62).

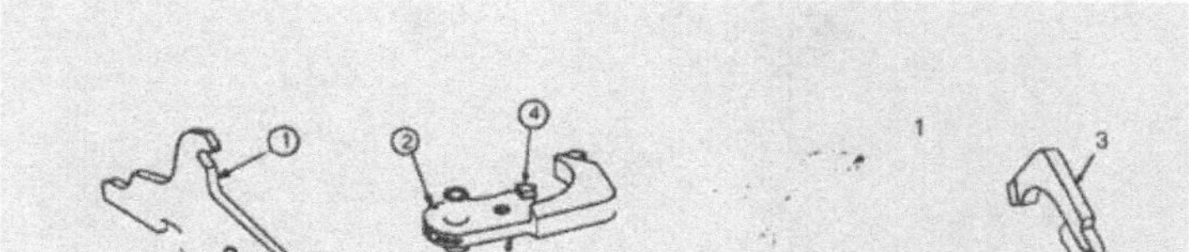

M16A2 and M4 M4A1

3-5. TROUBLESHOOTING PROCEDURES (CONT).

TROUBLESHOOTING (CONT)

MALFUNCTION
TEST OR INSPECTION
CORRECTIVE ACTION

13 FIRES TWO ROUNDS WITH ONE PULL OF TRIGGER WITH SELECTOR LEVER SET ON SEMI (DOUBLE FIRING) (CONT).

Step 4 Worn or broken trigger (5) (searing portion).
Replace trigger (5) (p 3-62).

Step 5 Worn trigger or hammer pin hole (6).
Gage trigger pin hole (7) and hammer pin hole (6) (p 3-62). If test falls, replace weapon.

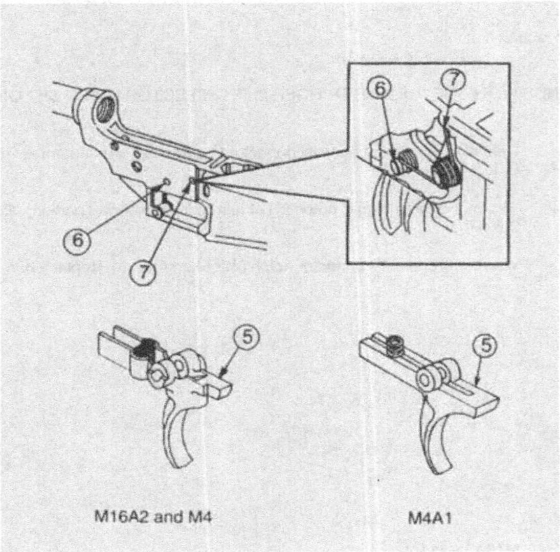

TROUBLESHOOTING (CONT.)

MALFUNCTION

TEST OR INSPECTION

CORRECTIVE ACTION

14. FIRES WITH SELECTOR LEVER ON SAFE OR WHEN TRIGGER IS RELEASED WITH SELECTOR LEVER ON SEMI.

Step 1. Defective selector lever (1).

Replace selector lever (1) (p 3-62).

Step 2. Worn or broken trigger (rear portion) (2).

Replace trigger (3) (p 3-62).

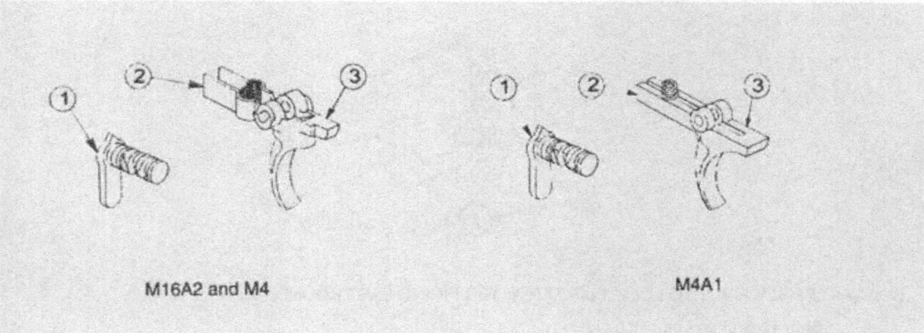

3-5. TROUBLESHOOTING PROCEDURES (CONT.).

TROUBLESHOOTING (CONT.)

MALFUNCTION

TEST OR INSPECTION

CORRECTIVE ACTION

15. HAMMER PIN "WALKS".

Hammer pin (1) "walks" or works loose during firing or hammer pin Is very easy to push out of receiver when hammer Is Installed.

Replace hammer assembly (p 3-62).

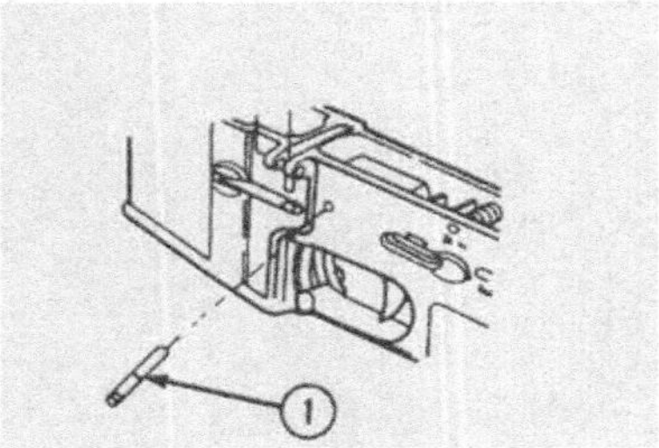

16. BOLT ASSEMBLY FAILS TO LOCK TO REAR AFTER FIRING LAST ROUND.

Step 1. Broken bolt catch (1).

Replace bolt catch (1) (p 3 62,.

Step 2. Weak or broken bolt catch spring (2).

Replace bolt catch spring (2) (p 3-62)

Step 3 Restricted movement of bolt catch (1).

Disassemble and clean.

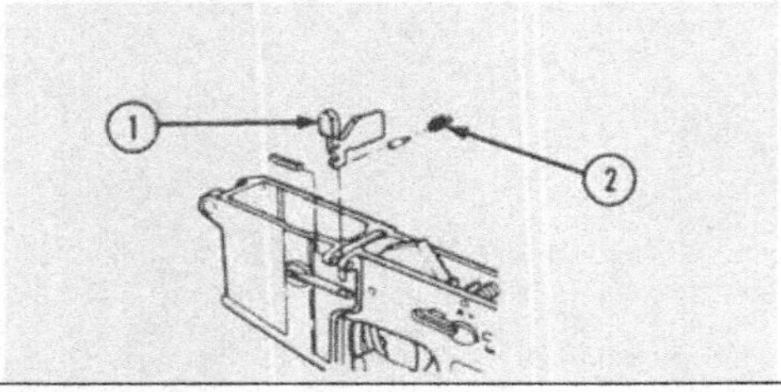

Section III. MAINTENANCE PROCEDURES FOR THE M16A2 RIFLE AND M4/M4A1 CARBINES

3-6. MAJOR COMPONENTS OF M16A2 RIFLE AND M4/M4A1 CARBINES.

This task covers disassembly.

INITIAL SETUP

Tools
(ARMY) Small Arms Repairman Tool Kit (item 3, app B)

References
TM 9-1005-319-10 (operator's manual)

Equipment Conditions
Weapon assembled

General Safety Instructions
Before starting an inspection, be sure to clear the weapon Do not pull the trigger until the weapon has been cleared.
Inspect the chamber and receiver to ensure no ammunition is present Do not keep live ammunition near work area.

DISASSEMBLY

a. Refer to operator's manual.

b. Remove magazine (1), sling (2), bolt carrier assembly (3), charging handle assembly (4), and upper receiver and barrel assembly (5), carrying handle assembly (5.1), from lower receiver and buttstock assembly (6).

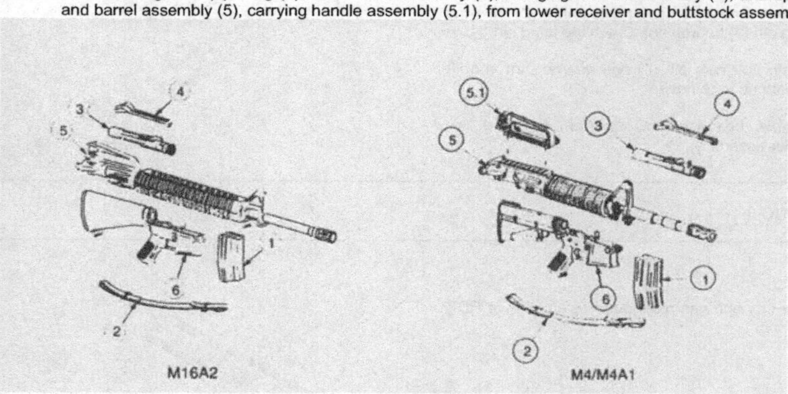

NOTE

Solid Film Lubricant (SFL) is the only authorized touchup for the M16A2 rifle and M4/M4A1 carbines and may be used on up to one third of the exterior finish of the weapon. **FOR ARMY CONUS USE ONLY AND AIR FORCE TRAINING WEAPONS ONLY**: SFL may be used as touchup without limitation on the upper receiver and barrel assembly. This is to say that units which DO NOT fall under the category of Divisional Combat Units or rapid deployment type units may have up to 100 percent of the exterior surface of the upper receiver and barrel assembly protected with SFL if necessary.

3-7. BOLT CARRIER ASSEMBLY.

This task covers:

a Disassembly	**d** Test
b Cleaning	**e** Repair
c Inspection	**f** Reassembly

INITIAL SETUP

Test Equipment
Tool and Gage Set (item 2, app B)

Tools
(ARMY) Small Arms Repairman Tool Kit (item 3, app B)

Materials/Parts
Cleaner, lubricant, and preservative (CLP) (item 9, app D)
Cleaning compound, rifle bore (RBC) (item 12, app D)
Pipe cleaner (item 11, app D)

Equipment Conditions
3-1 5 Bolt carrier assembly removed

a. DISASSEMBLY

1. Remove firing pin retaining pin (1).
2. Tip key and bolt carrier assembly (2) allowing firing pin (3) to drop out. Catch the firing pin.
3. Rotate bolt cam pin (4) one quarter turn and lift straight up to remove.
4. Remove bolt assembly (5) from key and bolt carrier assembly (2)

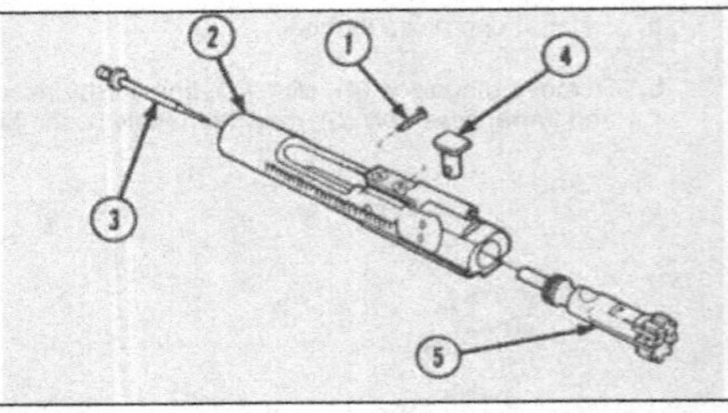

b. CLEANING

Clean carrier key and gas relief ports using CLP or RBC and pipe cleaner

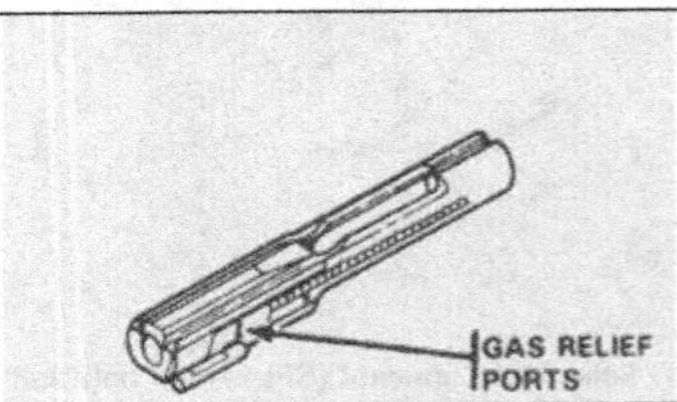

c. INSPECTION

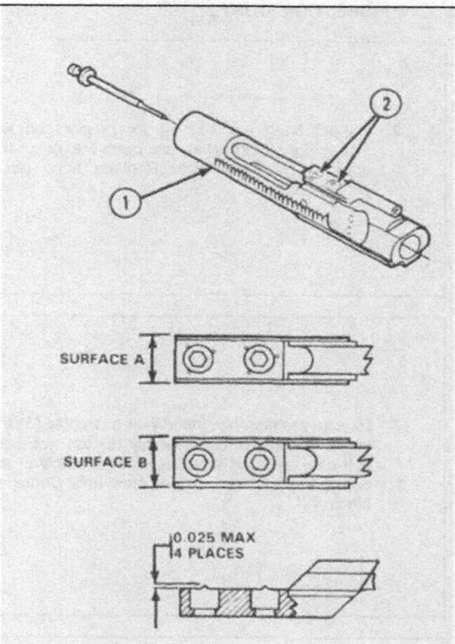

1. Inspect bolt carrier assembly using the following guidelines.

(a) Inspect bolt carrier assembly (1) for burrs, cracks, wear, and evidence of gas loss.

(b) Visually inspect the carrier and key screws (2) for looseness and proper staking as shown below.

NOTE

Do not attempt to retorque if there is no loosening of the screws indicated by the staking marks.

Surface "A" must not Indicate distortion or damage which Impairs parallelism.

NOTE

A maximum of 0.025 in. (0.064 cm) protrusion In an upward direction is permissible.

NOTE

There are bolts and bolt carriers on fielded rifles, some with chrome-plated exterior surface finishes and some with phosphate coating Both finishes are acceptable under certain operational requirements and or restrictions Phosphate-coated bolt carriers are required for divisional combat units Chrome plated bolt carriers are acceptable for divisional noncombat units and training center units. Chrome-plated and phosphate-coated bolt assemblies, bolt carrier assemblies, and repair parts for these assemblies may be intermixed In any combination, with the following exception:

Phosphate-coated bolt carriers are required for all deployable and deploying units Chrome-plated bolt carriers are acceptable for nondeployable and training center units.

3-7. BOLT CARRIER ASSEMBLY (CONT.).

c. INSPECTION (CONT.)

2. Inspect firing pin 131 tip for proper con tour Inspect for pitting, wear, and burrs Pits or wear in area (4) Is permissible Replace firing pin If defective

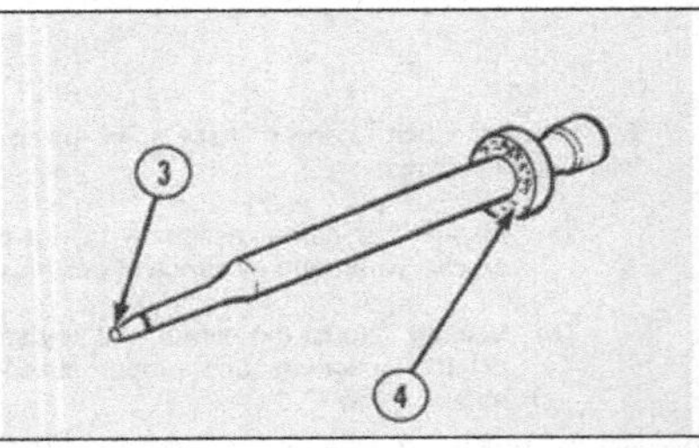

3. Prior to reassembly, insert bolt assembly (5) Into key and bolt carrier assembly (6) (do not insert bolt cam pin) and exercise bolt assembly in and out of key and bolt carrier assembly Check for binding

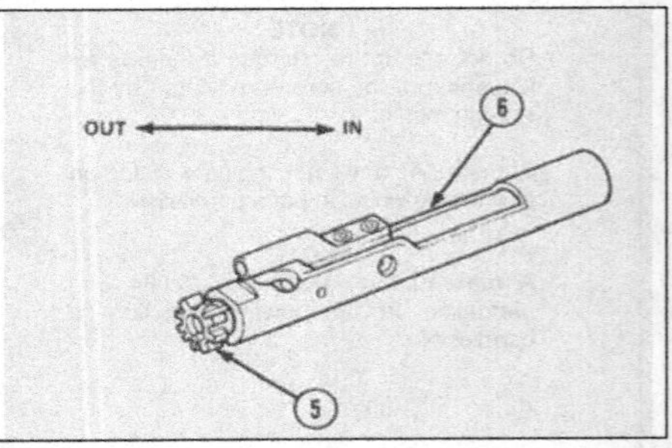

4. Check bolt assembly (5) for proper fit with bolt cam pin removed Turn key and bolt carrier assembly (61 and suspend so the bolt assembly is pointed down

NOTE

The bolt assembly must not drop out If weight of bolt assembly allows It to drop out of key and bolt carrier assembly, replace bolt rings (p 3-21)

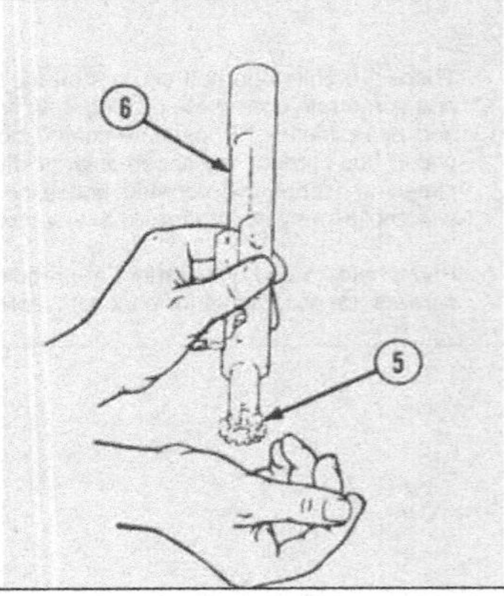

This page is blank. Procedures and figure were moved to page 2-36.

3-7. BOLT CARRIER ASSEMBLY (CONT.).

d. TEST

1. Insert firing pin (1) through bolt (2).
2. Position firing pin protrusion gage (3) PN 7799735 to check for proper firing pin (1) protrusion (minimum 0.028 in. (0 07 cm)-maximum 0 036 In. (0.09 cm))

NOTE
Firing pin should touch the gage on minimum but should not touch on maximum.

3. Replace a defective firing pin.

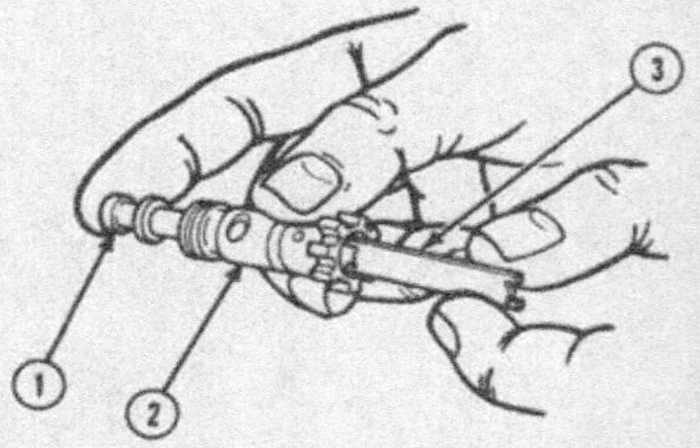

e. REPAIR

Replace all authorized unserviceable items. Retest all replaced parts.

f. REASSEMBLY

1. Install bolt assembly (1) into key and bolt carrier assembly (2).
2. Install bolt cam pin (3) and rotate one quarter turn to secure bolt assembly (1)
3. Hold key and bolt carrier assembly (2) with bolt assembly (1) down and drop In firing(pin (4)
4. Install firing pin retaining pin (5) from left side only to ensure proper Installation Check Installation by attempting to shake out firing pin
5. Reassemble rifle, refer to page 3-77

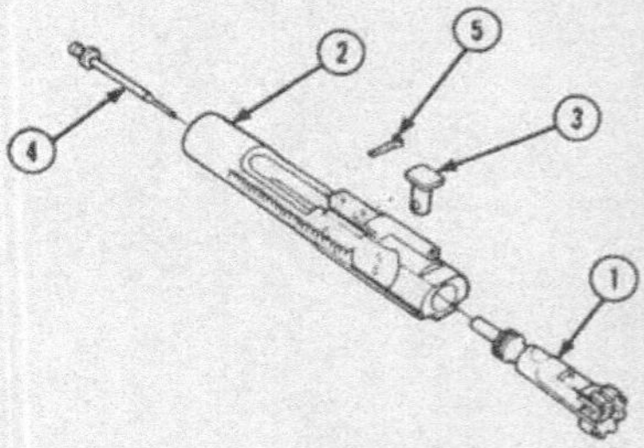

3-8. BOLT ASSEMBLY.

This task covers:
a. Disassembly
b. Inspection/Repair
c. Test
d. Reassembly

INITIAL SETUP

Test Equipment
Tool and Gage Set (Item 2, app B)

Tools
(ARMY) Small Arms Repairman Tool Kit (Item 3, app B)

Materials/Parts
Penetrant kit (item 25, app D)
Rag, wiping (item 26, app D)

Equipment Conditions
3-16 Bolt assembly removed

a. DISASSEMBLY

NOTE
Do not remove bolt rings unless they require replacement and three new replacement bolt rings are on hand.

Using small flat tip jeweler's screwdriver, remove the three bolt rings (1) from the bolt (2).

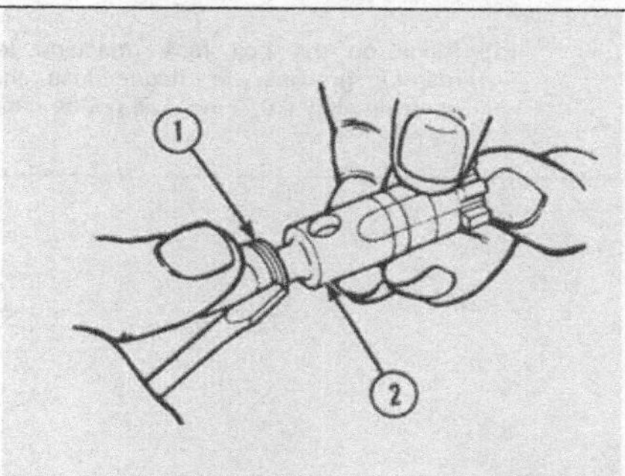

b. INSPECTION/REPAIR

1. Visually Inspect bolt rings for cracks, kinks, and bends. Replace all three bolt rings if one or more bolt rings is damaged. See page 3-16 for bolt ring wear check.

3-8. BOLT ASSEMBLY (CONT.).

b. INSPECTION/REPAIR (CONT.)

2. Inspect bolt for pits, burrs, and wear as follows.

(a) Bolt faces with a cluster of pits which are touching or tightly grouped, covering an area measuring approximately 1,8 Inch across, will be rejected and replaced.

(b) Bolts which contain individual pits or a scattered pattern will not be cause for rejection.

(c) Bolts that contain pits extending Into the firing pin hole will not be rejected unless firing pin hole gaging check determines excess wear.

(d) Rings on the bolt face (machine tool marks), grooves, or ridges less than approximately 0.010 inch will not be cause for rejection.

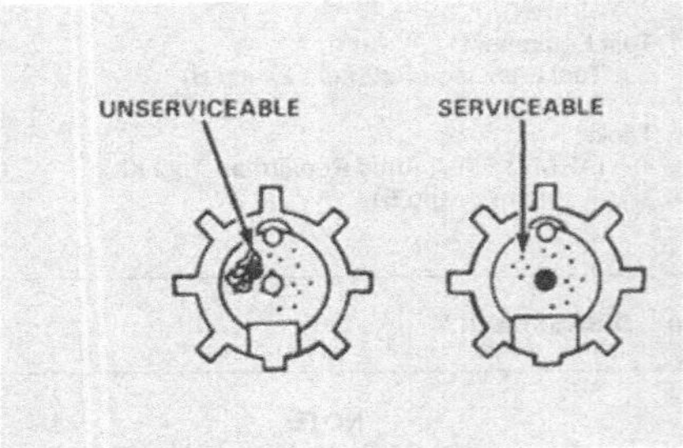

3. Inspect bolt for cracks In the locking lugs and the bolt cam pin hole area. Use black light If available; otherwise, use a glass of no more than 3X magnification or use a penetrant kit (item 25, app D). Pay close attention to the area where the locking lugs meet the body. Replace bolt assembly if bolt is defective.

WARNING

Dry cleaning solvent is flammable and toxic and should be used in a well-ventilated area. The use of rubber gloves is necessary to protect the skin when washing rifle parts.

4. Use penetrant kit (Item 25, app D) to check for cracks in bolt as follows:

(a) The area to be inspected must be clean, free of oil, etc. Spray a small amount of remover on the area to be inspected, let dry and wipe off with a wiping rag.

(b) Spray penetrant (only enough to wet the area) on the area of the bolt (1) to be inspected.

(c) Spray developer over the penetrant and let the developer work. Cracks will be indicated by a change in color where there is a crack. If there are cracks, the component 1' unserviceable.

(d) Pay close attention to the area where the locking lugs (2) meet the body.

(e) If there are no cracks, spray remover on the area; let dry and wipe off with a wiping rag. Oil the area to prevent corrosion.

(f) Replace bolt assembly If bolt (1) is defective.

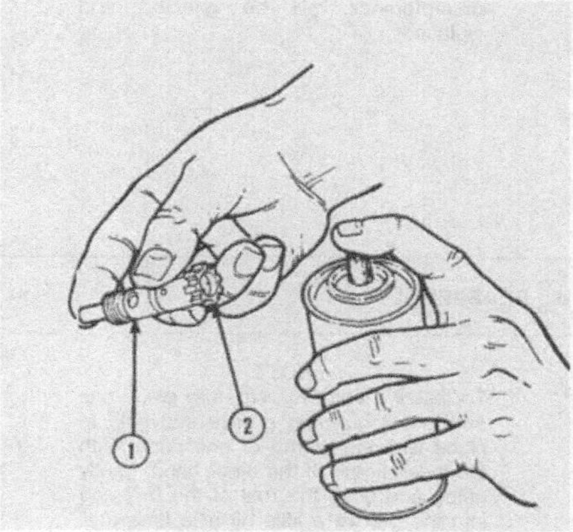

NOTE

Replacement of the bolt assembly will require that the headspace be tested (p 3-45, TEST).

3-8. BOLT ASSEMBLY (CONT.).

c. TEST

Test bolt (1) for elongated or oversized firing pin hole using special no-go plug gage (2) PN 12620101.

NOTE

Bolts with firing pin holes which permit the special no-go plug gage to fully penetrate at any position on the circumference will be rejected and replaced.

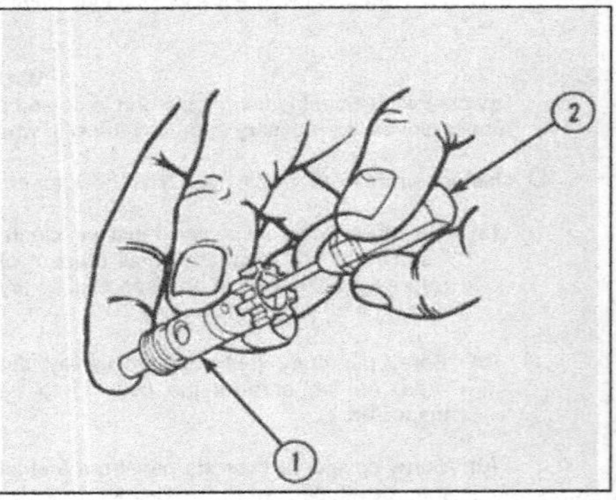

d. REASSEMBLY

NOTE

To install a bolt ring, carefully place one end in the bolt ring groove and hold In place with the thumb of one hand With the index finger of the other hand, gently guide and push the rest of the bolt ring into the groove a little bit at a time until the entire bolt ring is in place.

1. Install the three bolt rings (1) one at a time onto the bolt (2) using care not to bend or "spring" new bolt rings Stagger the bolt ring gaps (approximately ' turn apart).

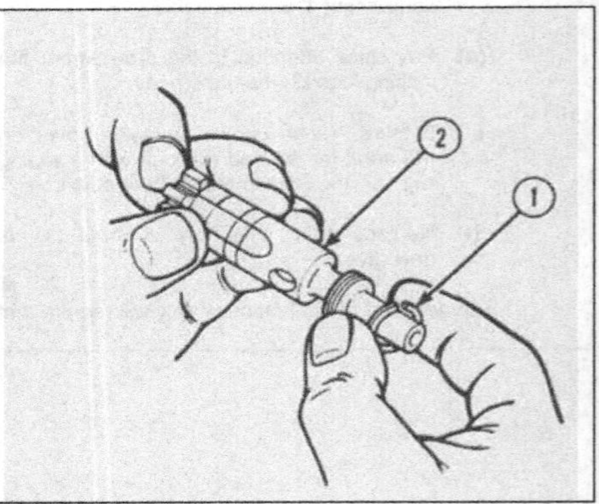

NOTE

Make certain bolt ring gaps are staggered to prevent loss of gas pressure. New bolt rings will make installing the bolt assembly difficult.

Lubricate inside key and bolt carrier assembly and use gentle pressure when installing.

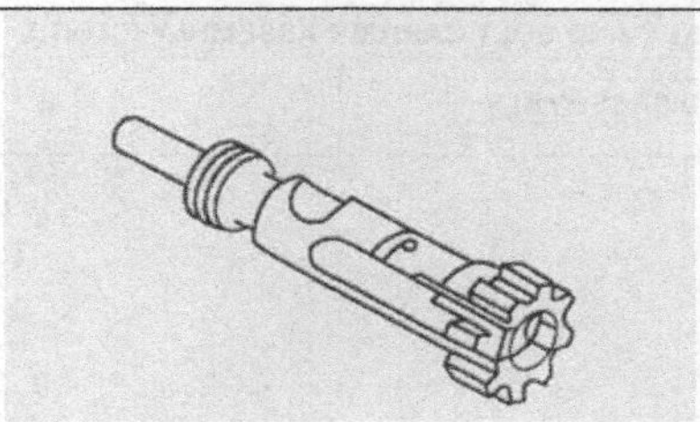

2. Reassemble rifle, refer to page 3-77.

3-9. KEY AND BOLT CARRIER ASSEMBLY.

This task covers:

a Disassembly
b Repair
c Reassembly

INITIAL SETUP

Tools
Field Maintenance Basic Less Power Small Arms Shop Set (item 1, app B)
(ARMY) Small Arms Repairman Tool Kit (item 3, app B)

Materials/Parts
Carrier and key screws (2) (8448508)

Equipment Conditions
3-16 Key and bolt carrier assembly removed

3-9. KEY AND BOLT CARRIER ASSEMBLY (CONT.).

a. DISASSEMBLY

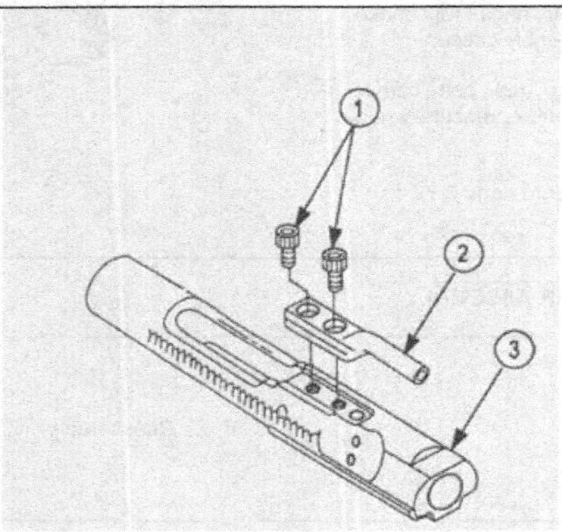

NOTE

Do not disassemble the key and bolt carrier assembly unless the bolt carrier key is defective as determined by inspection procedures described in section III, preceding.

1. Using socket wrench handle and tight fitting 1/8 inch socket head screw socket wrench attachment, remove two carrier and key screws (1).

NOTE

The heads and part of the bolt carrier key may be ground off in order to remove bolt carrier key from bolt carrier if carrier and key screws cannot otherwise be removed.

2. Remove bolt carrier key (2) from bolt carrier (3).

b. REPAIR

NOTE

Do not retorque carrier and key screws if staking marks do not indicate loosening screws.

Repair by replacing, torquing, and restaking carrier and key screws. Refer to the following reassembly procedures.

c. REASSEMBLY

NOTE
Do not reuse old carrier and key screws New carrier and key screws must be used at assembly.

1. If disassembled, place bolt carrier (1) In vise using vise law caps Install and position bolt carrier key (2) on bolt carrier (1).

2. Install two carrier and key screws (3). Always use new screws.

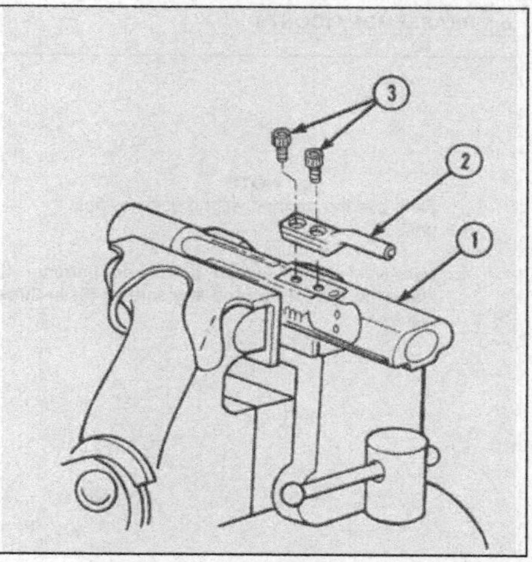

3. Use a tight-fitting 1 8 inch socket head screw wrench attachment and an Inch-pound torque wrench to torque the carrier and key screws (3) to 35 to 40 Inch-pounds (3 95 to 4 52 N m)

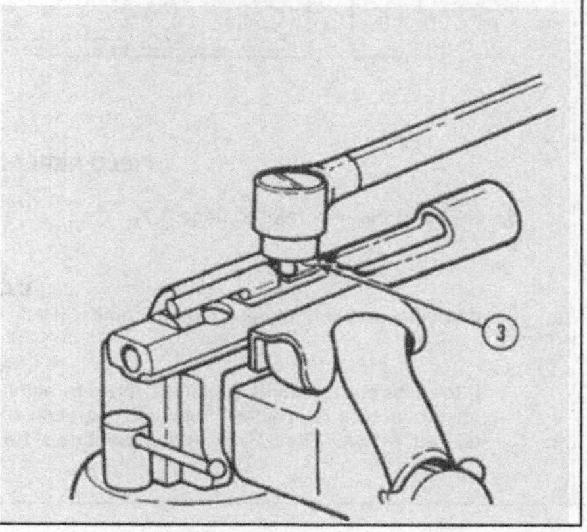

3-9. KEY AND BOLT CARRIER ASSEMBLY (CONT.).

c. REASSEMBLY (CONT.)

NOTE
Field staking method will be used by field units.

4. Use solid center punch and hand hammer to stake the two carrier and key screws (3) In three places.

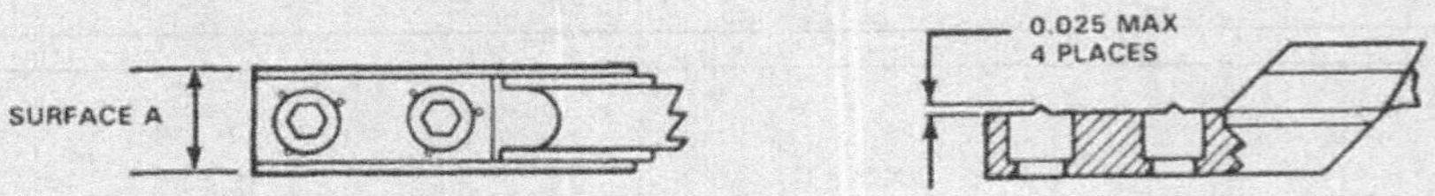

FIELD REPLACEMENT STAKING

5. Reassemble rifle, refer to page 3-77

CAUTION
If blanks are used, blank firing attachment (BFA) must be attached

NOTE
If the bolt carrier key is replaced, three to eight rounds of blank or ball ammunition must be fired to ensure a seal is created. Manual operation of the rifle may be required. If blank ammunition is utilized, M16A2 Blank Firing Attachment must be adapted

3-10. UPPER RECEIVER AND BARREL ASSEMBLY.

This task covers:

a Disassembly
b Inspection/Cleaning
c Repair
d Reassembly
e Test

INITIAL SETUP

Test Equipment
Tool and Gage Set (item 2, app B)

Tools
(ARMY) Small Arms Repairman Tool Kit (item 3, app B)
Field Maintenance Basic Less Power Small Arms Shop Set (item 1, app B)
Modified Needle Nose Pliers (fig E-7, app E)

Materials/Parts
Brush, cleaning, small (item 3, app D)
Carbon removing compound (item 8, app D)
Cloth, abrasive (item 13, app D)
Dichloromethane, technical (item 15, app D)
Dry cleaning solvent (item 16, app D)
Gloves, chemical and oil protective (item 18, app D)
Grease, molybdenum disulfide (item 19, app D)
Lubricant, solid film (item 21, app D)
Pan, wash (item 24, app D)
Polyethylene (item 32, app D)
Sealing compound (item 28, app D)
Target (item 31, app D)

References
FM 23-9
TM 9-1005-319-10

Equipment Conditions
3-15 Upper receiver and barrel assembly removed
2-43 Handguard assemblies removed

General Safety Instructions
To avoid injury to your eyes, use care when removing and installing spring-loaded parts.
When using solid film lubricant or dichloromethane, be sure the area is well ventilated.
When using carbon removing compound, avoid skin contact If carbon removing compound comes in contact with the skin, wash thoroughly with running water Using a good lanolin base cream after exposure to the compound is helpful Using gloves and protective equipment is required.

3-10. UPPER RECEIVER AND BARREL ASSEMBLY (CONT.).

a. DISASSEMBLY

1. Using ball-peen hammer and 5/64 inch diameter drive pin punch, drive spring pin (1) (which retains the gas tube) out of front sight assembly (2).

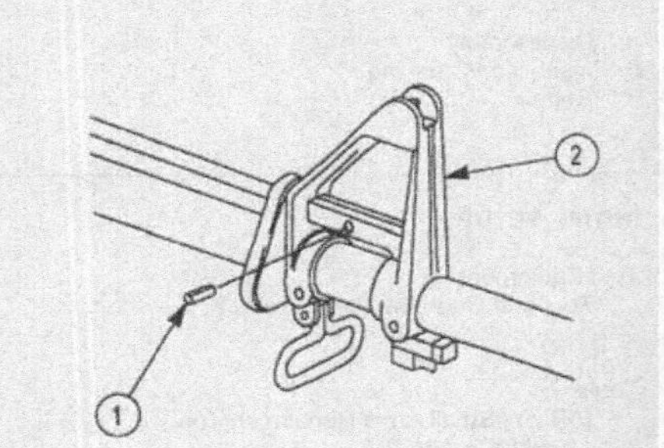

2. Slide gas tube (3) back into upper receiver assembly (4) to clear front sight assembly (2). Then lift slightly, pull forward, and remove gas tube (3).

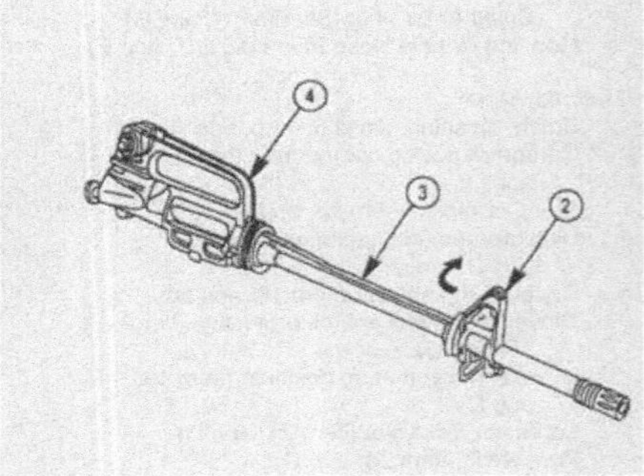

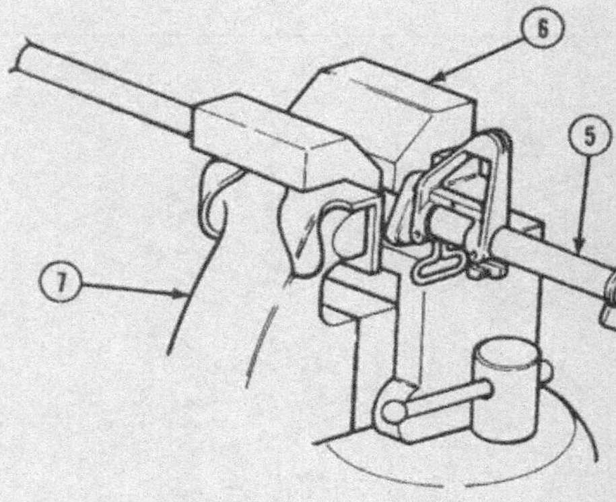

3. Position upper receiver and barrel assembly (5) in barrel removal fixture (6) and secure both In machinist's vise (7).

4. Using combination wrench (8) and 12 inch drive handle, remove compensator (9).

5. Remove peel washer (10) being careful not to lose or bend thin sections.

6. Remove upper receiver and barrel assembly (5) from barrel removal fixture and machinist's vise.

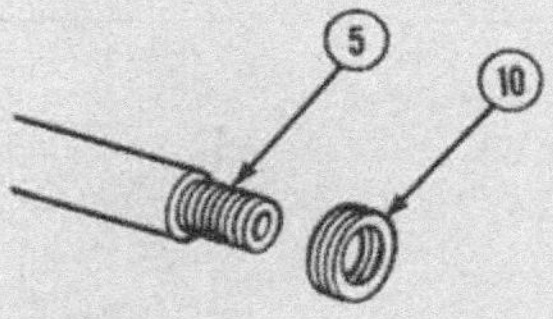

3-10. UPPER RECEIVER AND BARREL ASSEMBLY (CONT.).

a. DISASSEMBLY (CONT.)

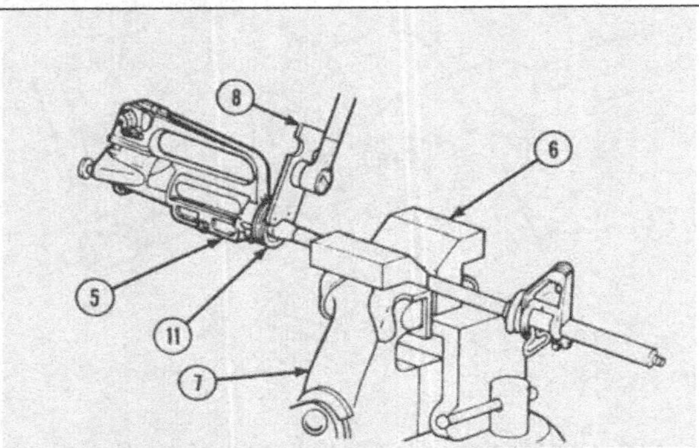

7. Place upper receiver and barrel assembly (5) Into barrel removal fixture (6) and clamp Into machinist's vise (7).

NOTE

Be sure all three drive pins on combination wrench are fully engaged with barrel nut assembly. Wrench must be pushed toward upper receiver assembly to compress the slip ring spring In barrel nut assembly. Do not use a torque wrench to loosen the barrel nut assembly.

8. Using '2 Inch drive handle and combination wrench (8), loosen barrel nut assembly (11)

9. Separate upper receiver assembly (4) from barrel assembly (12)

10. Remove barrel assembly (12) from machinist's vise and barrel removal fixture

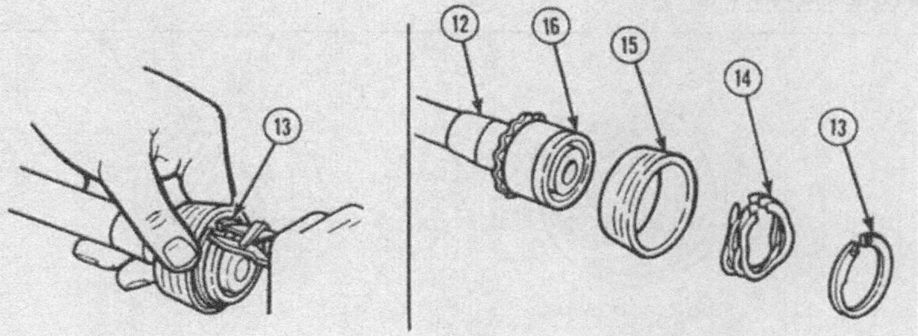

11. Remove retaining ring (13) using retaining ring pliers.

12. Remove slip ring spring (14) and handguard slip ring (15).

13. Do not remove barrel nut (16) from barrel assembly (12).

b. INSPECTION/CLEANING

WARNING

When using carbon removing compound, avoid skin contact. If carbon removing compound comes In contact with the skin, wash thoroughly with running water. Using a good lanolin base cream after exposure to compound is help ful. Using gloves and protective equipment is required.

1. Inspect gas tube for cracks. Replace if defective

2. Use carbon removing compound to remove carbon deposits from interior and exterior of gas tube. If a large amount of carbon is found and cannot be removed, replace the gas tube

NOTE

A small arms cleaning brush (bore) (item 4, app D) may be used to clean Interior of front sight assembly where gas tube is secured.

3-10. UPPER RECEIVER AND BARREL ASSEMBLY (CONT.).

b. INSPECTION/REPAIR (CONT.)

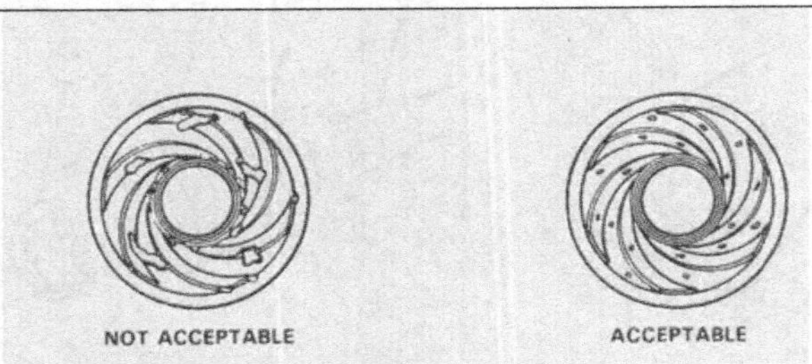

3. Inspect bore for burrs, cracks, rust, bulges, and pits using the following guidelines

(a) Pits no wider than a land or groove and no longer than 3'8 inch (0.95 cm) are al lowed in the bore.

(b) Lands that appear dark blue due to coating of gliding metal from projectiles are allowable.

(c) Definitely ringed bores or bores ringed sufficiently to bulge the outside surface of the barrel are cause for rejection. Replace barrel assembly if defective

4. If the upper receiver is separated from the barrel assembly, Inspect chamber for pits utilizing a flashlight. Pits 1'8 inch 10 32 cm) In length are cause for rejection Replace rifle barrel assembly If defective

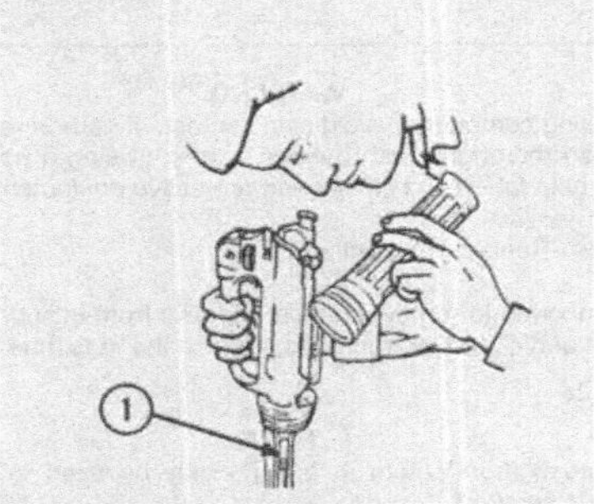

5. If upper receiver and barrel assembly is assembled, Inspect chamber using reflector tool (1) and flashlight. Pits 1 '8 inch (O 32 cm) In length are cause for rejection. Replace rifle barrel assembly If defective. If rifle barrel assembly Is replaced, Inspect headspace (p 3-45).

6. Inspect upper receiver assembly for cracks, corrosion, wear, or damage

(a) Small dents or gouges that do not affect functioning will not be cause for rejection.

(b) If upper receiver assembly contains cracks or holes, the upper receiver assembly will be replaced

7. Inspect all parts for damage and wear. Replace all defective parts.

NOTE

Damaged or missing teeth of the barrel nut is not cause for rejection provided the proper torque value can be obtained during installation using the tools depicted. If removal of the barrel is not possible with the combination tool, a pipe wrench or other such tool may be used during removal.

8. Inspect front sight guards for bends, if bent see page 3-37 for repair procedures.

c. REPAIR

1. Repair corroded upper receiver assembly surfaces as follows:

(a) Sand corroded area with abrasive cloth and make sure all corrosion has been removed.

WARNING

When using solid film lubricant or dichloromethane, be sure the area is well ventilated

(b) Wash area with technical dichloromethane (methylenechloride) to re move all dirt, grease, and foreign material

(c) Apply sealing compound, mixed In accordance with manufacturer's directions, to areas to be filled

(d) Spread sealing compound as smoothly as possible Into defective area using a putty knife or similar tool

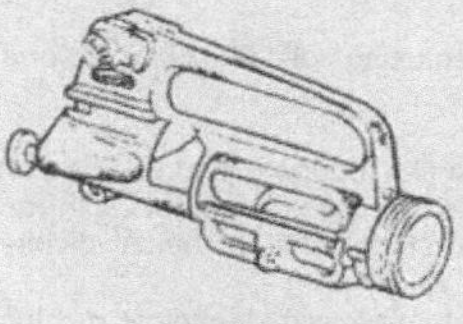

CORRODED
(REPARABLE)

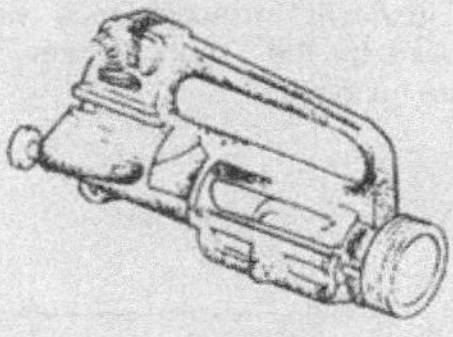

CORRODED
(NONREPARABLE)

3-10. UPPER RECEIVER AND BARREL ASSEMBLY (CONT.).

c. REPAIR (CONT.)

NOTE
Do not feather edges

(e) Place a sheet of polyethylene (Item 32, app D), cut to size, over filled area Rub by hand to smooth.

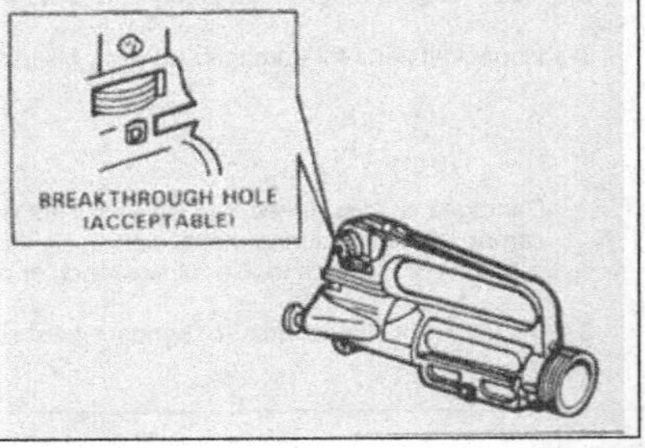

2. After curing, remove polyethylene sheet In accordance with Instructions by the manufacturer.

WARNING
When using solid film lubricant or dichloromethane, be sure the area is well ventilated.

CAUTION
Solid film lubricant is to be used only as an exterior surface protective finish and touchup If solid film lubricant comes in contact with recoiling parts or functional surfaces of the rifle, remove immediately by washing with technical dichloromethane.

3. Wash area with technical dichloromethane (methylenechloride) ; to remove dirt and grease, and foreign material.

4. Roughen area to be refinished with abrasive cloth and clean surface again Do not touch the area with fingers.

5. Repair shiny surfaces by spraying a coat of solid film lubricant in accordance with Instructions supplied by the manufacturer Dry 24 hours before handling.

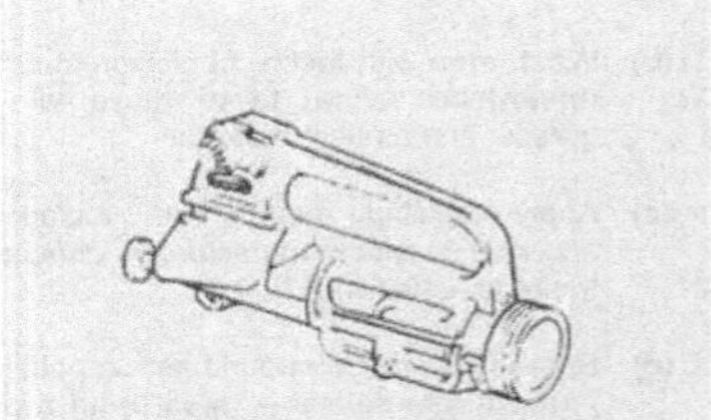

SHINY SURFACES (REPARABLE).

6. Straightening bent front sight guards (1) as follows:

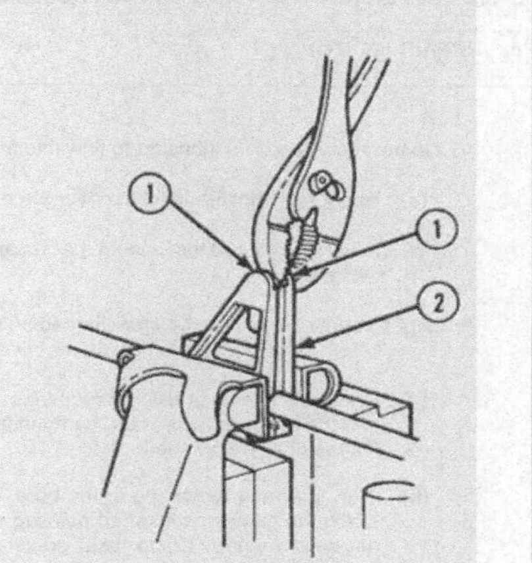

NOTE

Remove spring before heating. (Heat will damage spring.) The sight post and plunger may be reused unless damaged.

(a) Remove front sight post, detent, and helical spring (see p 2-43).

NOTE

Use copper or brass caps (jaw Inserts on bench vise to prevent damage to front sight base (2) during clamping.

(b) Place front sight base (2) in a bench vise.

(c) Heat front sight guards (1) and bend with pliers. The front sight guards (1) should be put back as nearly as possible to the original position. Allow front sight housing to air cool.

WARNING

Dry cleaning solvent is flammable and toxic and should be used in a well ventilated area. The use of rubber gloves is necessary to protect skin when washing rifle parts.

(d) Roughen any damaged surface of front sight guards with abrasive cloth and clean with dry cleaning solvent. Always wear rubber gloves when using dry cleaning solvent.

3-10. UPPER RECEIVER AND BARREL ASSEMBLY (CONT.).

c. REPAIR (CONT.)

CAUTION
Do not allow solid film lubricant to flow into front sight post threaded well.

(e) Apply solid film lubricant to cover the damaged finish.

(f) If front sight guards cannot be straightened utilizing the above procedures, reduce the rifle barrel assembly.

7. Slightly bent barrels may be straightened as follows'.

(a) Check straightness using straightness gage 8448202 (p 3 45) If the barrel falls the straightness test, and the gage remains in the barrel In the area of the front sight assembly, perform step (b) to determine If it may be straightened.

(b) With the gage remaining in the bore, hold the rifle in a vertical position with the end of the barrel into which the gage was inserted pointing up Ensure that if when the gage passes through the barrel it will not be damaged Using hand pressure' ONLY, flex the portion of the barrel between the front sight assembly and the compensator in all four directions (left, right, forward, and back) If the barrel is only slightly bent, the gage will drop through when the barrel is flexed in one of these directions Note the direction which allowed the gage to drop through the' barrel.

CAUTION
Remove the gage from the barrel before continuing

NOTE
If the gage does not pass through the barrel when it is flexed, replace the rifle barrel assembly.

(c) Place the barrel in a vise using appropriate protective laws . Clamp the barrel between the front sight assembly and the compensator approximately 1 inch (2.54 cm) from the front sight assembly The rifle barrel assembly should be in a horizontal position with the side noted in step (b) toward you.

CAUTION
Do not apply pressure to the receiver.

(d) Grasp the BARREL near the receiver so that when force is applied the barrel will flex In the same direction as noted In step (b).

(e) Give the barrel a sharp jerk of approximately 20 to 40 pounds of force (f) Remove the barrel from the vise and recheck straightness (step (al).

(g) If gage still will not pass through the barrel, perform step (b) to determine direction of bend If the barrel is still bent in the same direction as before, perform steps (c) through (f) using slightly more force. If the barrel Is now bent In the opposite direction, replace the rifle barrel assembly.

(h) If the gage passes freely through the barrel, the barrel shall be considered straight and continue In service.

(i) If the barrel has been straightened, the rifle must be targeted (p 3-45).

d. REASSEMBLY

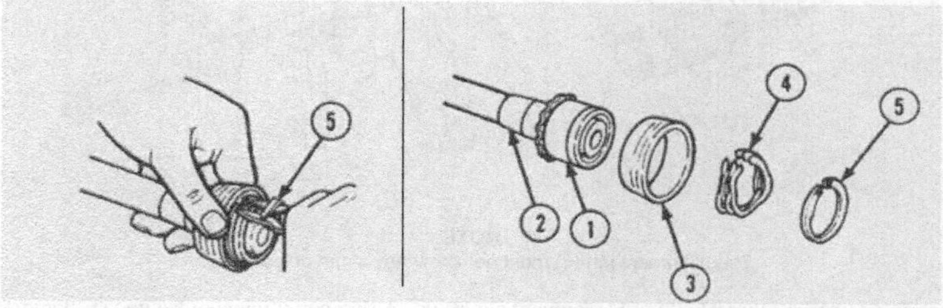

1. Position barrel nut (1) by sliding it to the rear of rifle barrel assembly (2) as far as possible
2. Slide handguard slip ring (3) over barrel nut (1).
3. Press slip ring spring (4) from both sides and insert It Into handguard slip ring (3)
4. Install retaining ring (5) against slip ring spring (4) using retaining ring pliers Snap retaining ring (5) to barrel nut (1).

NOTE
After cleaning, apply molybdenum disulfide grease to threads of barrel nut assembly before Installation.

3-10. UPPER RECEIVER AND BARREL ASSEMBLY (CONT).

d. REASSEMBLY (CONT)

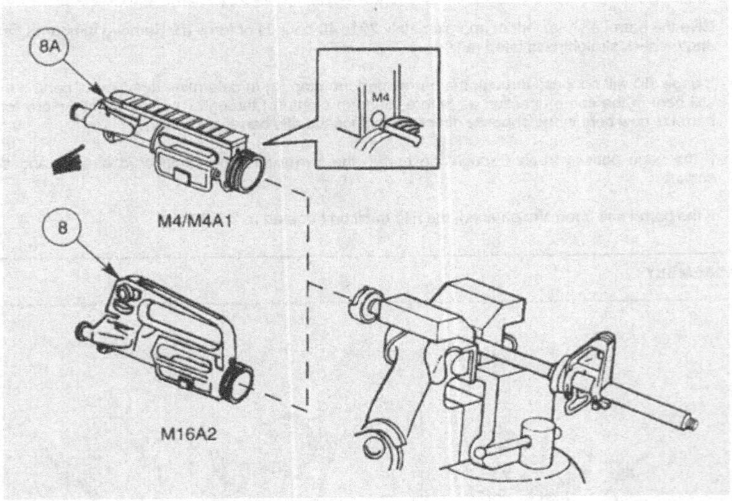

NOTE
The alignment pin (6) must not show any signs of looseness.

5. Position rifle barrel assembly (2) with alignment pin (6) up. Using barrel remover fixture (7), clamp rifle barrel assembly (2) in vise.

NOTE
The slot should fit the alignment pin perfectly with very little or no rotational play present. Note the play in a new rifle barrel assembly and new upper receiver assembly and use this as a guide.

6. Align upper receiver assembly (8) or (8A) using alignment pin (6) and the slot in upper receiver assembly (8) or (8A). Install over end of rifle barrel assembly (2).

7. Wipe upper receiver threads clean and ensure there are no burrs. Apply molybdenum disulfide grease to the threads before installation.

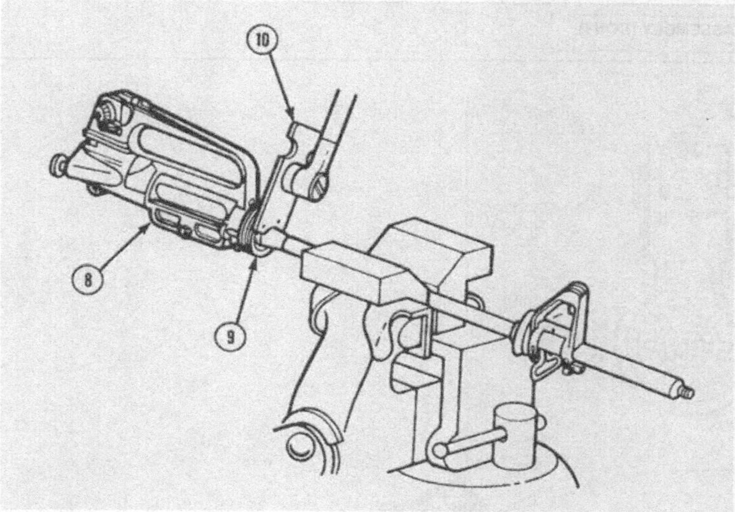

8. Engage threads of barrel nut assembly (9) with upper receiver assembly (8).

9. Using combination wrench (10) and torque wrench, torque barrel nut assembly (9) to 30 ft-lb (40.5 N-m). Torque is measured when both wrenches are used together.

NOTE

Three times torquing procedure provides for a better thread fit and prevents barrel nuts from becoming loose. Do not use the torque wrench for loosening.

10. Make certain all three drive pins on combination wrench are engaged with barrel nut assembly (91 Loosen and repeat torque operation. Then loosen the barrel nut again.

3-10. UPPER RECEIVER AND BARREL ASSEMBLY (CONT).

d. REASSEMBLY (CONT)

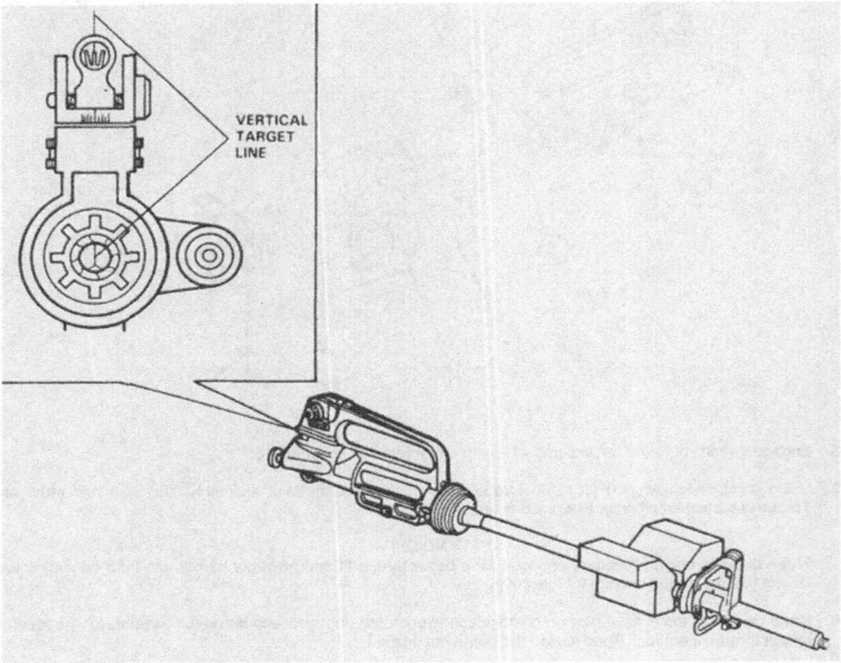

11. Front sight post must be Installed. Loosen the vise and align the bore on a distant vertical target Center the target In the bore from 12 o'clock through 6 o'clock The, front sight post should be on line and vertical with the target. Tighten vise Adjust the rear sight windage until a proper sight picture is obtained on the vertical target The rear sight aperture will be approximately In the center of the rear sight base If the rifle barrel assembly is properly aligned in the upper receiver assembly

NOTE

If rifle barrel assembly (usually new) is not properly aligned In the upper receiver assembly (usually an old part), excessive windage will be present and the upper receiver assembly will require replacement to obtain the proper fit between the alignment pin and slot.

CAUTION

Do not torque over 80 ft-lb (108 N-m) while tightening the barrel nut assembly to the next hole, to allow for proper alignment of gas tube.

NOTE

Do not attempt to hold the upper receiver assembly with a pry bar, however, If the rifle barrel assembly turns In the holding fixture, a pry bar may be used through the front sight assembly base to help prevent the rifle barrel assembly from turning In the holding fixture. Use care not to distort or bend front sight assembly or retaining pins. Use "buddy system" to hold pry bar.

12. Torque the barrel nut assembly again to 30 ft-lb (40.5 N-m) while maintaining sight alignment. The barrel nut assembly may be tightened beyond 30 ft-lb (40.5 N-m) to align the barrel nut assembly serrations for proper gas tube clearance Never loosen the barrel nut assembly to align for gas tube clearance.

13. Check alignment of barrel nut assembly (9) with upper receiver assembly (8). The front 8 inches (20.32 cm) of a gas tube may be used as an alignment tool (see Illustration). This Is Inserted Into the bolt carrier key and then Inserted into the rear of the receiver. If the parts of the barrel nut assembly are properly aligned, the tool will pass freely and lay top dead center along the top of the barrel. A number 15 twist drill (0.180 inch) may also be needed as an alignment tool. If necessary, tighten barrel nut assembly to next hole to allow proper alignment.

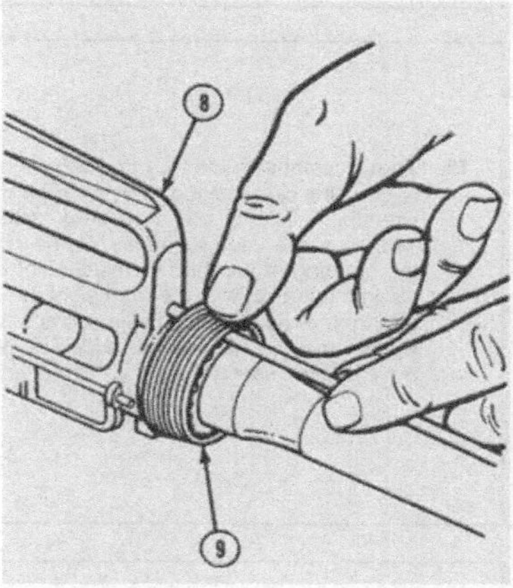

3-10. UPPER RECEIVER AND BARREL ASSEMBLY (CONT).

d. REASSEMBLY (CONT)

NOTE

The peel washer may be heated to remove thin sections. Always place thin section to rear.

14. Install peel washer (11) and compensator (12) on rifle barrel assembly (2).

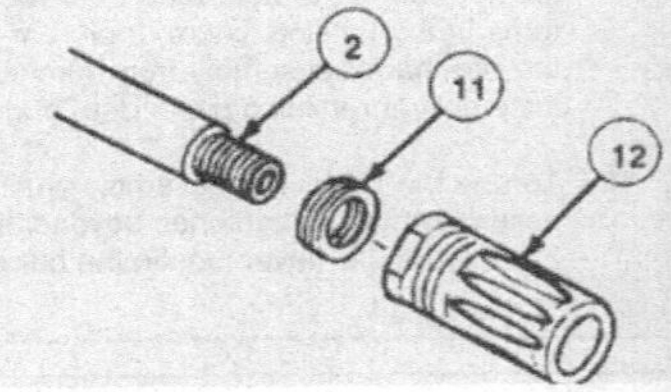

15. Torque compensator (12) to 15-20 ft-lb (20-27 N-m) using combination wrench (10) and torque wrench. Torque is measured when both wrenches are used together. Some portion of the openIng of the third or middle slot must be straight up (align with the front sight post) (12.1) at proper torque level. Thin sections of the peel washer may be removed or added as required. Save unused sections.

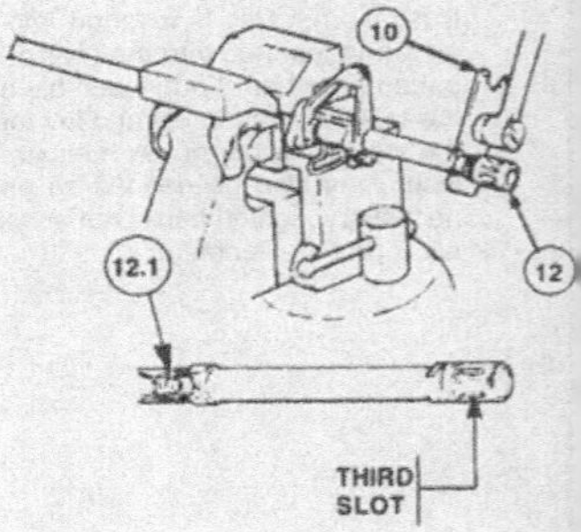

16. Slide gas tube (13) through the barrel nut assembly (9) and then slide forward, Inserting gas tube into hole In the front sight assembly base (14).

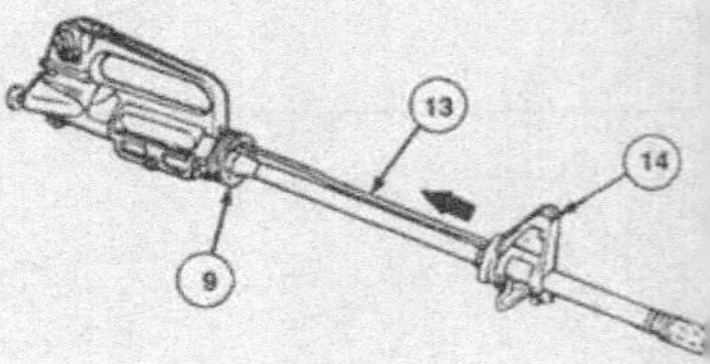

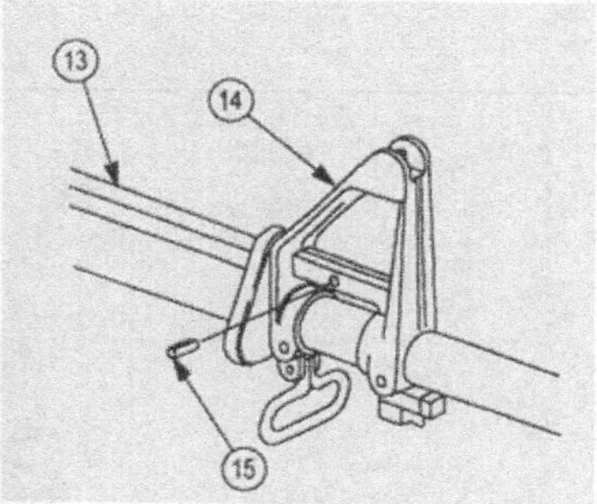

17. Align the holes in gas tube (13) and front sight assembly base (14).

NOTE

To assist in installing spring pin (1) modified nose pliers may be used.

18. Using ball-peen hammer and 5/64-inch diameter drive pin punch, drive spring pin (15) into front sight assembly base (14) to secure gas tube (13).

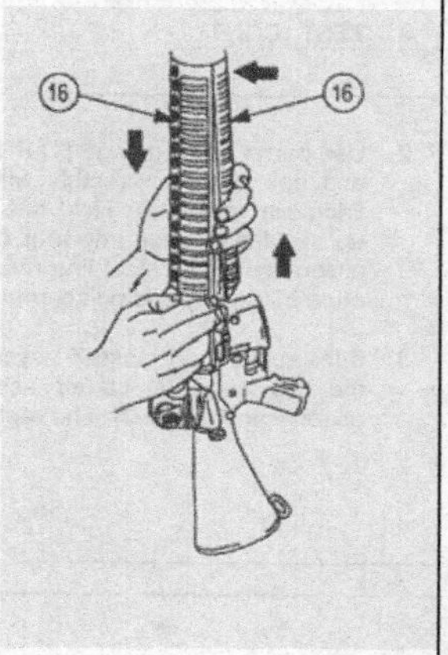

19. Install upper and lower handguard assemblies (16).

e. TEST

1. The following information pertains to the use of breech, bore, and other gages:

(a) All M1 6A2 barrels and chambers are chromed.

(b) Barrel erosion gage, PN 8448496 (normally used on M16A1 fully-chromed barrels), can be used to gage M16A2 barrels.

(c) The bore straightness gage, PN 8448202, is required for use on all barrels. The gage must pass through the barrel without being forced.

3-10. UPPER RECEIVER AND BARREL ASSEMBLY (CONT).

e. TEST (CONT)

2. Use barrel erosion gage (1) PN 8448496. Install key and bolt carrier assembly with bolt assembly and firing pin removed. Hold rifle vertical with receiver up. Insert gage into rear of key and bolt carrier assembly. The reject line must be read at the rear edge of the key and bolt carrier assembly.

3. If the reject mark passes beyond the rear surface of the key and bolt carrier assembly, the barrel is unserviceable and shall be replaced.

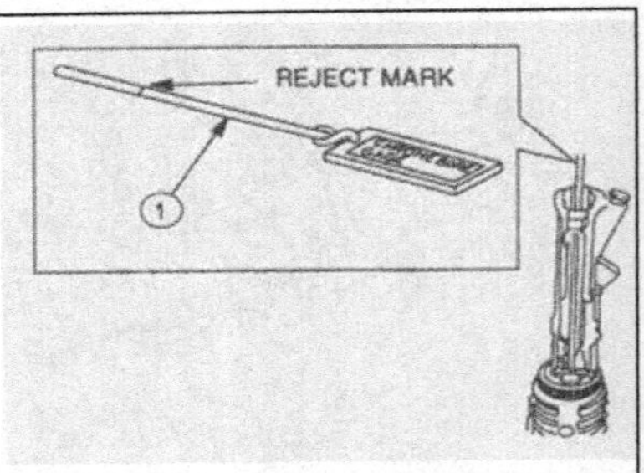

NOTE
Ensure barrel is clean prior to performing the following test.

4. Check straightness of bore using straightness gage (2) PN 8448202. Put gage in barrel. Tilt barrel and allow gage to fall through. Catch gage.

5. Gage must pass freely through barrel. If gage does not pass through barrel, recheck as follows. Hold upper receiver and barrel assembly (3) in vertical position with muzzle pointed down; insert gage into chamber end of barrel. Release gage and catch it as it exits muzzle end. If gage passes freely through the barrel, barrel is acceptable. If it does not, the barrel must be straightened or replaced. (See page 3-38 for straightening instructions/ procedures.)

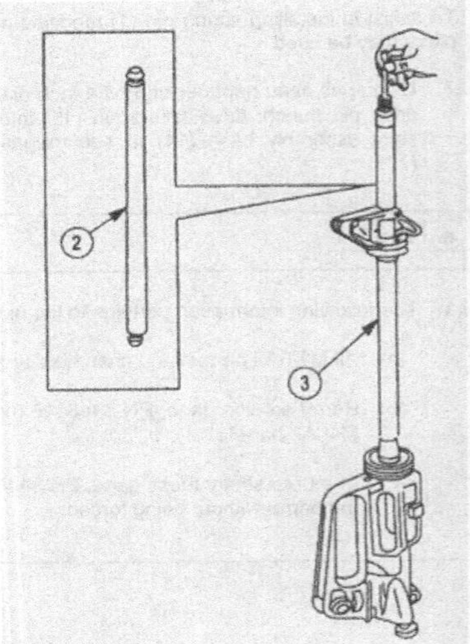

6. Assemble charging handle assembly (4), bolt assembly, and key and bolt carrier assembly (5) into upper receiver assembly (6).

7. Insert headspace gage (7) PN 7799734 in chamber.

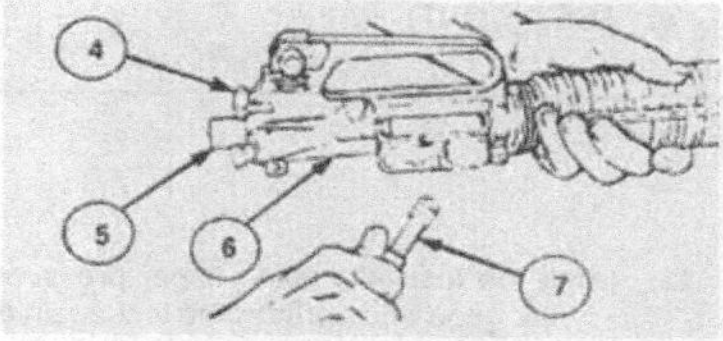

NOTE

For the purpose of this test "light finger pressure" is defined as 8 1/2 to 8 3/4 pounds.

8. Check headspace by pressing key and bolt carrier assembly (5) and charging handle assembly (4) forward using light finger pressure.

9. Bolt should not rotate to locked position. Key and bolt carrier assembly (5) must protrude from rear of upper receiver assembly (6) for proper headspace. If excessive headspace, first replace old bolt assembly with an unused bolt assembly and then recheck. If headspace is not corrected, replace rifle barrel assembly; then recheck with the original bolt assembly to determine if the bolt assembly is still good or if the bolt assembly should be replaced also.

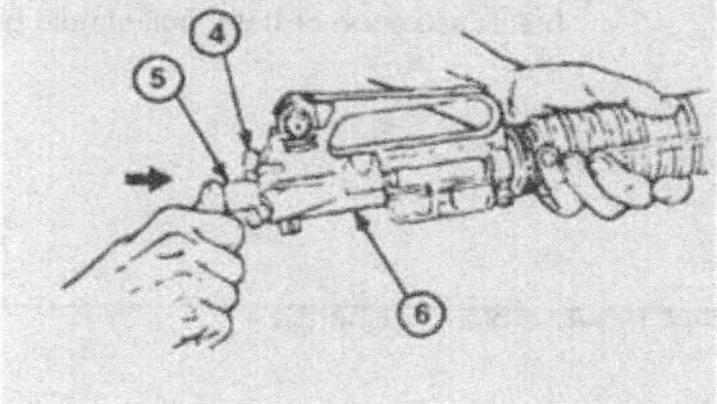

10. Remove key and bolt carrier assembly, bolt assembly, charging handle assembly, and headspace gage.

11. Reassemble rifle, refer to page 3-77.

NOTE

Rifles which have been rebarreled must be function-fired with seven rounds of 5.56mm ball ammunition. After rebarreling, the rifle must be targeted with three rounds of 5.56mm ball ammunition at 25 meter range using target. Refer to TM 9-1005-319-10 and FM 23-9

3-10. UPPER RECEIVER AND BARREL ASSEMBLY (CONT).

e. TEST (CONT)

12. (a) If the test fails using finger pressure, remove the gage and perform the test again as follows: With the muzzle down, stack 8 1/2 to 8 3/4 pounds of trigger weights (8) on a locally fabricated spacer/weight (9) on the bolt carrier assembly (10) Insert headspace gage (11) and test per above instructions.

(b) Remove trigger weights (8), spacer/weight (9), bolt carrier assembly (10), charging handle (12), and headspace gage (11).

(c) If excessive headspace, first replace bolt assembly and then recheck If headspace is not corrected, replace barrel assembly; then recheck with the original bolt to determine if the bolt is still good or If the bolt should be replaced also.

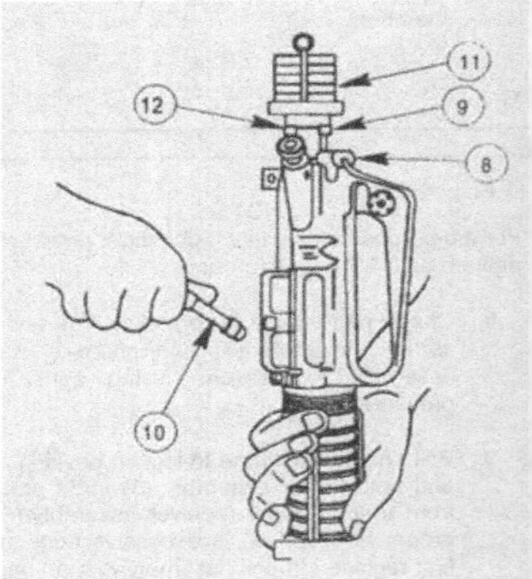

3-11. UPPER RECEIVER ASSEMBLY AND REAR SIGHT ASSEMBLY.

This task covers:

a. Disassembly
b. Inspection
c. Repair
d. Lubrication
e. Reassembly
f. Mechanical Zero Procedures (A.F. Only)

INITIAL SETUP:

Tools
(ARMY) Small Arms Repairman Tool Kit (item 3, app B)
Field Maintenance Basic Less Power Small Arms Shop Set (item 1, app B)

Materials/Parts
Cleaner, lubricant, and preservative (CLP) (item 9, app D)
Lubricant, solid film (item 21, app D)
Index screw (9349065)

Equipment Conditions
3-29 Upper receiver assembly removed.

General Safety Instructions
To avoid injury to your eyes, use care when removing and installing spring-loaded parts.
When using solid film lubricant or dichloromethane, be sure the area is well ventilated.

a. DISASSEMBLY

WARNING
To avoid injury to your eyes, use care when removing and installing springloaded parts.

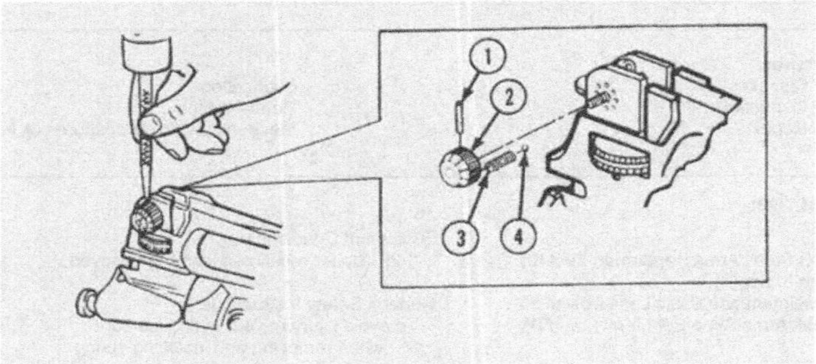

CAUTION

Be sure to catch small parts (1, 2, 3, and 4). AIR FORCE ONLY: Use magnet to keep from losing small parts (1, 3, and 4).

1. Drive out spring pin (1) using a hammer and 1/16 inch punch.

2. Catch rear sight assembly windage knob (2), helical spring (3), and ball bearing (4).

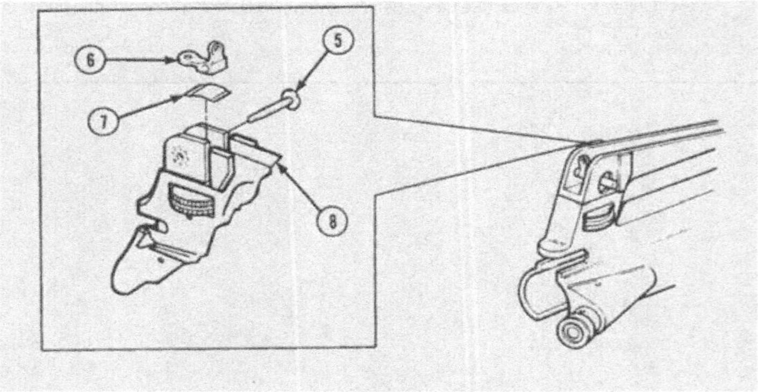

3. Using a flat-bladed screwdriver, remove rear sight assembly windage screw (5)

4. Remove sight aperture (6) and flat spring (7) from upper receiver (8).

3-11. UPPER RECEIVER ASSEMBLY AND REAR SIGHT ASSEMBLY (CONT).

a. DISASSEMBLY (CONT)

WARNING
To avoid injury to your eyes, use care when removing and installing spring-loaded parts.

5. Drive out spring pin (9) using a 3/32 Inch punch Catch helical spring (10) when punch is withdrawn.

6. Rotate elevation index (11) until rear sight assembly base (12) clears upper receiver (8). Catch ball bearing (13) and helical spring (14) as rear sight base clears.

7. Push elevation index (11) out with thumb using slight rotation motion. Catch ball bearing (15) and helical spring (16).

8. Use 1 16 Inch allen wrench to remove index screw (17). Discard Index screw (1 7) Separate elevation index (11) from elevation knob (18) by hand.

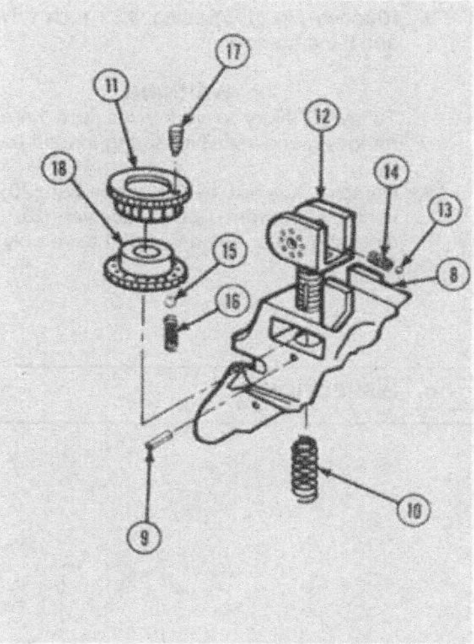

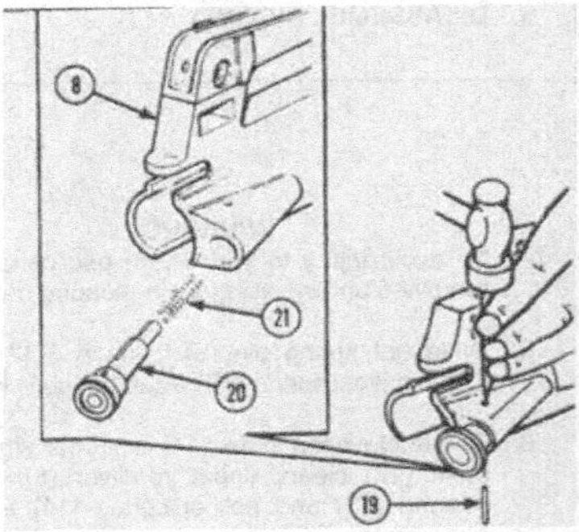

9. Remove pin (1 9) using 3,32 Inch drive pin punch and hand hammer .

WARNING

To avoid Injury to your eyes, use care when removing and installing spring-loaded parts.

10. Remove forward assist assembly (20) and helical spring (21) from upper receiver (8). For further disassembly of forward assist assembly see page 3-59.

b. INSPECTION

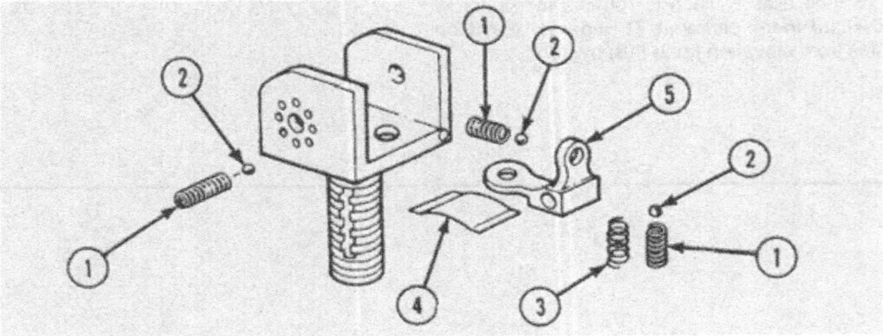

1. Check rear sight parts for serviceability. Inside of apertures should be round and distinct. Replace If defective

2. Visually inspect rear sight assembly helical springs (1) ball bearings (2), and helical spring (3) for breaks, bends, and missing parts Ball bearings should be smooth and round. Replace If defective

3. Check upper receiver for cracks, corrosion, and damage Clear drain hole with a piece of wire. Repair (p 3-29) or replace If defective.

4. Check that flat spring (4) retains sight aperture (5) firmly In either position. Replace flat spring (4) If sight aperture is not firm.

3-11. UPPER RECEIVER ASSEMBLY AND REAR SIGHT ASSEMBLY (CONT).

b. INSPECTION (CONT)

5. Check elevation Index 161 and windage knob (71 for legibility of markings Check underside of windage knobs for cracks. Detent indexing surfaces should be well formed.

6. Check rear sight assembly base (81 for serviceability Clear drain holes for springs Threaded portion of rear sight assembly base and elevation knob should be well formed.

7. Inspect rear sight guards for bends, if bent refer to the following page for repair procedures.

8. Inspect all parts for damage and wear Replace all defective parts.

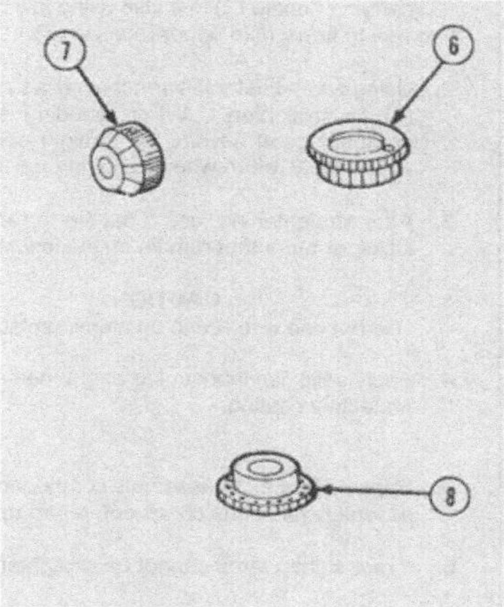

c. REPAIR

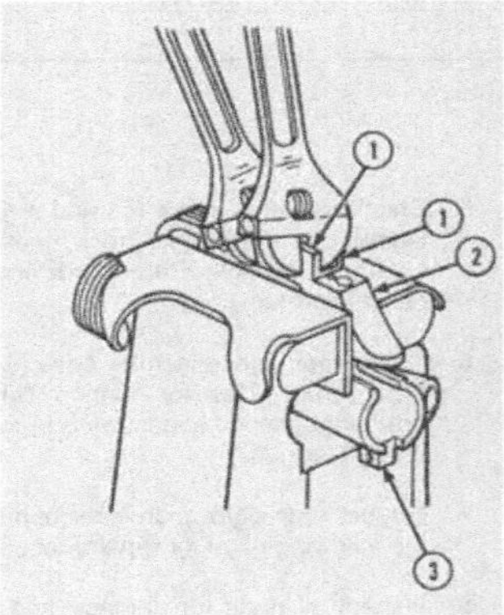

1. To straighten bent rear sight assembly guards (1), remove rear sight assembly components. Place carrying handle (2) In a vise using jaw clamps. Tighten vise to firmly hold upper receiver (3).

2. Using two adjustable wrenches, gradually bend guards (1) to straighten. When bending the guards (1), gradually bend beyond the straight point as the guard will partially return when bending pressure is stopped.

3. After straightening, use a flat file to remove any nicks, kinks, or burrs that remain on the inside of guards (1).

CAUTION
Do not use wire brush on aluminum surfaces

4. Apply solid film lubricant to brightened area for final protective coating.

5. Replace rear sight assembly components and check that rear sight assembly functions properly. If rear sight assembly functions check out, return upper receiver assembly to service.

6. If rear sight guards cannot be straightened utilizing the above procedures, replace the upper receiver

d. LUBRICATION

Lubricate upper receiver assembly and rear sight assembly. Apply CLP to helical springs and ball bearings (three each) and threaded portion of screws before installation. Lubricate helical springs and ball bearings through their respective drain holes.

3-11. UPPER RECEIVER ASSEMBLY AND REAR SIGHT ASSEMBLY (CONT).

e. REASSEMBLY

WARNING
To avoid injury to your eyes, use care when removing and Installing spring-loaded parts.

1. Apply CLP to helical spring (1) and forward assist assembly (2) and Install them into upper receiver (3).

2. Install spring pin (4) using 3, 32 Inch drive pin punch and hand hammer

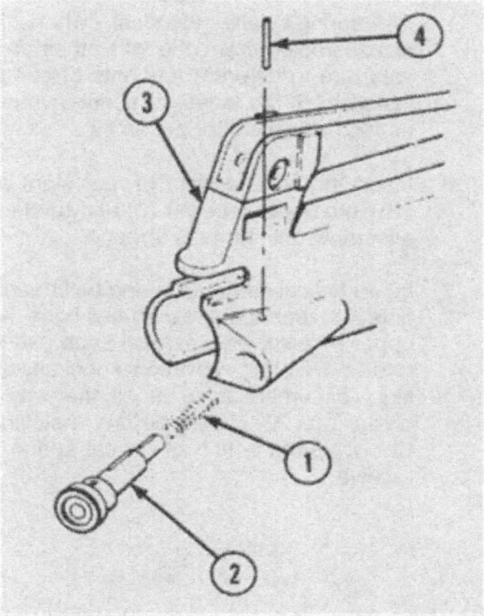

3. Assemble elevation knob (5), elevation index (6), and new index screw (7) using 1 16 Inch alien wrench. Do not overtighten Index screw as scale will require adjustment

4. Install ball bearing (8) and helical spring (9) using needle nose pliers or tweezers

5. Depress ball bearing (8) with a punch Inserted through access hole, and slide elevation knob assembly In upper receiver (3) from the side Center elevation knob assembly

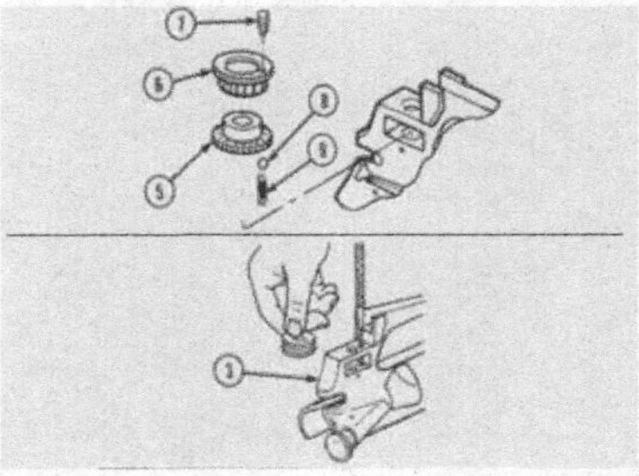

WARNING

To avoid injury to your eyes, use care when removing and installing spring-loaded parts.

NOTE

All springs are identical when new. Once disassembled from the rifle, their free length may vary due to different amounts of compression when installed. If the length of springs varies, use longest as #22, and the shortest as #9

6. Insert threaded portion of rear sight assembly base (10) into upper receiver (3) and rotate elevation knob assembly until threads engage.

7. Insert helical spring (11) and ball bearing (12) in their hole as rear sight assembly base is lowered into upper receiver as elevation knob assembly is further rotated. Rotate elevation knob assembly until rear sight assembly base is all the way down. Then come up 22 clicks before installing spring pin. Check spring action of helical spring (11) on upper receiver.

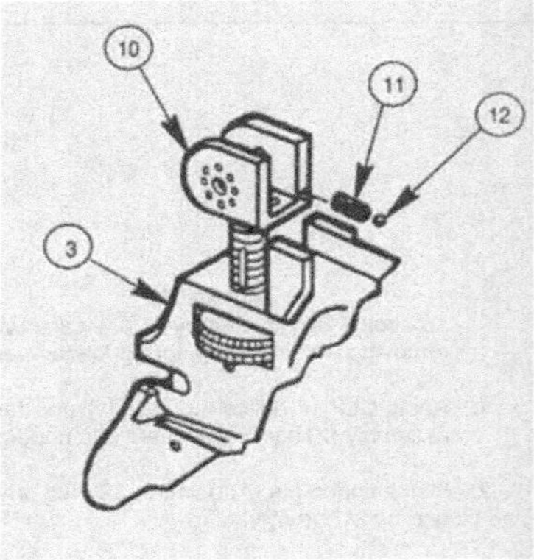

3-11. UPPER RECEIVER ASSEMBLY AND REAR SIGHT ASSEMBLY (CONT).

e. REASSEMBLY (CONT)

CAUTION

Ensure that spring pin (1 3) passes over helical spring (14), not through its coils

8. Insert helical spring (14) through underside of upper receiver (15). Compress helical spring with a small tip screwdriver (16) to install spring pin (1 3) Spring pin must pass over helical spring, not through its coils Rotate elevation knob (17) until rear sight base (18) Is all the way down.

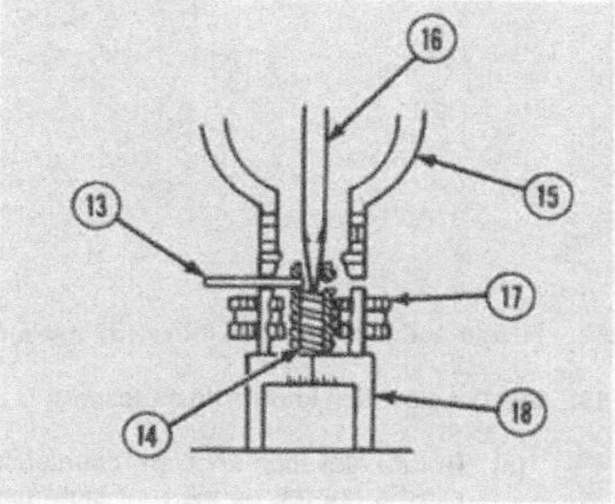

9. Install flat spring (19) and sight aperture (20) In rear sight assembly base (10). Install windage screw (21) with screwdriver.

10. Insert helical spring (22) and ball bearing 123) In windage knob (24)

NOTE

Tilt upper receiver toward wind age knob during positioning to prevent loss of ball bearing.

11. Position windage knob on shaft of windage screw (211 Align holes In windage knob with hole of shaft In windage screw and hold In alignment with a small screwdriver or punch Install spring pin (25).

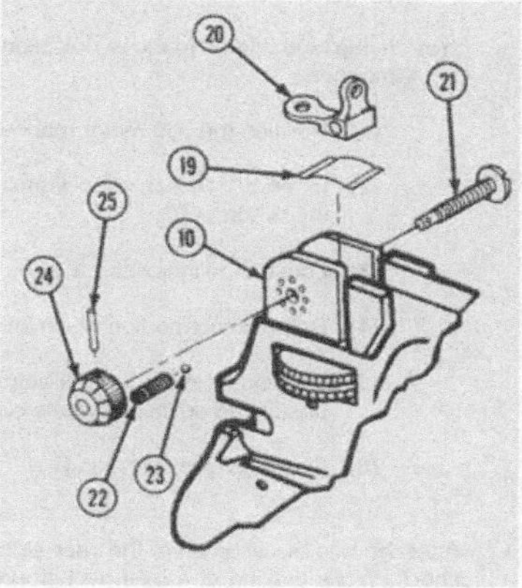

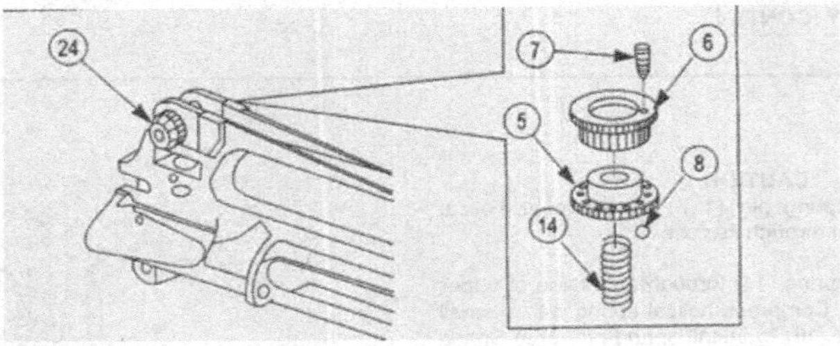

12. Rotate and test elevation index (6) and windage knob (24) for ease of functioning.

13. Inspect elevation knob zero as follows:

(a) Rotate elevation knob (5) counterclockwise until the rear sight assembly is all the way down. If a whole click is not felt as the rear sight assembly stops, the rear sight assembly has bottomed out and will not pivot freely.

(b) Position elevation knob back slightly to its last whole click so the rear sight assembly base is under tension of the ball bearing (8) and helical spring (14). The 300 meter mark should align with the mark on to receiver.

(c) If the 300 meter mark is not aligned with the mark on receiver, slip the range scale in the following manner:

(1) Position the 300 meter mark with the mark on the receiver.

(2) Insert a 1/16 inch allen wrench through the access hole of the rear sight assembly base and into the index screw (7).

(3) Loosen the index screw three turns and leave the wrench in place.

(4) Rotate lower portion of elevation knob

(5) counterclockwise until it stops (range scale should not have moved). Elevation knob should be positioned on its last whole click. (5) Tighten index screw (7) and remove wrench.

(6) Check for proper setting.

NOTE

After the rifle is assembled, the rear sight is centered, and placed at the 300 meter mark, perform the owing check: While looking at a light background, obtain good sight alignment. If the hole in the rear sight aperture appears oval Instead of round, the rear sight base or upper receiver should be replaced. To determine which part requires replacement, replace the rear sight base first. If this does not resolve the problem, replace the upper receiver.

14. Assemble rifle, refer to page 3-77

3-11. UPPER RECEIVER ASSEMBLY AND REAR SIGHT ASSEMBLY (CONT).

f. MEACHANICAL ZERO PROCEDURES (A.F. ONLY)

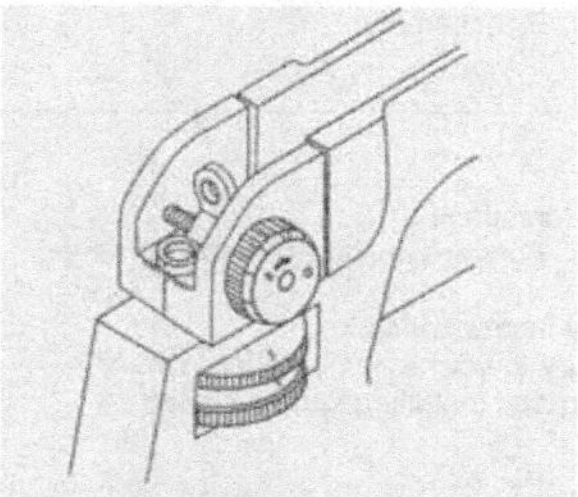

1. Center rear sight by moving windage knob in the appropriate direction.

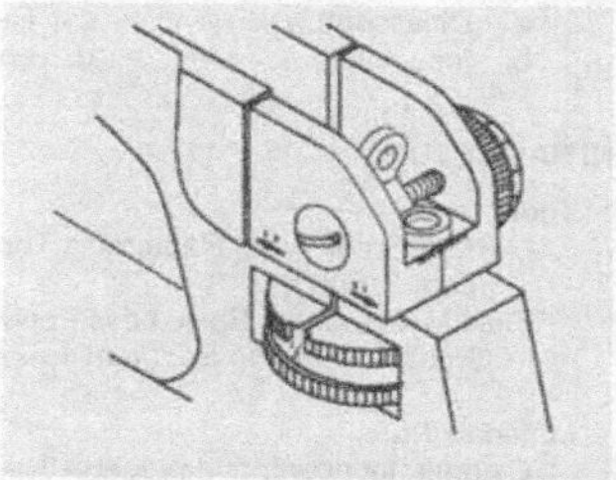

2. Always push in on windage screw head after making rear sight adjustments.

3. Visually check rear sight to ensure it is centered after making adjustments. Also, ensure the rear sight is set in the shortrange (unmarked aperture) position.

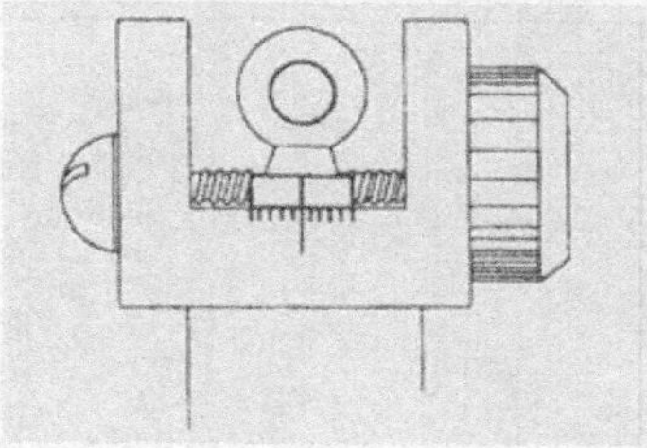

NOTE

This procedure, when used in conjunction with front sight mechanical zero adjustment (p 2-54), will give an approximate battle site zero to most M16A2 rifles. The above steps can also be used before firing a new or newly assigned rifle. Use the procedure to check rifles stored in preferred packaging during routine inspections. This will help ensure people armed with the rifles will stand a better chance of hitting an enemy if the rifles must be used before a live fire zero can be made. Whenever possible, zeroing of the rifle should be accomplished using ball ammunition on a 25 meter zeroing target using the "L" aperture.

After the rifle is assembled, the rear sight is centered, and placed at the 300 meter mark, perform the following check: While looking at a light background sight through the rear sight and obtain good sight alignment with the front sight placed in the center of the rear sight aperture. If while performing this check the hole in the rear sight aperture appears oval instead of round, the rear sight base or upper receiver should be replaced. To determine which part requires replacement, replace the rear sight base first. If this does not resolve the problem, replace the upper receiver.

3-12. FORWARD ASSIST ASSEMBLY.

This task covers:

- **a.** Disassembly
- **b.** Inspection
- **c.** Repair
- **d.** Lubrication
- **e.** Reassembly

INITIAL SETUP:

Tools
(ARMY) Small Arms Repairman Tool Kit (item 3, app B)
Field Maintenance Basic Less Power Small Arms Shop Set (item 1, app B)

Materials/Parts
Cleaner, lubricant, and preservative (CLP) (item 9, app D) | a

Equipment Conditions
3-48 Forward assist assembly removed

General Safety Instructions
To avoid injury to your eyes, use care when removing and installing spring-loaded parts

a. DISASSEMBLY

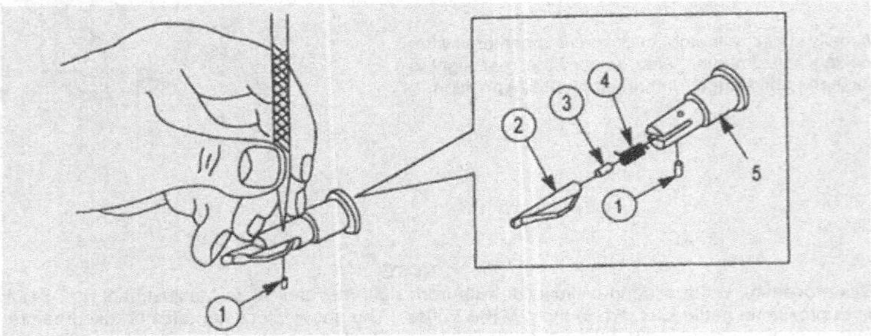

WARNING
To avoid injury to your eyes, use care when removing and installing spring-loaded parts.

1. Drive out spring pin (1) using 1/16 inch drive pin punch and hand hammer
2. Remove forward assist pawl (2), pawl detent (3), and helical spring (4) from plunger assembly (5)

3-12. FORWARD ASSIST ASSEMBLY (CONT)

b. INSPECTION

1. Inspect forward assist pawl (1) for burrs, chips, and cracks. Minor burrs may be removed using fine files or stones, as required. Replace forward assist pawl if defective.

2. Inspect pawl detent (2) for burrs and cracks. Minor burrs may be removed using fine files or stones, as required. Replace pawl detent if defective.

3. Inspect helical spring (3) for kinks, breaks, and wear. Replace helical spring if defective.

4. Inspect plunger assembly (4) for wear, burrs, chips, and breaks. Minor burrs may be removed using fine files or stones, as required. Replace forward assist assembly if defective.

5. Inspect spring pin (5) for wear. Replace if defective.

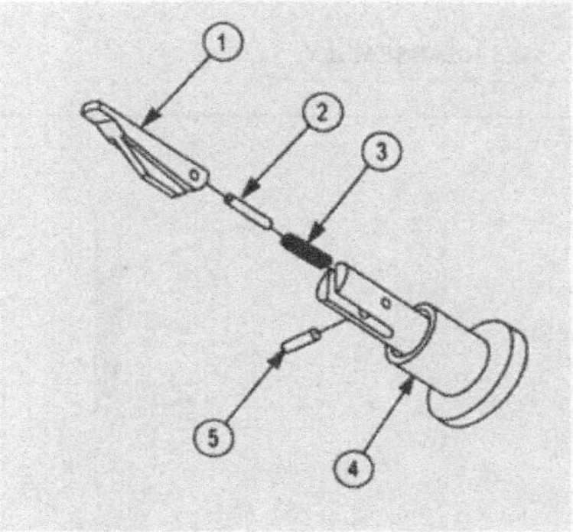

c. REPAIR

1. Repair forward assist pawl using fine files or stones, as required, to smooth burrs. Do not deform forward assist pawl.

2. Repair pawl detent using fine files or stones, as required, to smooth burrs. Do not deform pawl detent.

3. Repair plunger assembly using fine files or stones, as required, to smooth burrs. Do not deform plunger assembly.

d. LUBRICATION

Lubricate helical spring, pawl detent, and forward assist pawl with CLP (p 2-33) before installation.

e. REASSEMBLY

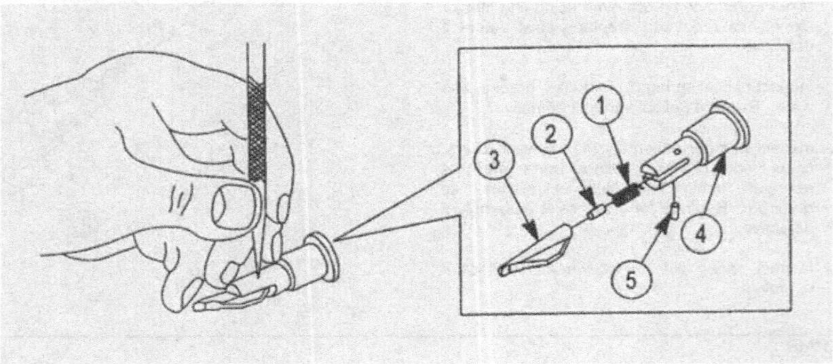

WARNING
To avoid injury to your eyes, use care when removing and installing spring-loaded parts.

1. Install helical spring (1), pawl detent (2), and forward assist pawl (3) into plunger assembly (4).

2. Align holes and install spring pin (5) using 1/16 inch drive pin punch and hand hammer. Spring pin must be flush or slightly below flush after reassembly.

3. Assemble rifle, refer to page 3-77.

3-13. LOWER RECEIVER AND BUTTSTOCK ASSEMBLY.

This task covers:

a. Disassembly
b. Inspection
c. Repair
d. Test
e. Reassembly

INITIAL SETUP

Test Equipment
Tool and Gage Set (item 2, app B)

Tools
(ARMY) Small Arms Repairman Tool Kit (item 3, app B)
Field Maintenance Basic Less Power Small Arms Shop Set (item 1, app B)
Pivot pin removal tool (fig E-3, app E)
Slave pin (fig E-5, app E)

Materials/Parts
Cleaner, lubricant, and preservative (CLP) (item 9, app D)
Dichloromehtane, technical (item 15, app D)
Lubricant, solid film (item 21, app D)
Screw, self-locking (item 6, p C-11)

Equipment Conditions
3-15 Lower receiver and buttstock assembly removed
2-57 Buttstock assembly and pistol grip removed

General Safety Instructions
To avoid injury to your eyes, use care when removing and installing spring-loaded parts.
When using solid film lubricant or dichloro methane, be sure the area is well ventilated.

a. DISASSEMBLY

WARNING
To avoid injury to your eyes, use care when removing and installing spring-loaded parts.

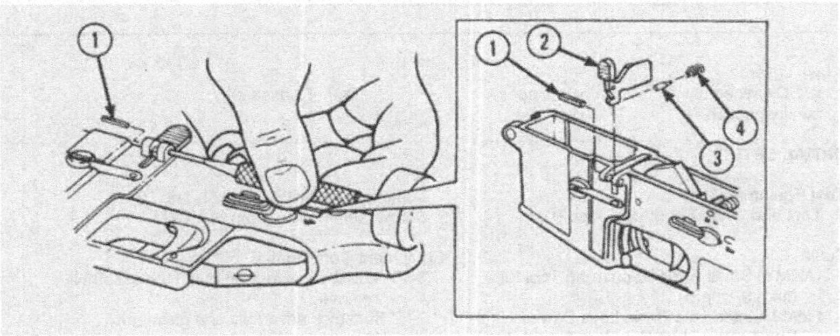

1. Remove spring pin (1) using 3/ 32 inch drive pin punch and hand hammer.

2. Remove bolt catch (2), bolt catch plunger (3), and bolt catch spring (4).

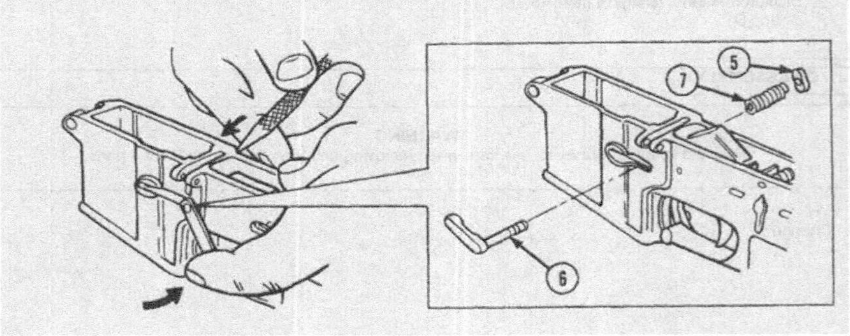

WARNING

To avoid injury to your eyes, use care when removing and installing spring-loaded parts.

3. Using drive pin punch, press in on magazine button (5) and turn magazine catch 16) counterclockwise to unscrew and remove.

4. Remove magazine button (5) and magazine catch spring (7).

3-13. LOWER RECEIVER AND BUTTSTOCK ASSEMBLY (CONT).

a. DISASSEMBLY (CONT)

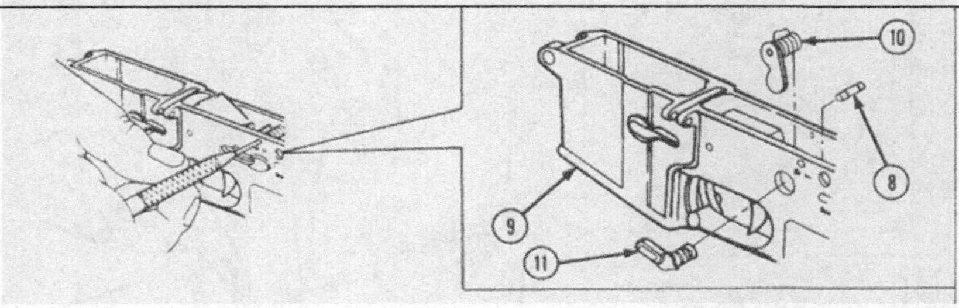

5. Use a rounded tip drive pin punch to push automatic sear pin (8) out of lower receiver (9).

NOTE
To remove automatic sear, selector lever must be positioned to BURST (if installed).

6. Remove automatic sear (10) and selector lever (11).

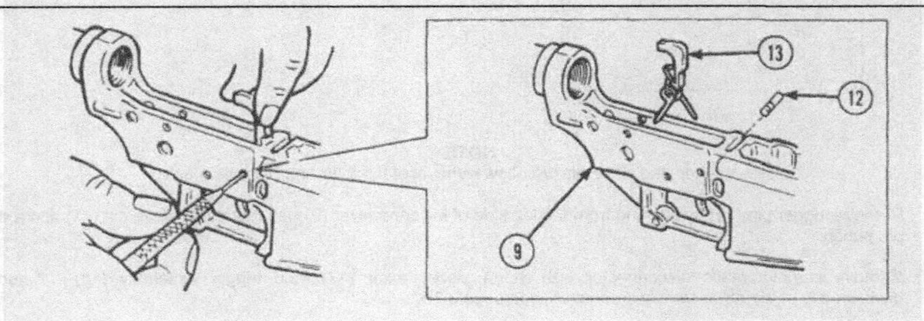

NOTE
To remove (hammer assembly should be forward), place selector lever (if installed) to SEMI position.

7. Use drive pin punch to push hammer pin (12) from lower receiver (9).

8. Remove hammer assembly (13). If further disassembly is required, see page 3-73.

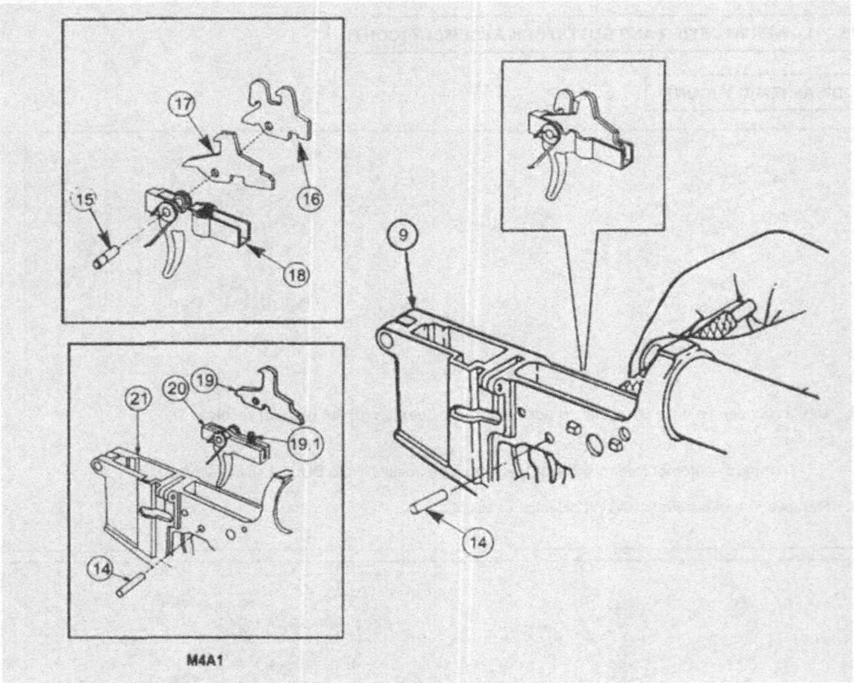

NOTE

Use of fabricated slave pin will allow removal of the following parts as a unit.

9. Remove trigger pin (14) by pushing from the left side of lower receiver (9) with fabricated slave pin (15) and a drive pin punch.

10. Remove semiautomatic disconnector (16), burst disconnector (17), and trigger assembly (18). If further disassembly of trigger assembly is required, see page 3-75.

M4A1 ONLY

11. Remove disconnector (19), disconnector spring (19.1), and trigger assembly (20) from lower receiver and extension subassembly (21).

3-13. LOWER RECEIVER AND BUTTSTOCK ASSEMBLY (CONT).

b. INSPECTION

NOTE
Refer to page 3-35 for repair of corroded surfaces.

Inspect lower receiver and buttstock assembly for corrosion in the lower receiver lobes of the pivot area or hinge pin area. If extensive corrosion appears in these areas, the receiver will not be repaired; turn rifle in for replacement.

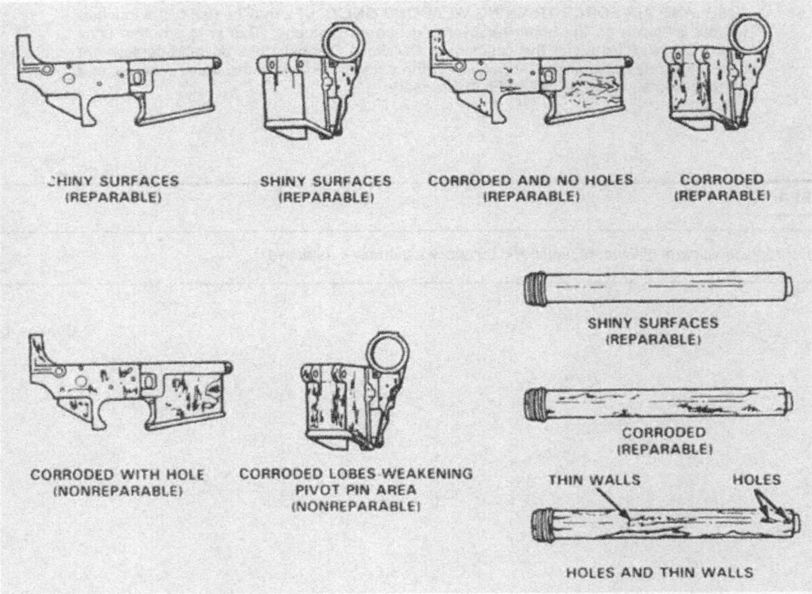

WARNING

When using solid film lubricant or dichloromethane, be sure the area is well ventilated.

CAUTION

Do not use a wire brush on aluminum surfaces.

NOTE

If a weapons lower receiver is missing one third or more of its exterior protective finish, resulting in an unprotected, light reflecting surface, it is candidate for overhaul. This missing finish will be considered a shortcoming. This shortcoming requires action to obtain a replacement weapon. Once a replacement has been received, evacuate the original weapon to depot for overhaul.

Solid Film Lubricant (SFL) is the only authorized touchup for the weapon and may be used on up to one third of the exterior finish of the weapon. **FOR ARMY CONUS USE ONLY AND AIR FORCE TRAINING WEAPONS ONLY:** SFL may be used as a touchup without limitation on the upper receiver and barrel assembly. This is to say that units which DO NOT fall under the category of Divisional Combat Units or rapid deployment type units may have up to 100 percent of the exterior surface of the upper receiver and barrel assembly protected with SFL if necessary.

c. REPAIR

Repair or replace all parts of lower receiver and buttstock assembly if defective.

3-13. LOWER RECEIVER AND BUTTSTOCK ASSEMBLY (CONT).

d. TEST

1. With the upper receiver attached to the lower receiver, and the pivot pin and takedown pins in place, perform the following test:

(a) Apply hand pressure to push the upper receiver as far to one side as possible.

(b) Attempt to insert a 0.020 inch thickness gage between the pivot pin lugs of the upper and lower receivers.

(c) If the thickness gage penetrates to the pivot pin at all accessible locations, repair by replacement of the upper receiver (see (b) below) or replacement of rifle is required.

2. If the rifle fails the above test, remove the upper receiver and install a "NEW" upper receiver and perform the test again.

3. If the rifle now passes the above test, it shall be considered serviceable and continue in use.

4. If the rifle fails the test with a new upper receiver, this failure shall be considered a shortcoming. This shortcoming requires action to obtain a replacement rifle. Once a replacement has been received, evacuate the original rifle to depot for overhaul.

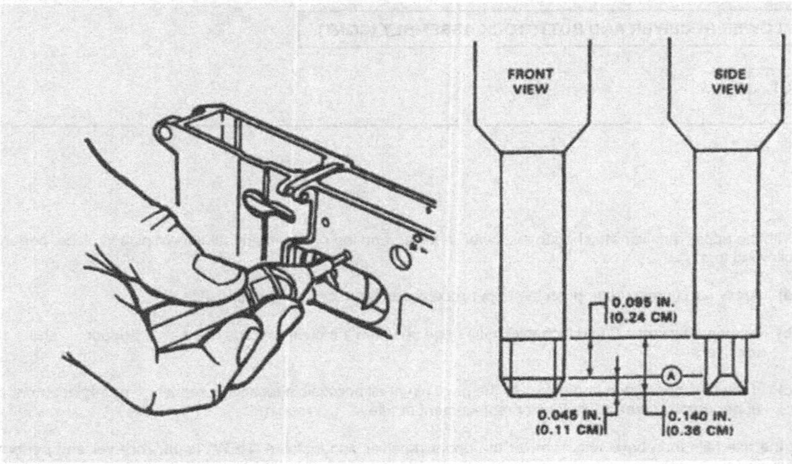

NOTE

If the lower receiver is not disassembled, visually inspect for broken or damaged parts, and to ensure that the hammer and trigger springs are correctly installed before beginning this test. It is not necessary to disassemble the lower receiver for the sole purpose of this visual Inspection. If broken or damaged parts are found, disassemble (p 3-62) and repair as authorized.

5. Test two hammer pin holes and two trigger pin holes using not-go plug gage PN 12006472. This test may be conducted by disassembly of the lower receiver (p 3-62) or by pushing the pin far enough to disengage the end of the pin from the side of the receiver which is being tested. If the lower receiver is not disassembled and the not-go plug gage enters any hole to first shoulder (A), the lower receiver must be disassembled and all four holes must be tested again.

6. Gently insert the not-go plug gage and rotate it 180 degrees. If the not-go plug gage passes through any one of the four pin holes, the rifle is unserviceable and will be turned in for replacement. The gage must extend through the wall thickness to be unserviceable.

7. After completion of gaging operation, visually inspect hammer and trigger springs to ensure proper location of spring legs.

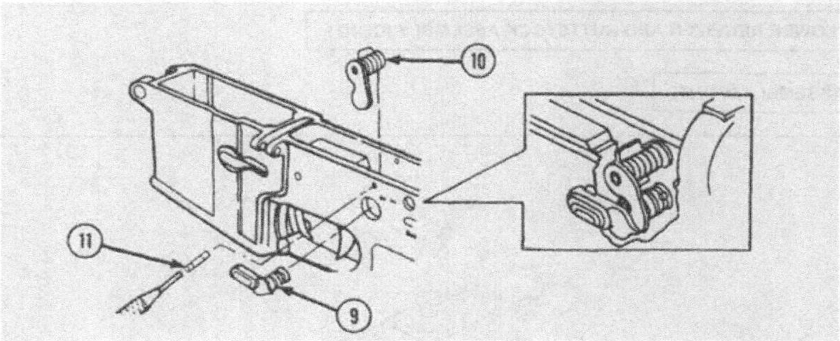

NOTE

Hammer assembly should be cocked prior to installing the selector lever.

Selector lever (9), if installed, must be positioned to BURST. Long leg of automatic sear (10) spring must rest on top of selector lever.

5. Install selector lever (9) and automatic sear (10).

6. Install automatic sear pin (11) (install on the right side) into receiver using drive pin punch. Push in until flush.

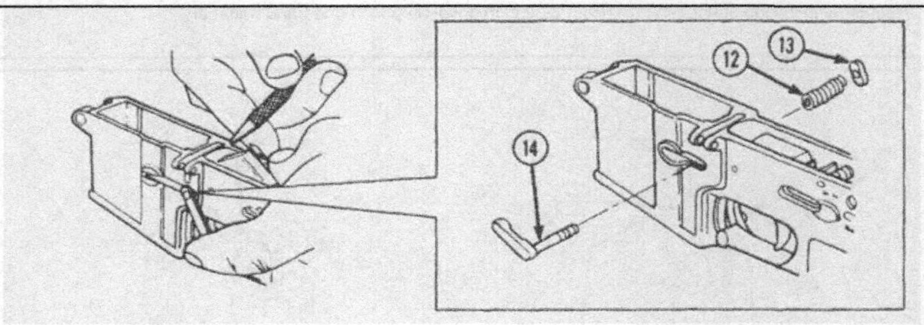

7. Install magazine catch spring (12) and magazine button (13).

NOTE

Drive pin punch should be larger than hole in magazine button.

8. Install magazine catch (14). Push in on magazine button (13) using a drive pin punch and turn magazine catch (14) clockwise until threaded end of magazine catch is flush with magazine button head.

3-13. LOWER RECEIVER AND BUTTSTOCK ASSEMBLY (CONT).

e. REASSEMBLY (CONT)

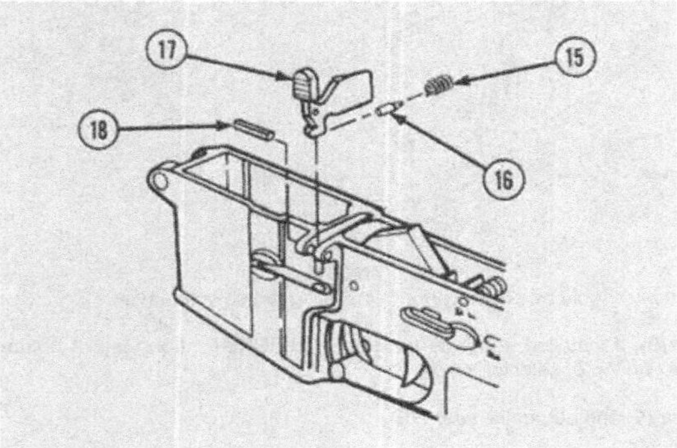

9. Install bolt catch spring (15), bolt catch plunger (16), and bolt catch (17).

10. Secure by installing spring pin (18) using 3/32 inch drive pin punch and hand hammer.

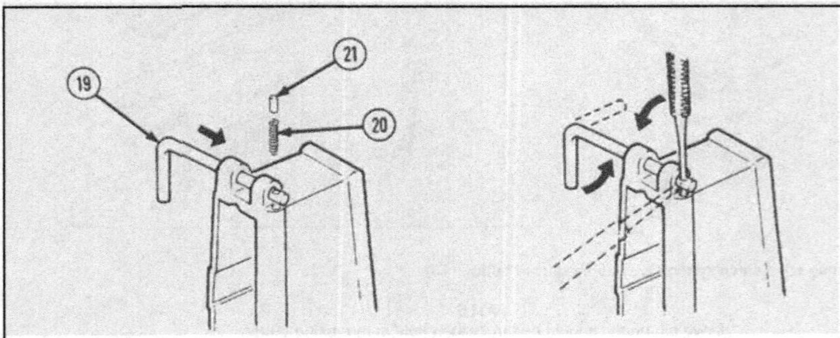

11. Install fabricated pivot pin installation tool (19). Insert helical spring (20) and pivot pin detent 21). Compress pivot pin detent in recess with punch and rotate tool. Remove punch.

WARNING
To avoid injury to your eyes, use care when removing or installing spring-loaded parts.

NOTE
Rounded end of pivot pin detent must be in the groove of the pivot pin (22) when assembly is complete.

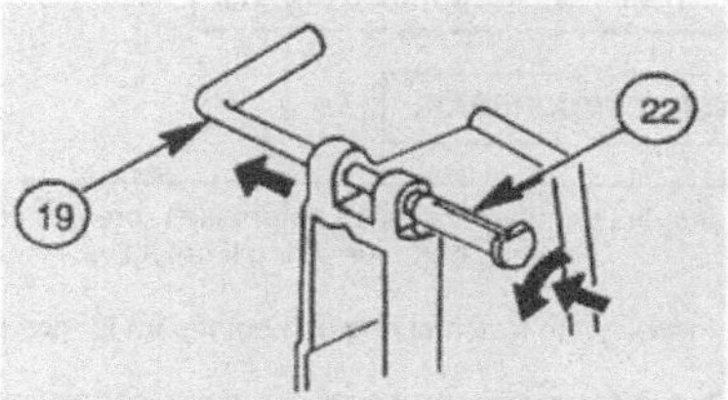

12. Position pivot pin (22) and removing fabricated pivot pin installation tool (19) while maintaining pressure, slide pivot pin (22) into hole. Rotate pivot pin to receive pivot pin detent.

13. Assemble rifle, refer to page 3-82.

3-14. HAMMER ASSEMBLY

This task covers:
a. Disassembly
b. Inspection/Repair
c. Reassembly

INITIAL SETUP

Equipment Conditions
3-62 Hammer assembly removed

a. DISASSEMBLY

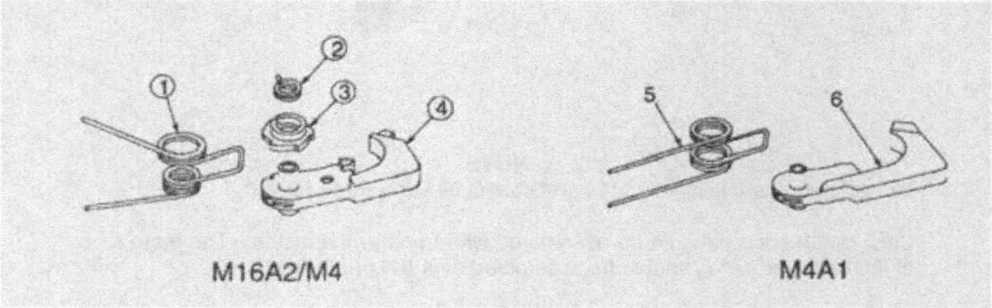

NOTE
M16A2 burst cam is black. M4 burst cam is nickel colored (shiny).

(M16A2/M4) remove hammer spring (1), cam clutch spring (2), and burst cam (3) from hammer and hammer pin retainer assembly (4). **(M4A1)** remove hammer spring (5) from hammer (6).

3-14. HAMMER ASSEMBLY (CONT).

b. INSPECTION/REPAIR

1. Inspect hammer spring for deformities, breaks, and bends. Pay special attention to the large coil. Replace hammer spring if defective.

2. Inspect cam clutch spring and burst cam for deformities, breaks, and bends; replace if defective.

3. Inspect hammer and hammer pin retainer assembly for chips and breaks. Hammer pin should click home unde strong finger pressure. Install hammer pin into hole in hammer to check spring retention of the hammer pir Replace hammer and hammer pin retainer assembly if defective.

c. REASSEMBLY

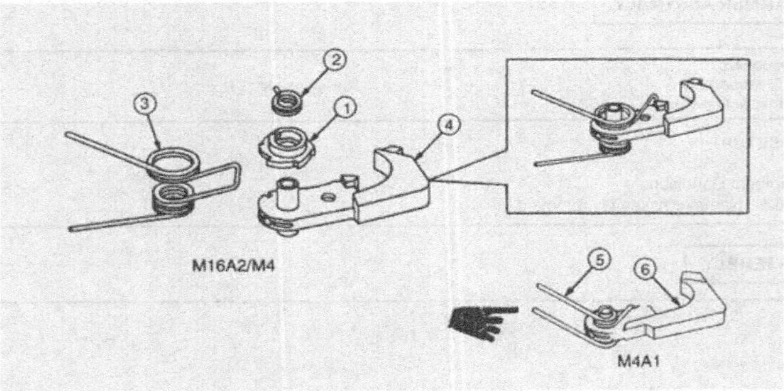

NOTE

M16A2 burst cam is black. M4 burst cam is nickel colored (shiny).

Cam clutch spring should be assembled with bend to the inside. The large loop of the hammer spring should be assembled over the burst cam.

(M16A2/M4) install burst cam (1), cam clutch spring (2), and hammer spring (3) on hammer and hammer pin retainer assembly (4). **(M4A1)** install hammer spring (5) on hammer (6).

3-15. TRIGGER ASSEMBLY AND TRIGGER SUBASSEMBLY (CONT).

b. INSPECTION/REPAIR

1. Inspect trigger spring for kinks, deformities, and weakness. Replace if defective.
2. Inspect disconnector springs for deformities, bends, breaks, and weakness. Replace if defective.
3. Inspect trigger for chips, wear, and cracks. Inspect for damaged searing surface on the trigger nose. Replace if defective.

c. REASSEMBLY

M16A2 AND M4 ONLY
NOTE
Use bolt carrier key tool to install disconnector springs.

1. Use the following procedures to install two disconnector springs (1) using the bolt carrier key tool:

(a) Secure the trigger (2) in a soft vise jaws or similar device.

(b) Place one spring (1) firmly on the tool with large diameter coils outward.

(c) Press spring (1) into recess to solid height.

(d) Hold spring at solid height and slide spring into one of the holes until the punch is flush and perpendicular with the recess wall.

(e) Turn the spring one to two turns opposite of coil winding of the spring.

(f) Discontinue winding when an audible click or snap is heard or felt. This indicates that the spring is seated.

(g) Hold the spring (1) in place when removing the tool to avoid unseating or damaging the spring.

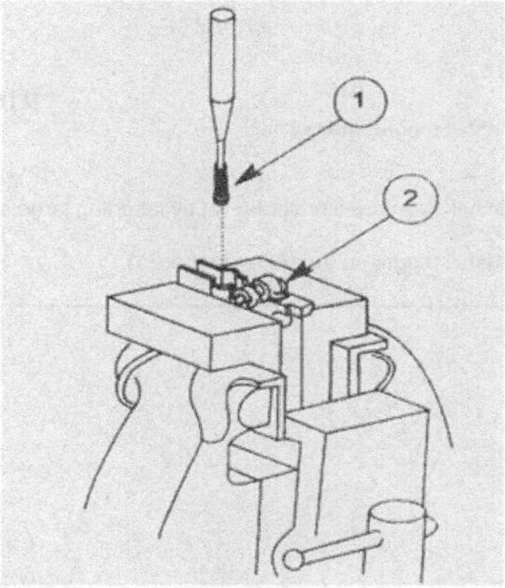

M4 CARBINE ONLY
NOTE
The semi and burst disconnector springs are not the same. The Semi disconnector spring (left side) (figure C-15, item 1B) is black while the burst disconnector spring (right side) (figure C-15, item 1A) is nickel (shiny). Ensure that the correct spring is installed on each side for proper functioning.

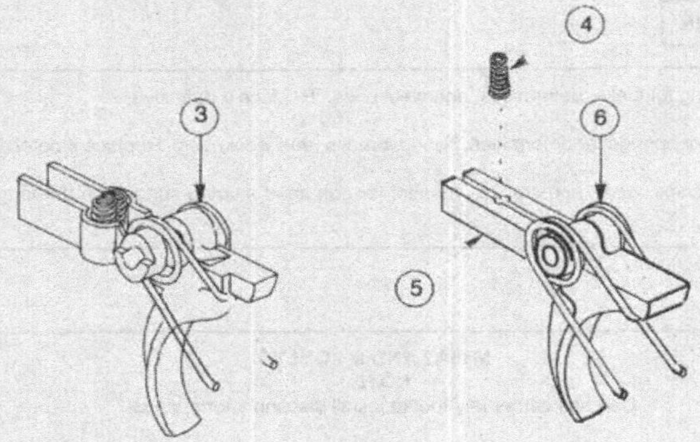

M16A2/M4 M4A1

M16A2 and M4

2. Install trigger spring (3).

M4A1 ONLY

3. Install disconnector spring (4) by inserting large end of spring into trigger (5).

4. Install trigger spring (6) on trigger (5).

3-16. LOWER RECEIVER AND RECEIVER EXTENSION ASSEMBLY.

This task covers:

a. Disassembly
b. Inspection
c. Repair/Modify
d. Reassembly

INITIAL SETUP

Tool Equipment
Tool and Gage Set (item 2, app B)

Tools
(ARMY) Small Arms Repairman Tool Kit (item 3, app B)
Field Maintenance Basic Less Power Small Arms Shop Set (item 1, app B)

Materials/Parts
Cloth, abrasive (item 13, app D)
Grease, molybdenum disulfide (item 19, app D)
Lubricant, solid film (item 21, app D)

Equipment Conditions
3-62 Lower receiver and receiver extension assembly removed

General Safety Instructions
To avoid injury to your eyes, use care when removing and installing spring-loaded parts.
When using solid film lubricant or dichloromethane, be sure the area is well ventilated.

a. DISASSEMBLY

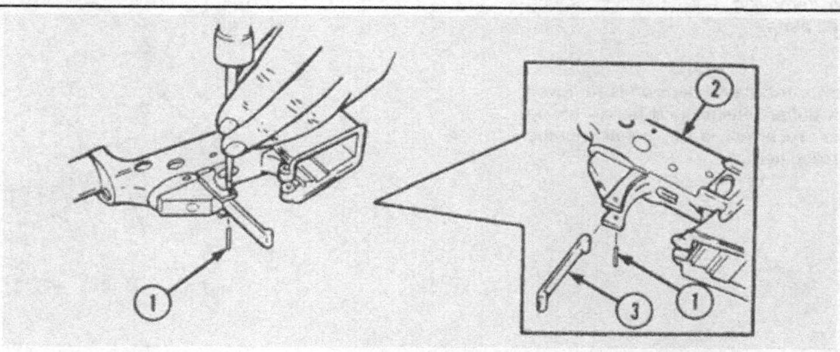

1. Remove spring pin (1) from lower receiver (2) using 1/8 inch drive pin punch and hand hammer.

2. Remove trigger guard (3).

3-16. LOWER RECEIVER AND RECEIVER EXTENSION ASSEMBLY (CONT).

a. DISASSEMBLY (CONT)

3. Use padding between lower receiver and brass vise laws. Grip the solid portion of the lower receiver with brass vise jaws which conform to the shape of the lower receiver in this area.

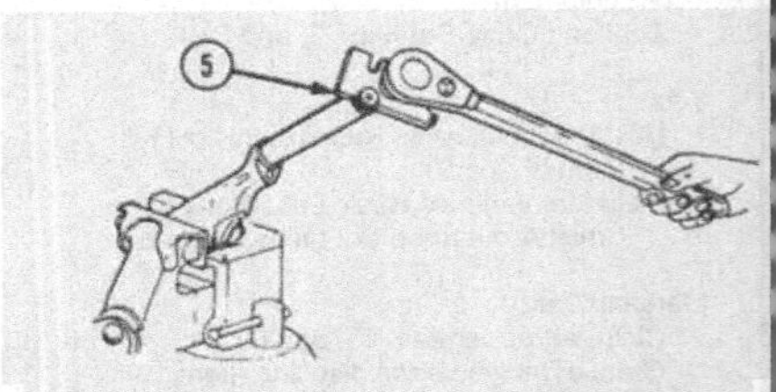

RIFLE ONLY

4. Clamp lower receiver (2) in a machinist's vise using vise jaw caps and tighten on solid portion just tight enough to hold.

WARNING

To avoid injury to your eyes, use care when removing and installing spring-loaded parts.

NOTE

As lower receiver extension is removed, catch buffer retainer and helical spring. Lower receiver is a serial number controlled item.

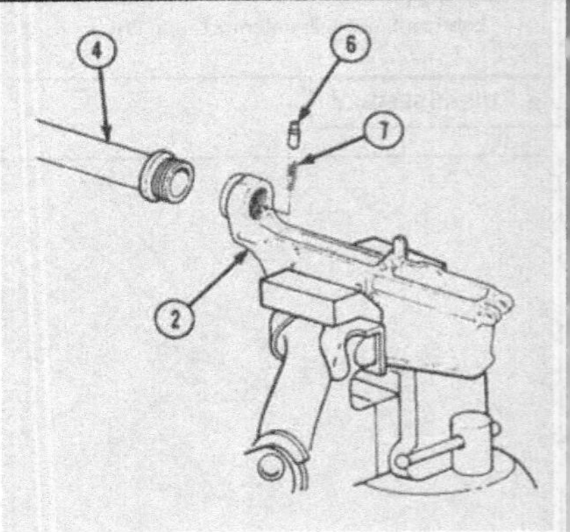

5. Remove lower receiver extension (4) from lower receiver (2) using combination wrench (5) and socket wrench handle. Catch buffer retainer (6) and helical spring (7).

CARBINE ONLY

6. Clamp lower receiver (8) in vise and tighten on solid portion just tight enough to hold.

NOTE

Use wooden vise jaws in place of brass vise jaw caps.

7. Remove the lower receiver extension (9), by loosening the extension locking nut (10) using the special tool (item 12, app C). Catch buffer retainer and spring (11).

CAUTION

While performing the following step, care should be taken to restrain the pivot pin spring and detent.

8. Loosen the locking nut (10) to allow the receiver end plate (12) to disengage from the lower receiver. Hold the buffer retainer and spring (11) in position with your index finger and unscrew the lower receiver extension (9) from the lower receiver (8).

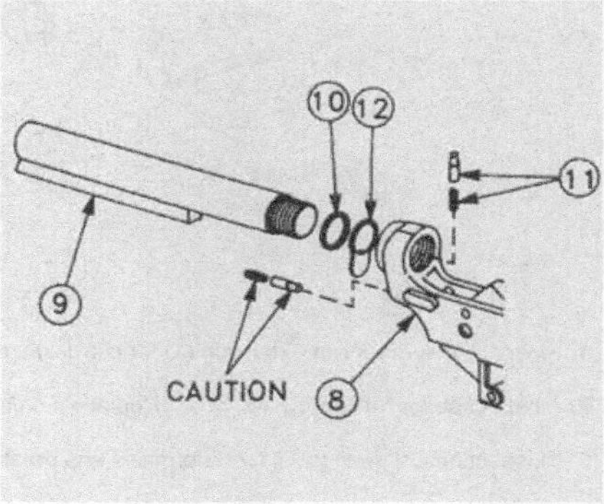

b. INSPECTION

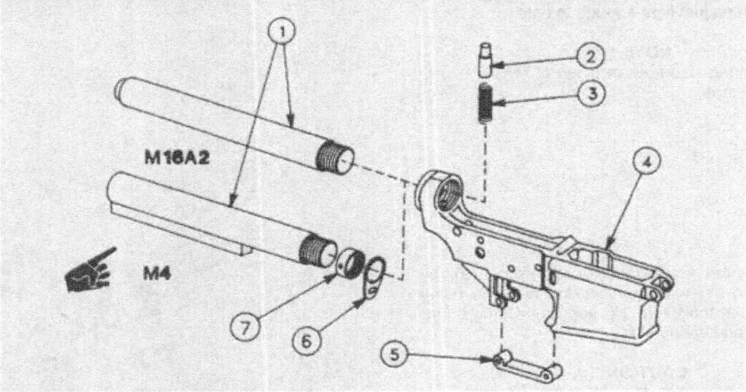

1. Inspect lower receiver extension (1) for corrosion, dents, and wear. Repair or replace if defective.
2. Inspect buffer retainer (2) for wear. Replace if defective.
3. Inspect helical springs (3) for deformities and breaks. Replace if defective.
4. Inspect lower receiver (4). See page 3-66, INSPECTION.
5. Inspect trigger guard (5) for deformities and check operation of plunger and spring. Replace trigger guard if defective.
6. **CARBINE ONLY:** Inspect receiver end plate (6) and locking nut (7) for damage. Replace if damaged.

c. REPAIR/MODIFY

1. Repair lower receiver extension by using abrasive cloth to remove light corrosion. Retouch using solid film lubricant.

NOTE

AIR FORCE ONLY: Only depot maintenance is authorized to restamp the serial number on rifle.

ARMY ONLY: Only direct support level is authorized to restamp serial number.

3-16. LOWER RECEIVER AND RECEIVER EXTENSION ASSEMBLY (CONT).

c. REPAIR/MODIFY (CONT)

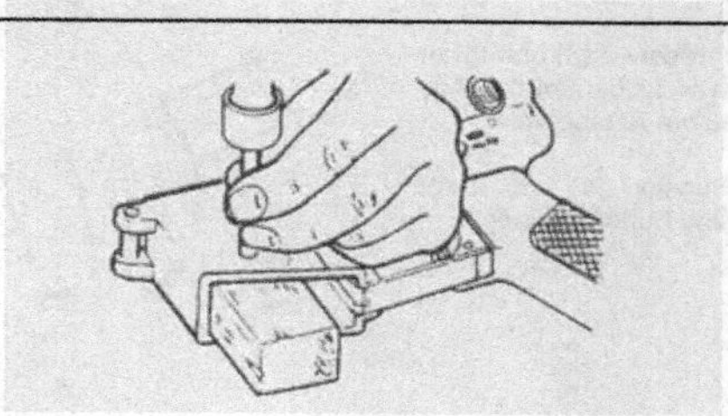

2. **ARMY ONLY:** If serial number is hard to read on rifle, restamp as follows:

   (a) Support the receiver In the stamping area to prevent bending and distortion of the receiver.

   (b) Exercise extreme care to restamp the same serial number as the original.

   (c) Restamp the serial number the same size as the original serial number.

NOTE

Most rifle serial numbers are 1/8 inch (0.31 cm) in height, or close enough that this size is acceptable for such restamping. In the event that a rifle has a serial number that cannot be reproduced by the use of the die sets contained In the Set D Field Maintenance Post, Camp, and Station Small Arms Shop Set, local purchase of an appropriate size die set is authorized.

d. REASSEMBLY

RIFLE ONLY

WARNING

To avoid injury to your eyes, use care when removing and installing spring-loaded parts.

1. Install helical spring (1) and buffer retainer (2) into lower receiver (3).

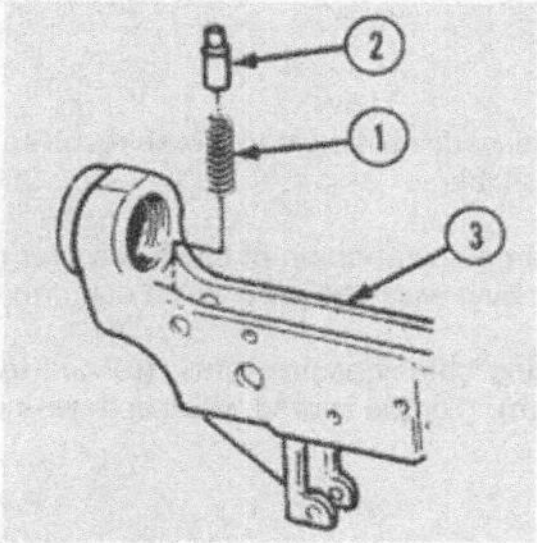

RIFLE ONLY

2. Lubricate threads of lower receiver (3) and lower receiver extension (4) with molybdenum disulfide grease (item 19, app D) before reassembly.

3. Install lower receiver extension (4) into lower receiver (3) while depressing buffer retainer.

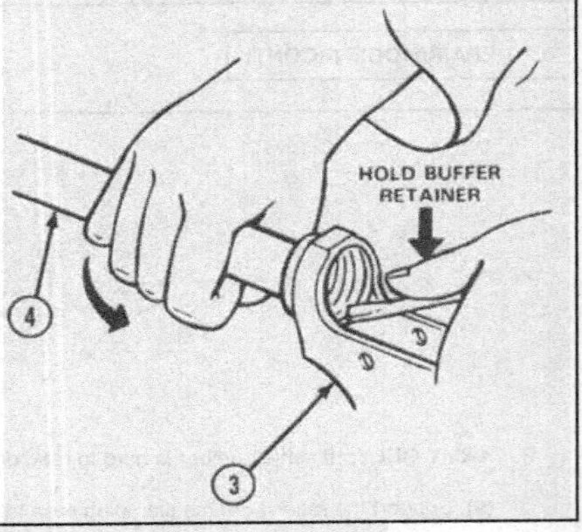

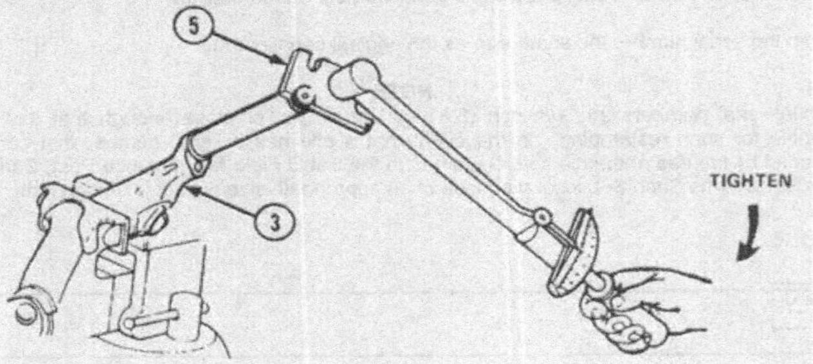

RIFLE ONLY

4. Use padding between lower receiver and brass vise jaws. Use vise jaws in vise and brass vise jaw caps, if available.

5. Clamp solid portion of lower receiver (3) in a machinist's vise using vise jaws. Grip the solid portion of the lower receiver with vise jaws which conform to the shape of the lower receiver in this area.

6. Using combination wrench (5) and torque wrench, torque lower receiver extension to 35-39 ft-lb (47 25 52 65 N•m). Torque is read when both wrenches are used to gether.

3-16. LOWER RECEIVER AND RECEIVER EXTENSION ASSEMBLY (CONT).

d. REASSEMBLY (CONT)

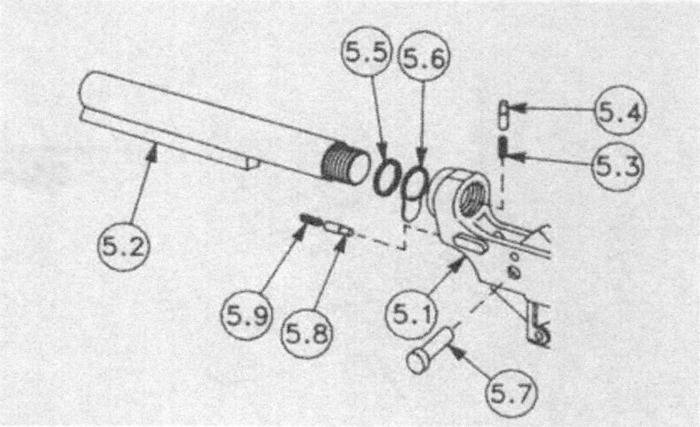

CARBINE ONLY

6A. Lubricate threads of lower receiver (5.1) and lower receiver extension (5.2) with molybdenum disulfide grease (item 19, app D) before reassembly.

6B. Pre-position the spring (5.3) and buffer retainer (5.4) into the retaining hole of the lower receiver (5.1). Screw the locking nut (5.5) onto the lower receiver extension (5.1) with the three notches on the locking ring (5.5) facing forward.

6C. Align the receiver end plate (5.6) onto the lower receiver extension (5.2) with the lug of the receiver end plate (5.6) facing forward.

6D. Pre-position the takedown pin (5.7), detent (5 8), and spring (5.9) in lower receiver assembly (5.1).

6E. Push down on the buffer retainer (5.4) and spring (5.3) and at the same time, screw the lower receiver extension (5.2) in until it retains the buffer retainer (5 4) in position.

6F. Align the lug of the receiver end plate (5.6) into the rear of the lower receiver (5.1). Screw the locking nut (5.5) forward until it contacts the receiver end plate (5.6).

6G. Using the special tool (Item 12, app C) tighten the locking nut (5.5) until snug.

6H. Using the special tool (item 12, app C), and torque wrench, torque locking nut (5.5) to 40 ±2 inch pounds.

6J. Stake the receiver end plate (5.6) in 2 places across from the notches in the locking nut (5.5).

3-16. LOWER RECEIVER AND RECEIVER EXTENSION ASSEMBLY (CONT).

d. REASSEMBLY (CONT)

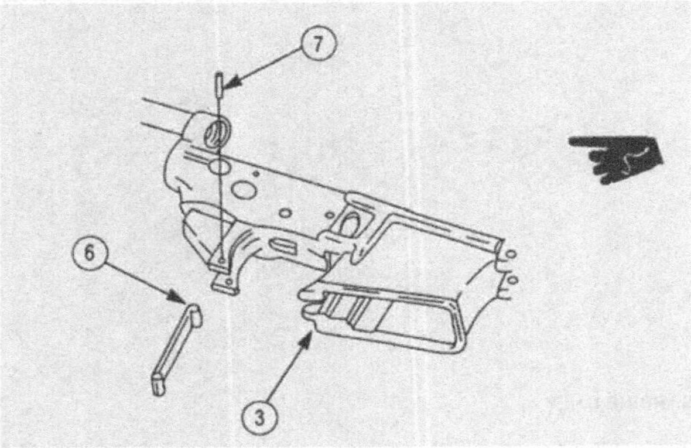

7. Install trigger guard (6) into lower receiver (3).

8. Install spring pin (7) using 1/8 inch drive pin punch and hand hammer.

9. Deleted.

3-17. MAJOR COMPONENTS OF M16A2 RIFLE

This task covers:

a. Reassembly b. Test c. Inspection

INITIAL SETUP

Test Equipment
Tool and Gage Set (item 2, app B)

Tools
(ARMY) Small Arms Repairman Tool Kit (item 3, app B)

Reference
TM 9-1005-319-10

General Safety Instructions
To avoid injury to your eyes, use care when removing and installing spring-loaded parts.
Live ammunition must not be near the work area.

a. REASSEMBLY

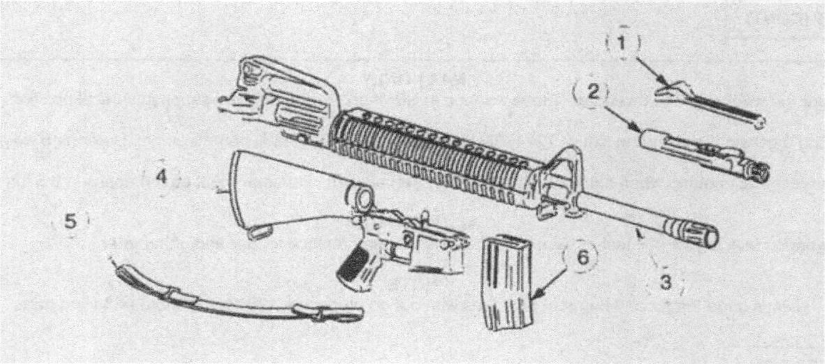

1. Refer to TM 9-1005-319-10.

2. Install charging handle assembly (1) and bolt carrier assembly (2) into upper receiver and barrel assembly (3). Join upper receiver and barrel assembly (3) and lower receiver and buttstock assembly (4).

3. Snap on sling (5) and install magazine (6).

b. TEST

Test trigger pull as follows:

M16A2 and M4 ONLY

(a) Clear the weapon. Place selector to BURST. Pull the trigger and hold it to the rear. Pull the charging handle to the rear and return to the bolt closed position three times. (This will place the BURST disconnector in the deep notch of the BURST cam.)

(b) Release the trigger. Place the selector to SEMI. Hold the rifle in the vertical position. Using trigger pull measuring fixture, 7274758, add weights until hammer trips. Determine weight applied.

(c) Hammer must not trip when 5.5 lbs (2.49 kg) have been applied, hammer must trip on applying 9.5 lbs (4.31 kg).

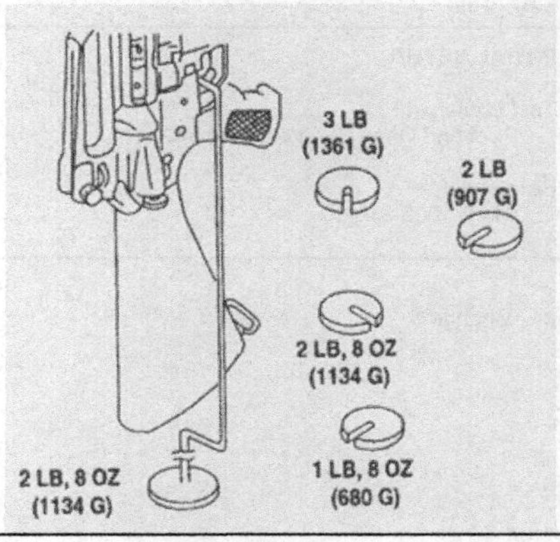

3-17. MAJOR COMPONENTS OF M16A2 RIFLE (CONT)

b. TEST (CONT)

M4A1 ONLY

(d) Clear the weapon. Cock weapon. Place weapon in SEMI position and hold weapon in vertical position.

(e) Using trigger pull measuring fixture 7274758 as shown, add weights until hammer trips. Determine weight applied

(f) Hammer must not trip when 5.5 lbs (2.49 kg) have been applied. Hammer must trip on applying 8.5 lbs (3.86 kg).

ALL WEAPONS

(g) If weapon fails trigger pull test or excessive creep is present, replace trigger and/or hammer.

NOTE

Always gage trigger and hammer pin holes with not go plug gage 12006472 before replacing parts.

c. INSPECTION

Perform final inspection procedures below.

3-18. M16A2 RIFLE FINAL INSPECTION.

This task covers:

a. Final inspection
b. Test
c. Functional theory of three-round burst control

INITIAL SETUP

Test Equipment
Tool and Gage Set (item 2, app B)

References
TM 9-1005-319-10

General Safety Instructions
Live ammunition must not be near the work area.

Air Force rifles will be inspected in accordance with AFM 36-2227, Vol. 1.

a. FINAL INSPECTION I

1 Visually inspect general appearance of rifle should look almost new. All metal surfaces are to have a dull, rustor corrosion-resistant finish with no burrs or deep scratches.

2 Visually inspect barrel for serviceability. Barrels must be straight, clean, free of rust, powder fouling, and free of bulges and rings. Fine pitting is allowable.

Using moderate hand pressure, check for rotational movement of the front sight in relation to the barrel. Also, using moderate hand pressure, check for rotational movement of the barrel in relation to the upper receiver. If movement between the front sight and the barrel exist, the barrel assembly must be replaced. If movement between the barrel and the upper receiver exist, the barrel assembly must be aligned and tightened (page 3-29).

3 Visually inspect rifle for missing parts. All parts must be attached and all modifications must be applies. Steel parts must be rust free. Spring pins must be secure and screws must be tight.

4 Functionally inspect key and bolt carrier assembly and gas tube alignment. Refer to TM 9-1005-319-10 and use the following procedures:

(a) Disengage the takedown pin and open the receiver.

(b) Remove bolt carrier assembly.

(c) Remove bolt assembly from bolt carrier assembly.

(d) Insert key and bolt carrier assembly into upper receiver and barrel assembly. The bolt assembly must not be installed while performing test.

(e) Slide key and bolt carrier assembly forward to detect binding between key and bolt carrier assembly and gas tube by feel. Badly bent gas tube could cause damage to both the key and bolt carrier assembly and the gas tube. A slightly bent gas tube will cause unnecessary wear of the key and bolt carrier assembly and gas tube.

(f) Correct slight binding by removing handguard assemblies and slightly bending gas tube in the handguard area while repeating step (e) above until no binding is detected. Badly bent gas tubes will be replaced and realigned.

(g) Remove key and bolt carrier assembly from upper receiver and barrel assembly.

(h) Reassemble bolt assembly into key and bolt carrier assembly.

(i) Reinstall key and bolt carrier assembly into upper receiver and barrel assembly.

5 Make a functional check on an assembled weapon with selector lever in SAFE, SEMI, and BURST/AUTO positions Any portion of this check may be used alone to determine the operational condition of any specific firing position selected

3-18 M16A2 RIFLE FINAL INSPECTION (CONT)

a. FINAL INSPECTION (CONT)I

6 Check rear sight assembly as follows:

(a) Rotate elevation knob counterclockwise until the rear sight assembly is all the way down. If a whole click Is not felt as the rear sight assembly stops, the rear sight assembly has bottomed out and will not pivot freely.

(b) Position elevation knob back slightly to its last whole click as the rear sight assembly base is under tension of the ball bearing and helical spring. The 300 meter mark should align with the mark on the receiver (c) If the 300 meter mark is not aligned with the mark on receiver, slip the range scale in the following manner:

(1) Position the 300 meter mark with the mark on the receiver.

(2) Insert a 1/16 inch alien wrench through the access hole of the rear sight assembly base and into the index screw.

(3) Loosen the index screw three turns and leave the wrench in place.

(4) Rotate lower portion of elevation knob counterclockwise until it stops (range scale should not have moved). Elevation knob should be positioned on its last whole click.

(5) Tighten index screw and remove wrench.

(6) Check for proper setting.

7 Pull charging handle assembly to rear. Check that chamber is clear. Leave hammer in cocked position.

8. Place selector lever is SAFE position and pull trigger. Hammer should not fall.

WARNING

If rifle falls any of the following inspections, continued use of the rifle could result In injury to, or death of, personnel.

NOTE

For the purpose of the following tests, "SLOW" is defined as 1/4 to 1/2 the normal rate of trigger release

9. Place selector lever in SEMI position.

0 Pull trigger. Hammer should fall.

1 Hold trigger to the rear, charge rifle and release the trigger with a slow, smooth motion without hesitations or stops, until the trigger is fully forward; an audible click should be heard. Hammer should not fall.

2 Repeat the above selector lever SEMI position test (steps 10 and 11) five times. The rifle must not malfunction during any of these five repetitions. If the rifle malfunctions during any of these five tests, refer to page 3-7.

M16A2 and M4 ONLY

3 Place selector lever in BURST position. Charge rifle and pull trigger. Hammer should fall.

4 While holding the trigger to the rear, pull charging handle assembly to the rear and release it three times. Hammer should not fall the third time the charging handle assembly is released. It may or may not fall the first and second time the charging handle assembly is released. When the burst disconnector reached the deep notch of the burst cam, the burst disconnector should have held the hammer to the rear while the trigger was in the pulled position. Release the trigger with a slow, smooth motion without hesitations or stops, until the trigger is fully forward. The hammer should not fall.

5 Pull trigger. Hammer should fall. This would be the first round of a three-round burst.

NOTE

A detailed explanation of the three-round burst control can be found in the following pages.

M4A1 ONLY

5.1 Place selector lever in AUTO position. Charge weapon and pull trigger. Hammer should fall.

5.2 Hold trigger to the rear, charge weapon, and release trigger. Pull trigger. Hammer should not fall. Automatic assembly sear should have released the hammer as the bolt closed.

ALL WEAPONS

6 With the hammer in the forward position, using moderate finger/thumb pressure, attempt to place the selector lever in the SAFE position. Selector lever should not go in the SAFE position.

7 Perform the following additional functional checks and adjustments on assembled weapon:

(a) Press magazine catch button. Make sure it functions properly.

(b) Press bolt catch. Make certain it operates smoothly and holds bolt in open position.

(c) Inspect front sight and rear sight assembly. Make certain they can be adjusted properly.

(d) Actuate forward assist assembly It must work freely.

3-18 M16A2 RIFLE FINAL INSPECTION (CONT)

a. FINAL INSPECTION (CONT)

(e) Inspect upper receiver and barrel assembly. Barrel assembly should not rotate within upper receiver assembly.

(f) Check that third or middle slot of compensator is straight up (TDC).

b. TEST

1 Check headspace using headspace gage PN 7799734. See page 3-45, TEST.

2 Check firing pin protrusion using firing pin protrusion gage PN 7799735. See page 3-45, TEST.

3 Check extent of barrel erosion using barrel erosion gage PN 8448496. See page 3-45, TEST.

4 Check barrel straightness using barrel straightness gage PN 8448202. See page 3-45.

c. FUNCTIONAL THEORY OF THREE-ROUND BURST CONTROL (M16A2 AND M4 ONLY).

NOTE

First become familiar with the functioning of the firing mechanism especially when in the SAFE and SEMI positions. You should also understand the role that the automatic sear plays when firing in the BURST position. Functioning of the mechanism is explained below In a step by step manner. This actually will seem to complicate something that is really very simple and happens in less than 1 second. The diagrams below and on the following pages do not show the associated springs for the sake of simplicity The positioning of the burst cam is shown in detail

Functional check of three-round burst is as follows:

NOTE

Assume the rifle is fully loaded with a live round in the chamber and the selector lever on BURST.

(a) Hammer is cocked

(b) Front hook of burst lever is in stop notch

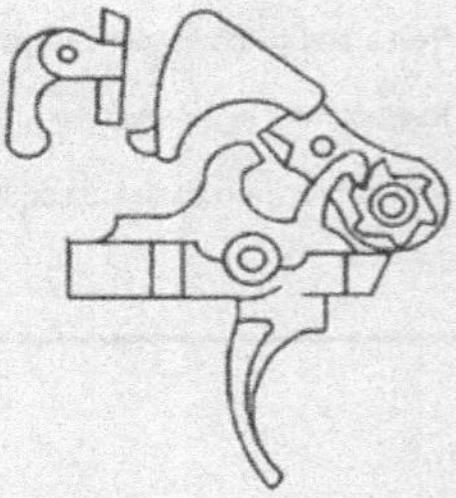

(c) Trigger Is pulled.

(d) Trigger nose drops and hammer falls firing e FIRST ROUND.

(e) Front hook of burst disconnector holds urst cam In place as hammer falls.

NOTE

Anytime the hammer falls forward, the clutch spring releases the burst cam and allows the front hook of the burst disconnector to keep It In place.

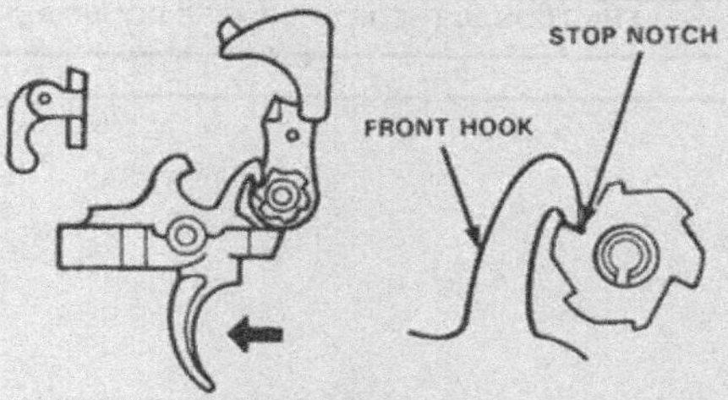

(f) As the key and bolt carrier assembly oves to the rear, the hammer is forced to the rear.

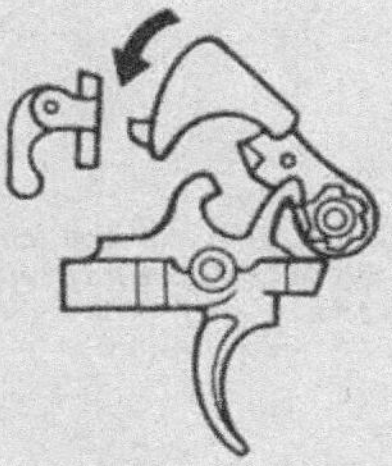

(g) The clutch spring of the burst cam lutches the burst cam and causes It to rotate one notch s the hammer is forced back.

(h) When hammer is fully to the rear, the utomatic sear catches It.

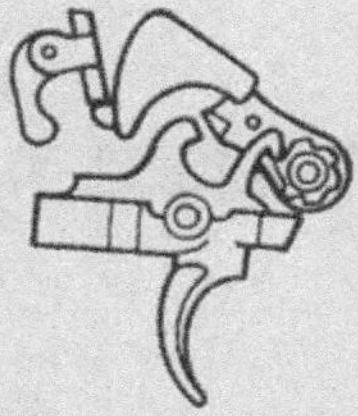

3-18 M16A2 RIFLE FINAL INSPECTION (CONT).

c. FUNCTIONAL THEORY OF THREE-ROUND BURST CONTROL (CONT)

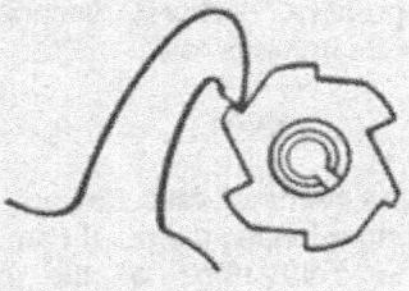

(i) The front hook of the burst disconnector is now fully In the second notch.

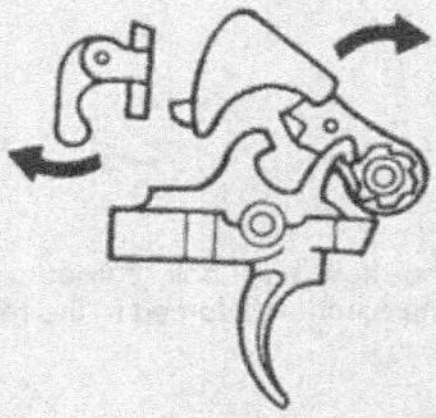

(j) As the key and bolt carrier assembly travels forward, the automatic sear releases the hammer and the hammer falls.

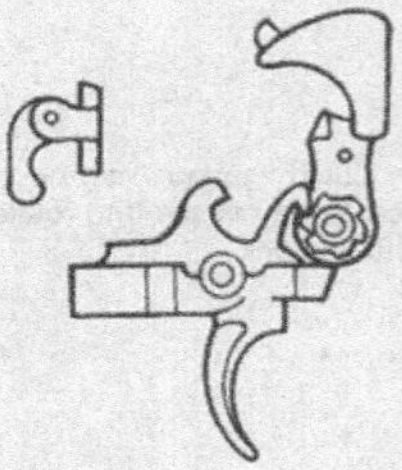

(k) When the hammer falls, the SECOND ROUND is fired.

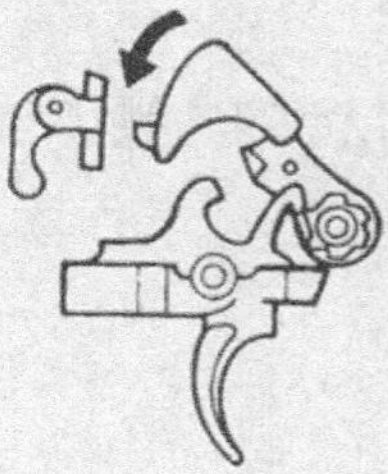

(l) As the key and bolt carrier assembly moves to the rear, the hammer is forced back to the rear.

(m) The clutch spring of the burst cam clutches against the burst cam and causes It to rotate one notch as the haiimmer is forced back

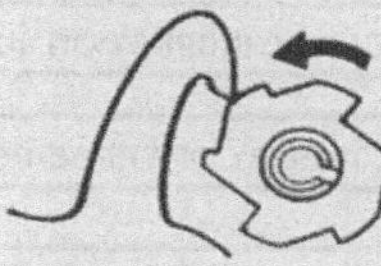

(n) When the hammer is fully to the rear, the automatic sear catches It

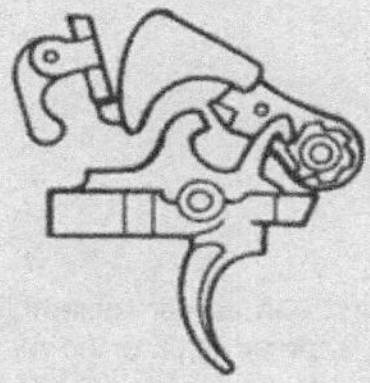

(o) The front hook of the burst disconnector IS now fully In the third notch

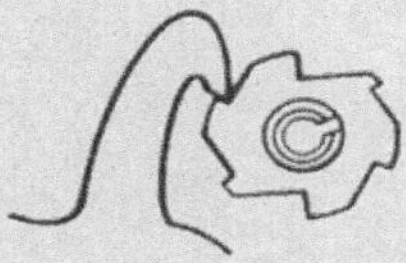

(p) As the key and bolt carrier assembly) travels forward, the automatic sear releases the hammer and the hammer falls

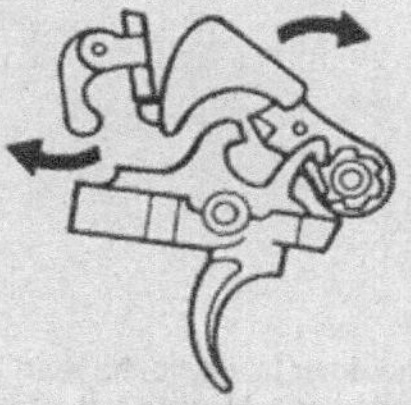

3-18. M16A2 RIFLE FINAL INSPECTION (CONT).

c. FUNCTIONAL THEORY OF THREE-ROUND BURST CONTROL (CONT)

(q) When the hammer falls, the THIRD ROUND Is fired

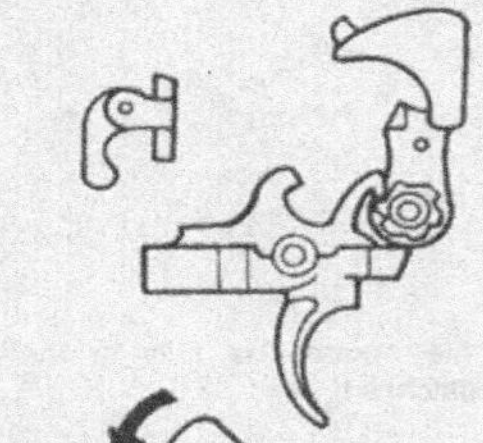

(r) As the key and bolt carrier assembly moves to the rear, the hammer is forced back to the rear

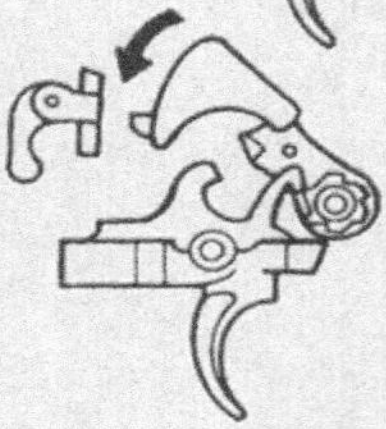

(s) The clutch spring of the burst cam clutches against the burst cam and causes it to rotate one notch as the hammer is forced back.

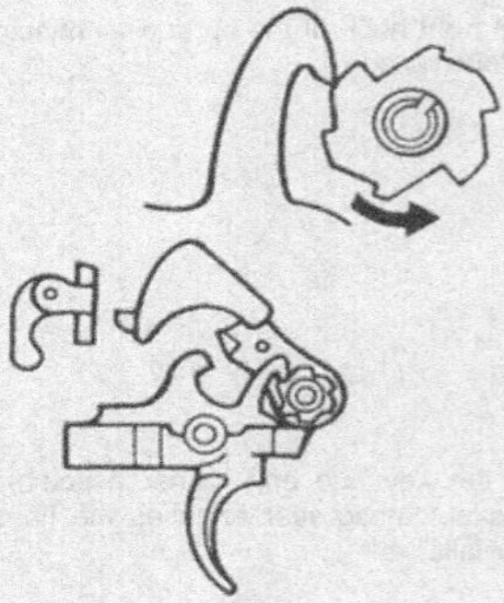

(t) When the hammer Is fully to the rear, It Is initially caught the, ;automatic sear However, the front hook of the first disconnector Is now fully in the next stop notch which is deeper than the others

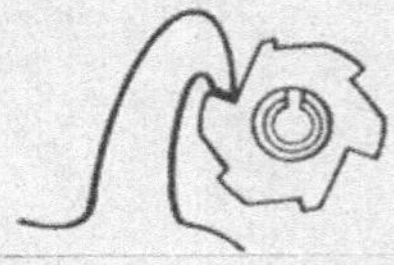

(u) Because a stop notch is deeper than the others, it allows the front hook of the burst disconnector further forward than before. This allows the rear hook of the burst disconnector to latch on the rear hammer notch. This holds the hammer fully to the rear even though the trigger is still to the rear. This happens when the burst Is over and the firing Is stopped.

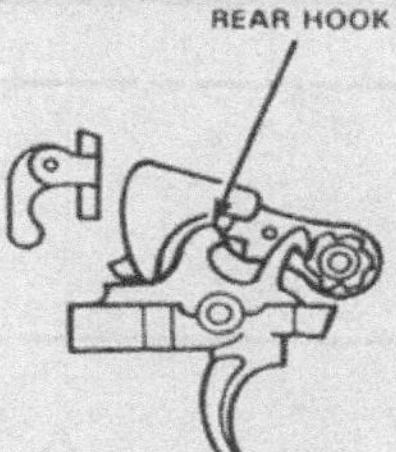

(v) Once the trigger is released, the trigger nose comes up and holds the hammer back.

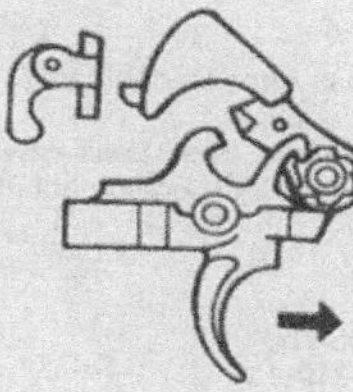

NOTE

Pulling the trigger to the rear and holding it back again will fire another three round burst This will continue until the magazine is empty However the trigger must be released between each burst

3-19 MI 6A2 RIFLE AND M4/M4A1 CARBINE ANNUAL GAGING REQUIREMENTS.

This task covers.

a Inspection

b Gaging

INITIAL SETUP

Test Equipment
Tool and Gage Set (item 2, app B)

Tools
(ARMY) Small Arms Repairman Tool Kit (item 3, app B)
Field Maintenance Basic Less Power Small Arms Shop Equipment (item 1, app B)

References
DA PAM 738-750
TM 9-1005-319-10
AFR 50-36
AFTO Form 105

General Safety Instructions
- To avoid injury to your eyes, use care when removing and installing spring-loaded parts
- Initial gaging is required 1 year from receipt of the weapons.
- All rifles and carbines must be gaged at least once annually for safety.
- All Army Reserve and Army National Guard M16A2 rifles must be inspected and gaged at least once every 2 years, after the initial inspection/gaging procedures have been accomplished. This 2 year interval may be maintained unless preventive maintenance checks and services (PMCS), or other physical evidence, indicates that an individual unit's rifles require inspection/ gaging at a more frequent interval. If it is determined that a yearly inspection is necessary for an individual unit, only that unit will be affected. This will not affect other units in regard to the interval of inspection.
- It is recommended that training units inspect/ gage all rifles at the end of each training cycle. Training units will inspect/gage all rifles at least once annually.
- Air Force M16 rifles will be inspected in accordance with AFR 50-36, Vol 1, Chapter 5.

a. INSPECTION

1 Visually inspect general appearance of weapon Overall appearance will be approximately that of a new weapon. F(inspection criteria refer to final inspection, page 3-84 All visual and functional Inspection requirements must be met.

NOTE

To perform the following tests, disassemble weapon only as far as allowed in chapter 2 (Unit Maintenance Instructions). unless a deficiency is uncovered

2 Perform a general inspection of weapon per section III of this chapter Repair as required and authorized

b. GAGING

Gage bolt carrier assembly for firing pin protrusion using firing pin protrusion gage PN 7799735. See page 3-20, TEST.

Gage bolt carrier assembly for firing pin hole wear using not-go plug gage PN 12620101. See page 3-24, TEST.

Inspect chamber in upper receiver and barrel assembly using chamber reflector tool PN 8448201 See page 3-33, INSPECTION/CLEANING.

Gage barrel in upper receiver and barrel assembly using barrel erosion gage PN 8448496 and bore straightness gage PN 8448202. See page 3-45, TEST.

Check headspace in upper receiver and barrel assembly by inserting headspace gage PN 7799734 in chamber. See page 3-45, TEST.

Gage pivot pin lug area clearance in lower receiver assembly using 0.020 thickness gage. See page 3-68.

Gage hammer and trigger pin holes in lower receiver assembly using taper plug gage PN 12006472. See page 3-68, TEST.

Gage trigger pull using trigger pull measuring fixture PN 7274758. See page 3-83, TEST.

Document inspection with DA PAM 738-750, AFTO Form 105, or NAVMC 11003 when completed.

Section IV. PREEMBARKATION INSPECTION OF MATERIEL IN UNITS ALERTED FOR OVERSEAS MOVEMENT

NOTE

This section refers to all the weapons (MI 16A2 rifle, M4, and M4A1) unless otherwise stated.

20 PURPOSE: This section establishes standards for overseas shipment (pre-embarkation in section criteria) for all eapons. These standards are provided to ensure that the user is furnished equipment which will perform its mission thout early failure or major maintenance problems.

3-21 SCOPE.

a. The standards prescribed provide for a high percentage of remaining life in affected rifles; therefore, rifle designated for overseas shipment must qualify under the standards contained in the following paragraph, table, and i referenced DA publications, before they can be approved for shipping action.

b. Provisions of this standard apply to all US Army agencies/activities selecting or preparing rifles for shipment to U troops overseas. It also applies to CONUS troops preparing rifles for shipment overseas. Provisions do not apply either t rifles being prepared for shipment to MAP/MAS recipients unless specifically prescribed by MAP/MAS transaction for th materiel or to rifle being returned to CONUS from overseas. The maintenance instructions and standards containe herein do not apply to rifles once the material has arrived at the overseas destination. At that time, maintenanc instructions contained in the applicable TM's will be used.

c. This applies to rifles which are the logistic responsibility of the US Army Armament, Munitions and Chemic Command.

3-22. GENERAL.

a. Only rifles which have been classified as serviceable condition code A, B, or C under AR 725-50 will b considered for overseas shipment. All items of equipment for which equipment serviceability criteria have been publishe must, as a minimum, be rated green under the ESC as a prerequisite to overseas shipment. In addition to the conditio code standard, as enumerated above, and the required ESC rating prescribed herein, the rifle being considered fo overseas shipment must meet the requirements of this section. The ESC will be discontinued as new operator manual are revised which will be used to determine serviceability condition of rifle.

b. Waivers to provisions can only be granted by the gaining command of any particular end Item being considered fo issue, deployment, or shipment. The issuing services may recommend issue or shipment of rifles not meeting th provisions when all the following conditions exist:

(1) Repair parts in required quantities cannot be obtained from the supply system prior to delivery of the en item.

(2) The gaining command concurs in the receipt of the end item for storage until required repair parts becom available. The gaining command must also state that capability, facilities, and funds are available to perform th necessary work when parts become available.

(3) Department of the Army approval is obtained on a case-by-case basis.

(4) Required repair parts are requisitioned by the issuing command for delivery to the Framing command.

c. All Department of the Army MWO's applicable to the specific rifle being considered for shipment overseas must have been applied.

d. Refer to SB 746-1 for pertinent publications relating to equipment processing and marking information

e. Refer to AMC-P 310-9 for publications containing applicable overhaul standards.

3-23. SHIPMENT OR ISSUE.

a. Organizational Repair Parts, Tools, and Equipment. Rifles must be complete with all Items required by applicable Department of the Army publications, including those In the basic Issue items list of the appropriate operator's manual.

b. Publications. Operator publications applicable to the equipment log book must accompany the equipment. All log book entries must be complete and up-to-date including those covering any repairs, replacements, or adjustments made to the rifle In complying with this section.

c. Documentation. Prepare DA Form 2408-9 (Equipment Control Record) at time of overseas shipment or Issue to another stock record or property book account, in accordance with the provisions of DA PAM 738-750.

d. Preparation. Process rifles for shipment as required by shipping documents and pertinent regulations.

3-24. DISPOSITION. Disqualified rifles which do not qualify for shipment will either be redistributed within the camp, post, or station, be repaired, or become candidates for overhaul, cannibalization, or other disposition as required by existing regulations.

3-25. GENERAL INSPECTION CRITERIA.

WARNING

Before starting an Inspection, be sure to clear the rifle Do not actuate the trigger until the rifle has been cleared. Inspect the bore and chamber to ensure that It is empty and free from obstructions, and check to see that no ammunition Is In position to be Introduced.

a. Before Inspection, the materiel must be thoroughly cleaned of all grease, dirt, or other foreign matter that might Interfere with Its proper function or the use of gages and tools during inspection.

b. Materiel must be free of burrs, rust, or corrosion on functional surfaces.

3-25. GENERAL INSPECTION CRITERIA (CONT).

c. Parts must not be cracked, bent, distorted, or damaged and must be free of detrimental wear or looseness

d. Minor defects In metal components do not normally affect their acceptability For example, scratches and tool marks are ordinarily of no Importance

e. Inspect finish of metal surface

(1) General. Satisfactory metal surfaces for rifles range from black to light gray. A worn shiny metal surface is objectionable only when It is capable of reflecting light No rifle will be reelected unless exterior parts have a shine All rear sights must have a dull gray or black finish of all surfaces that would cause a glare

(2) M16A2 Rifle. Minor loss of finish (shiny spots, nicks, scratches) on exterior surfaces of the barrel and flash suppressor shall not be cause for rejection of M16A2 rifles located In hands of troops at training centers Large shiny surfaces, nicks, scratches, etc., can be restored by the use of solid film lubricant (Item 21, app D) Rifles (small arms) missing in excess of one-third or more of the exterior finish resulting In an unprotected, light reflecting surface, are considered candidates for overhaul The only authorized level of maintenance to phosphate finish small arms Is depot.

f. Plastic components must not be cracked or damaged In such a way as to Interfere with their structural strength Surface cracks, bruises, or dents that do not affect their strength will not be cause for rejection. Cracks will be cause for rejection. Criteria for determining which cracks are repairable are on page 2-1 4.

g. Barrels must be clean and free of corrosion such as that caused by moisture and powder fouling Standards of serviceability are indicated In (1) through (9).

(1) Pits In the chamber are allowable If they do not cause extraction difficulties

(2) Pits as wide as a land and 3'8 Inch (O 95 cm) or less in length are allowable for 5 56-mm barrels Pits not greater than the width of a land and less than 3 8-inch (O 95 cm) long are permissible

(3) Scattered or uniformly fine pits, or fine pits In a densely pitted area are allowable

(4) Tool marks are acceptable regardless of length They will appear as lines running laterally In the grooves, or may run spirally across the top of lands

(5) Ringed bores or bores ringed sufficiently to bulge the outside surface of the barrel are cause for rejection. However, faint rings or shadowy depressions do not indicate an unserviceable barrel and will not be cause for rejection. Gap in lined barrels will not be classified as a ringed bore.

(6) Lands that appear dark due to coating of gilding metal from projectiles will not be cause for rejection.

(7) Breech bore diameter will be checked on unlined barrels using the appropriate breech bore gage.

(8) Barrel erosion gages are provided for lined barrels. Bore wear will be checked using barrel erosion gage for the M16A2. For detailed instructions in the use of the above gage and for serviceability limits, refer to page 3-46.

(9) Flaking or checking (fine cracks) of chromium plate in barrels or chambers will not be cause for rejection, unless accompanied by pitting to the degree that extraction difficulties are encountered or accuracy is unacceptable.

h. Springs must be free of distortion and broken coils. Springs must have sufficient tension to perform their intended function.

i. Screw heads must be in serviceable condition and threads must not be stripped. Internal threads must not be stripped.

j. The sear, hammer, and/or cocking notches must be in good condition. Chipped engaging corners will be cause for rejection. Slight wear on functional surfaces, including engaging corners, shall be acceptable, providing the minimum trigger pull requirements and selector lever checks are met in accordance with instructions on page 3-83.

k. Chips, flat spots, or bent striker points on firing pins will be cause for rejection.

l. The cartridge engaging surfaces on extractors must not be chipped or deformed.

m. Evidence of any damage to sights will be cause for a sight alignment check. Rear sight bases should have no movement.

n. Rear sight elevating and windage mechanisms must operate with distinct clicks, without binding. Sights must have sufficient tension to retain their setting during firing. Graduations and numerals must be legible. Graduation filler Is not required.

o. Safeties must positively position in both the ON and OFF position. When in the ON or safe position, the rifle must not fire when the trigger is squeezed; when in the OFF or fire position, the rifle must fire when the trigger is squeezed.

3-25. GENERAL INSPECTION CRITERIA (CONT).

p. All locking devices such as latches, magazine latches, or detents must be positive in action and must not become disengaged due to normal handling and firing. Retaining pins and similar devices must not be subject to accidental loss during use or transportation.

q. Each rifle must be hand functioned to check for unusual binding, positive cocking action, and general operation Dummy ammunition must be used to assure positive feeding, chambering, extraction, and ejection action.

Table 3-1. 5.56MM Rifle M16A2

Item	Standard
RIFLE:	
General	Clear rifle of any ammunition and Inspect In accordance with paragraph 3-25.
Barrel and barrel extension	Check barrel erosion. Use barrel erosion gage 8448496 for chrome lined barrels. Stripping of lands and grooves shall not be cause for rejection unless so determined by barrel erosion gage.
	Visually inspect, using chamber reflector tool 8448201.
	Pits 1/8 inch (0.31 cm) In length and those pits large enough to extend from the body of the chamber into the shoulder stop area and forcing cone area are cause for rejection. Large pits are defined as those 1/8 inch (0.31 cm) or more in diameter as determined by visual Inspection. Only closed flash suppressors are acceptable.
	Check barrel for straightness using bore straightness gage 8448202. Gage must pass freely through the bore to be acceptable, either dropped from the muzzle or chamber end.
Front sight and gas tube	Inspect gas tube for proper alignment with carrier key. Gas tube must not bind when mating with the key.
	Evidence of gas leaks around the front sight connection of the gas tube shall be cause for rejection until rifle has been function fired to determine if the loss of gas is sufficient to cause malfunction.
	If function firing malfunctions occur, repairs are necessary.
	Inspect front sight for damage.

Table 3-1. 5.56MM Rifle M16A2, M4/M4A1 Carbine - Cont

Item	Standard
olt carrier group	Inspect bolt for elongated or oversized firing pin hole using plain cylinder gage 12620101. Firing pinholes which permit the plain cylinder plug gage to fully penetrate at any position on the circumference will be rejected. Bolt face with a cluster of pits which are touching or tightly grouped, covering an area measuring approximately 1/8 inch (0.31 cm) across will be rejected. Bolts which contain pits extending into the firing pinhole will not be rejected unless firing pin hole gaging check determines rejection. Bolts which contain individual pits or scattered pits will not be cause for rejection. Only phosphated bolt carriers are acceptable. Both phosphated and chrome plated bolts are acceptable.
olt locking lugs and bolt cam pin hole	Inspect for cracks in the locking lugs and cam pin hole area. Use a black light, if available; otherwise, use a glass of no more than 3X magnification or use inspection penetrant (item 25, app D). Use instructions contained in kit for application. If cracks are detected, the bolts will be replaced. **NOTE** Particular attention must be given to the area where the lugs meet the bolt body and around the side walls of the cam pin hole. Bolt rings must not be broken. Ring gaps must be properly spaced approximately 1/3 turn apart and not in line. Firing pin protrusion must be not less than 0.028 inch (0.071 cm) or more than 0.036 inch (0.091 cm). (Use firing pin protrusion gage 7799735.) Socket head capscrews must be staked. Carrier key must not be dented where end mates with gas tube. Repair or replace damaged carrier keys.
eadspace	Inspect headspace using headspace gage 7799734. Excessive headspace will be cause for rejection.
rigger pull	Inspect trigger pull using trigger measuring fixture 7274758. Trigger pull must be minimum 5.5 pounds (2.49 kg), maximum for M16A2 and M4 is 9.5 pounds (4.31 kg) and maximum for M4A1 is 8.5 pounds (3.86 kg). Test trigger pull, refer to page 3-83, steps a, b, c, and d.

3-25. GENERAL INSPECTION CRITERIA (CONT).

Table 3-1. 5.56MM Rifle M16A2, M4/M4A1 Carbine - Cont

Item	Standard
Lower receiver group	Inspect hammer and trigger pin holes using plain cylinder plug gage 12006472 Penetration of the gage in any one or more of the four holes will be cause for rejection. Inspect for cracks, corrosion, or mutilation which would affect functioning. Small dents or gouges will not be cause for rejection. Inspect receiver for corrosion in the lobes of the pivot or hinge pin area. Width between lobes shall not exceed 0 515 inch (1.30 cm). Inspect receiver for break through of metal. Inspect receiver and receiver extension for initial loss of protective coating.
Action spring	Free length of spring shall be between 11-3/4 and 13-1/2 inches (29.84 and 34.29 cm) **M16A2 ONLY** and 10-1/16 and 11-1/4 inches (25.56 and 28.58 cm) **M4/M4A1 ONLY**.
Handguard	Inspect handguard assembly for breaks, separations, and cracks. Breaks and separations of material which prevent proper retention or interfere with functioning of the weapon will be cause for handguard rejection and replacement. Cracks up to 1 inch (2.54 cm) in length are acceptable provided they do not extend into the handguard retaining flange (1) (critical area). **M16A2 ONLY** Each handguard assembly may have up to two of the three front retaining tabs (2) missing. If all three front tabs are missing, the handguard assembly must be replaced. **M4/M4A1 ONLY** No tabs can be missing. **ALL WEAPONS** Replace severely cracked handguards. Handguards that have a heat shield loose enough to rattle when installed on the weapon must be replaced.
Stock assembly	Inspect buttstock assembly for dents, cracks, and chips. Check for breaks and separation of material which could prevent proper functioning of weapon. M4/M4A1 inspect for proper functioning. Repair as required

Table 3-1. 5.56MM Rifle M16A2, M4/M4A1 Carbine - Cont

Item	Standard
	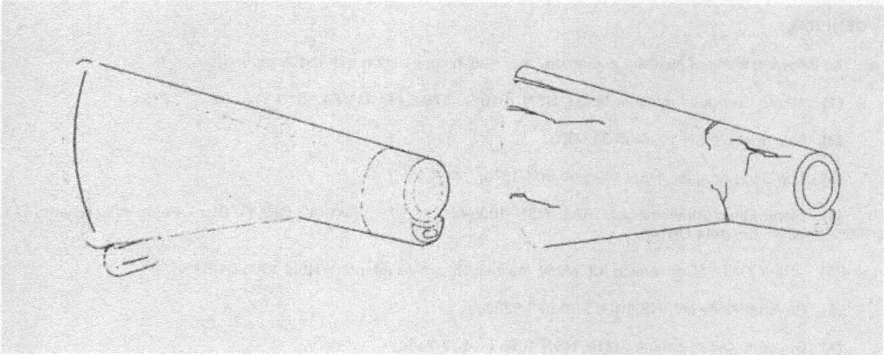
	M16A2 ONLY Under the following conditions, hairline cracks originating from buttplate end of buttstock are acceptable. No chipped away material is allowed. a. One hairline crack, not to exceed 1 inch (2.54 cm) in length, per side of buttstock. b. Two additional hairline cracks up to 0.22 inch (0.55 cm) in length, per side of buttstock. Buttstocks with unauthorized markings stamped into their surfaces will be replaced. Unauthorized markings, scratched, etched, carved, etc., are acceptable if they do not extend into the fiber of the buttstock which may weaken it. These marks may lie at any location on the buttstock. Cracks in the critical area at the front end of the buttstock are not acceptable and these buttstocks must be replaced.

CHAPTER 4

MAINTENANCE OF AUXILIARY EQUIPMENT

CHAPTER OVERVIEW

This chapter contains information and instructions to keep auxiliary equipment used with your weapon in good repair.

Section I. AUXILIARY EQUIPMENT REPAIR

4-1 GENERAL.

a The following items of auxiliary equipment are used in conjunction with the Weapons:

(1) 40mm Grenade Launcher M203, NSN 1010-00-179-6447, (M16A2 ONLY)

(2) Lock Plate, NSN 1005-00-233-9031.

(3) Top Sling Adapter, NSN 1005-00-406-1570.

(4) Blank Firing Attachment M15A2, NSN 1005-00-118-6192, (M16A2 ONLY), Blank Firing Attachment M23, NSN 1005-01-361-8208 (M4 ONLY)

Change 4**(5)** ARMY ONLY: Conversion Kit, M261 (caliber .22 rimfire adapter), NSN 1005-01-010-1561.

(6) Bayonet-Knife M7, NSN 1005-00-073-9238.

(7) Bayonet-Knife Scabbard M10, NSN 1095-00-223-7164.

(8) M9 Multi-Purpose Bayonet System, NSN 1005-01-227-1739.

(9) Night Vision Sight, Individual Served Weapon, AN/PVS-4, NSN 5855-00-629-5334 (M16A2 ONLY)

(10) M2 Practice Bolt, NSN 1005-01-184-4041 (M16A2 ONLY)

b Refer to TM 9-1010-221-23&P for unit maintenance for the Grenade Launcher M203.

c ARMY ONLY: Refer to TM 9-6920-363-12&P for unit maintenance for the M261 Conversion Kit (caliber rimfire adapter) M16, M16A1, and M16A2 Rifles.

d Refer to TM 9-1005-237-23&P for repair instructions and repair parts for Bayonet-Knife M7 and Bayonet Knife Scabbard M10 and M9 Multi-Purpose Bayonet System.

e Refer to TM 11-5855-213-23&P for maintenance for the Night Vision Sight, Individual Served Weapon, AN/PVS-4.

f Refer to TM 9-6920-746-12&P for unit maintenance for the M2 Practice Bolt.

4-2. LOCK PLATE

This task covers:

a Installation
b Removal
c Inspection

INITIAL SETUP

Tools
(ARMY) Small Arms Repairman Tool Kit (item 3, app B)

WARNING
The lock plate prevents the selector lever from being placed in BURST and will be installed at the discretion of the unit commander. It is mandatory for use in civil disturbance (riot control). Do not keep live ammunition near work area.

a. INSTALLATION

1 Using a screwdriver, reach inside pistol grip (1) and remove screw (2) and lockwasher (3).

WARNING
To avoid injury to your eyes, use care when removing and installing spring-loaded parts.

2 Carefully remove pistol grip (1). Hold detent helical spring (4) in place.

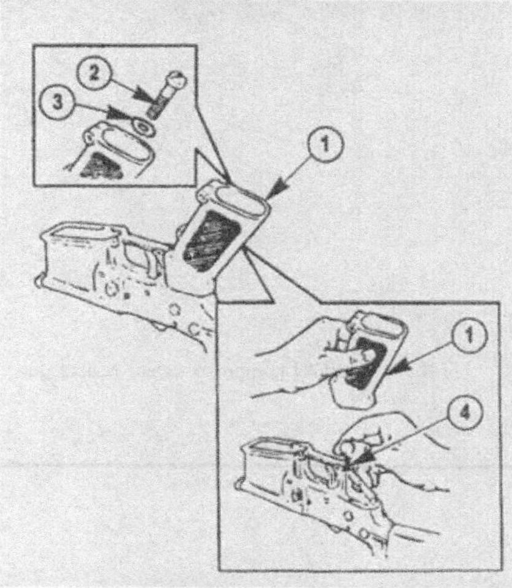

3 Install lock plate (5) with the detent helical spring (4) passing through the hole In the right side of the lock plate and the arm on the outside of the receiver The selector lever just point to the SAFE position

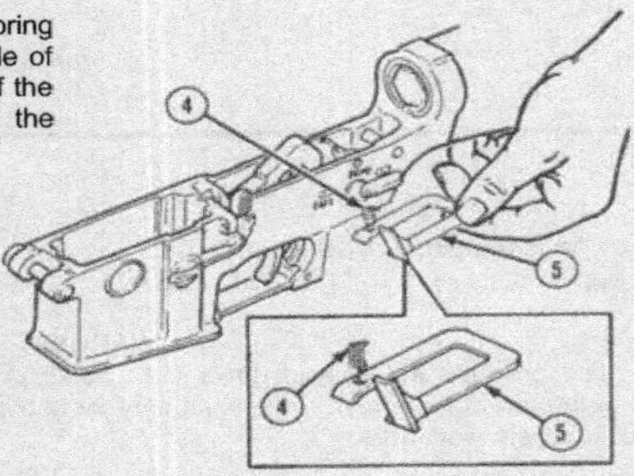

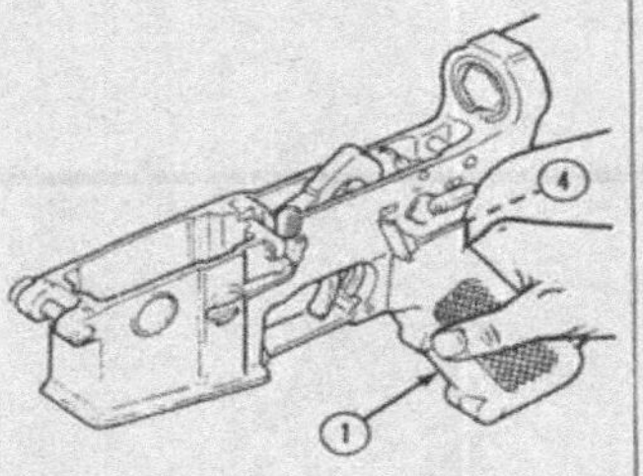

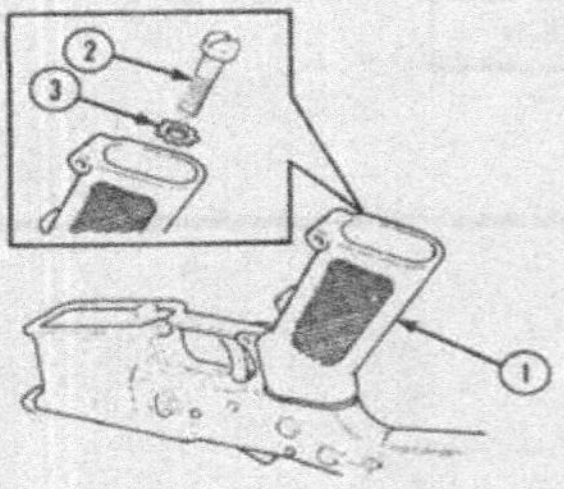

4 Carefully compress detent helical grip (1)

5. Using a screwdriver secure pistol by Installing lockwasher 131 spring (4) and position pistol grip (1) and screw 12)

-2. LOCK PLATE (CONT).

b REMOVAL

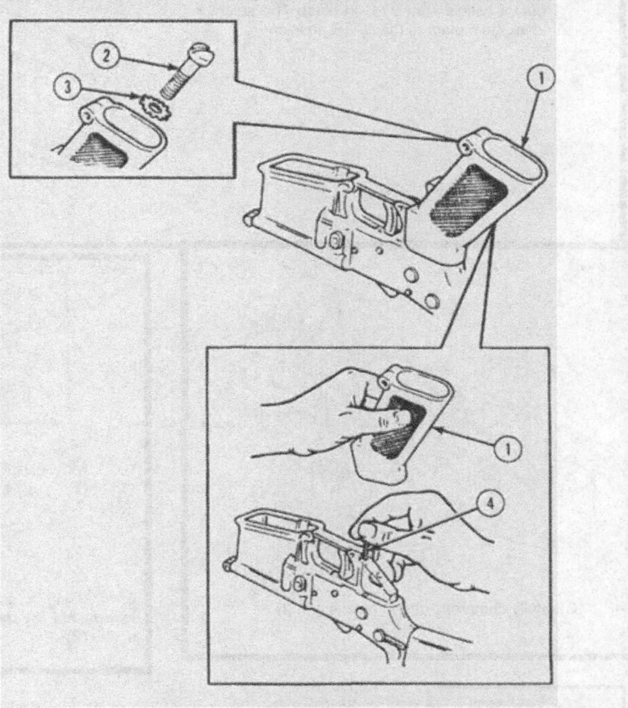

1. Using a screwdriver, reach inside pistol grip (l) and remove screw (2) and lockwasher (3)

WARNING

To avoid Injury to your eyes, use care when removing and installing spring loaded parts

2. Carefully remove pistol grip (1) Hold detent helical spring (4) In place

3 Remove lock plate (51 while continuing to hold detent helical spring (4) In place The selector lever must point to the SAFE position

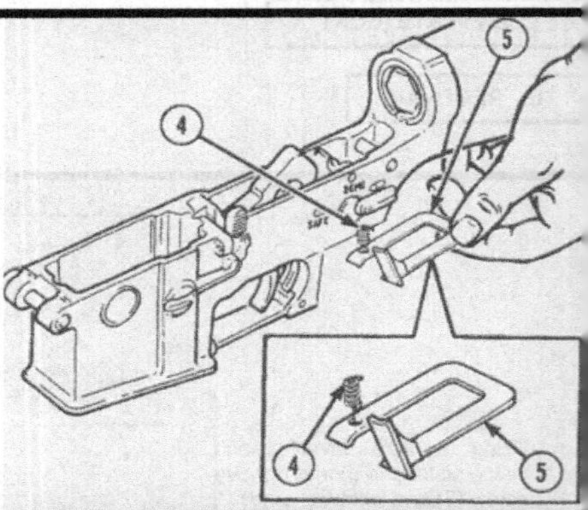

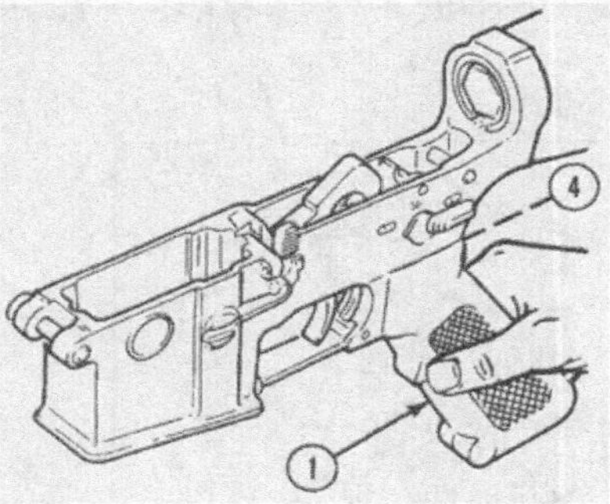

4 Carefully compress detent helical grip (l)

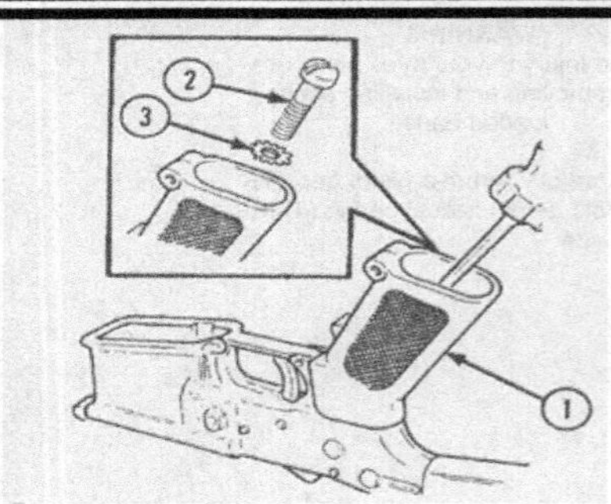

5. Using a screwdriver, secure pistol by installing lockwasher (3) spring (4) and position pistol grip (1) and screw (2)

c. INSPECTION

Inspect lock plate for serviceability and broken arm Replace If unserviceable or if arm is broken off

4-3. TOP SLING ADAPTER

This task covers:

a Installation
b Removal
c Inspection

INITIAL SETUP

Materials/Parts
Top sling adapter kit PN 8448471

References
TM 9-1005-319-10

a. INSTALLATION

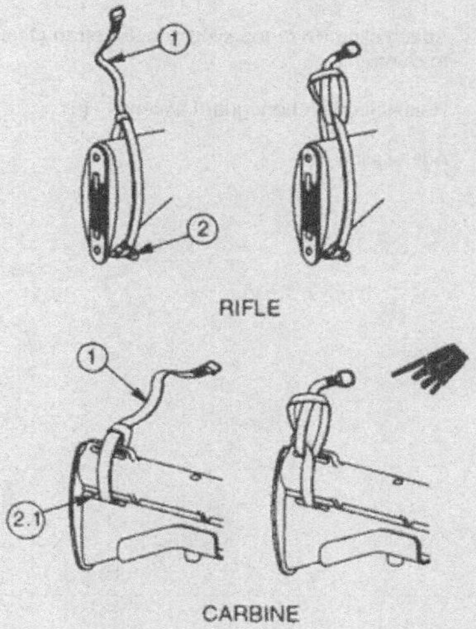

1 Refer to TM 9-1005-319-10 to remove sling.

RIFLE ONLY

2 Install top sling adapter strap (1) through sling swivel (2) and tie.

CARBINE ONLY

2.1. Install top sling adapter strap (1) through sling opening (2.1) and tie.

3 Remove upper handguard assembly.

4 Use pliers to install clamp (3) on front sight base (4).

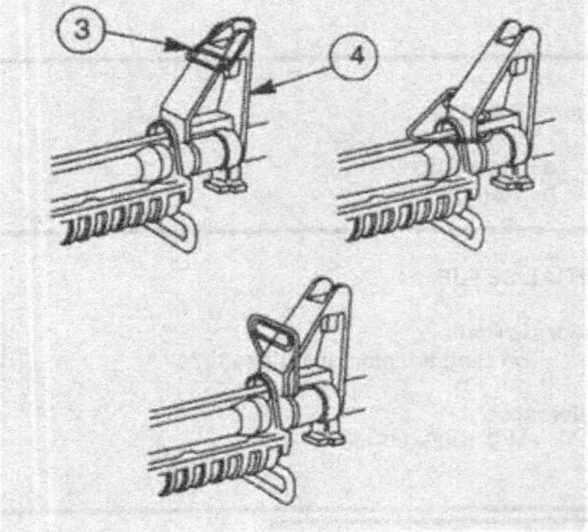

5 Attach sling (5) to top sling adapter strap (1) and to clamp (3).

6 Reinstall upper handguard assembly (6).

7 Adjust sling (5).

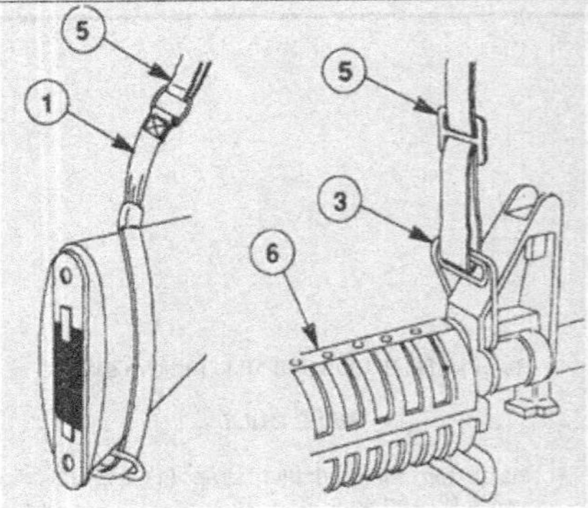

4-3. TOP SLING ADAPTER (CONT).

b. REMOVAL

1 Remove upper handguard assembly.

2 Remove sling (1) from clamp (2) and top sling adapter strap (3).

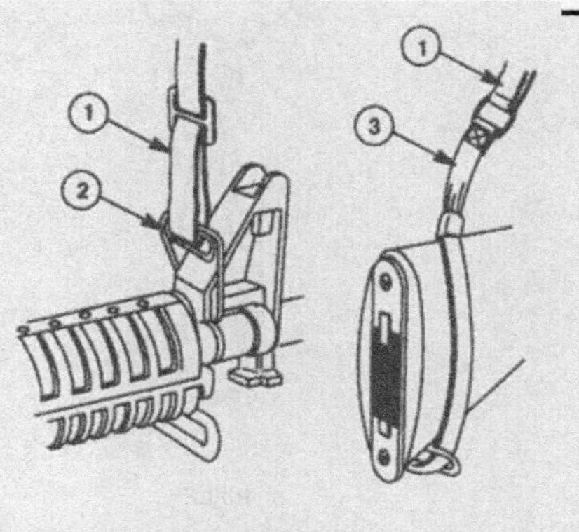

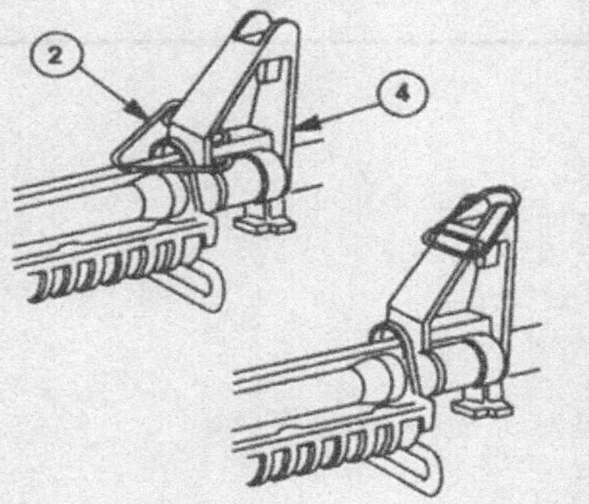

3 Use pliers to remove clamp (2) from front sight base (4). (5).

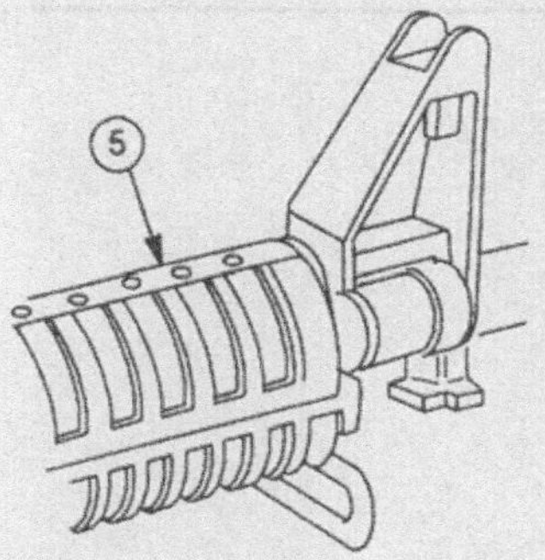

4 Reinstall upper handguard assembly

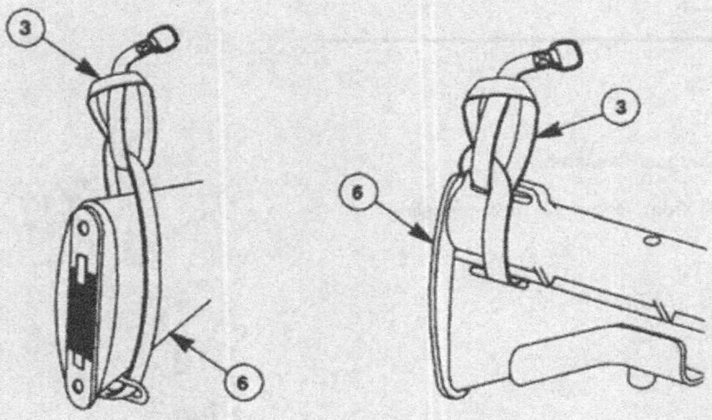

RIFLE

CARBINE

5 Untie top sling adapter strap (3) and remove from buttstock (6).

6 Refer to TM 9-1005-319-10 and install sling.

4-3. TOP SLING ADAPTER (CONT)

c. INSPECTION

Visually Inspect top sling adapter strap and replace If it is badly worn or damaged

4-4. BLANK FIRING ATTACHMENTS M15A2 (M16A2) AND M23 (M4/M4A1).

This task covers:

- a Installation
- b Removal
- c Cleaning
- d Inspection
- e Repainting
- f Replacement

INITIAL SETUP

Materials/Parts
Cleaner, lubricant, and preservative (CLP) (item 9, app D)
Coating compound, enamel (Red - M16A2) (item 23.2, app D)
Coating compound, enamel (Yellow - M4/M4A1) (item 23.3, app D)

General Safety Instructions
Do not keep live ammunition near the work area.

Only blank cartridge M200 is to be used when the blank firing attachment is attached to the weapon.
Do not fire blank ammunition at a representative enemy at distances of less than 20 feet (6.10m).
The unburned propellant grains can cause injury within this distance.

a. INSTALLATION

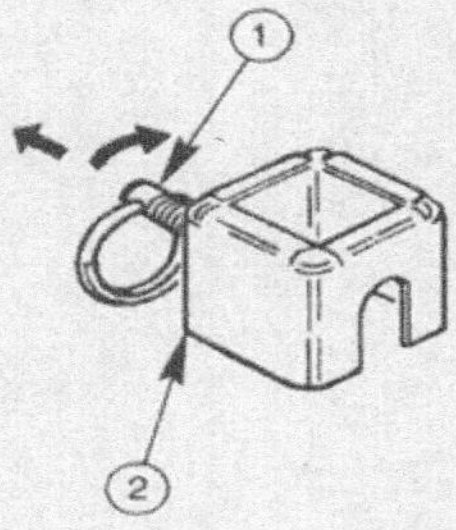

NOTE
M23 blank firing attachment is stamped "M4 Carbine Only" painted yellow and may be used on the M4 and M4A1 carbines. M15A2 BFA is painted red and is used on the M16A2 rifle.

1 Unscrew and pull slide (1) all the way out on blank firing attachment (2)

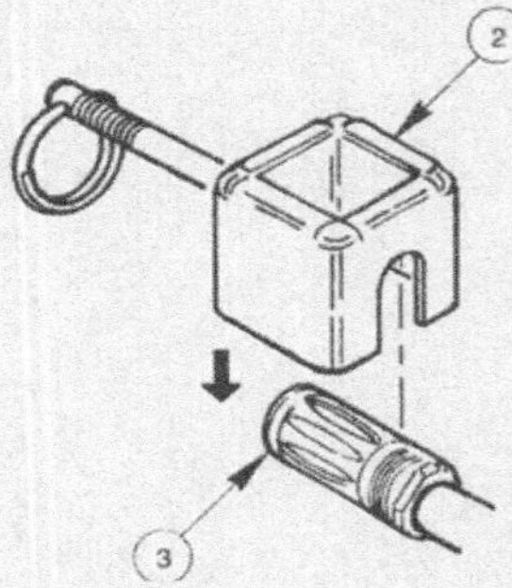

2 Hook blank firing attachment (2) behind the first groove of the compensator (3).

4-4 BLANK FIRING ATTACHMENTS M15A2 (M16A2) AND M23 (M4/M4A1)(CONT).

a. INSTALLATION (CONT)

CAUTION

Do not use tools to tighten the blank firing attachment. **USE HANDS ONLY.**

3 Push slide (2) into compensator (3) and hand tighten.

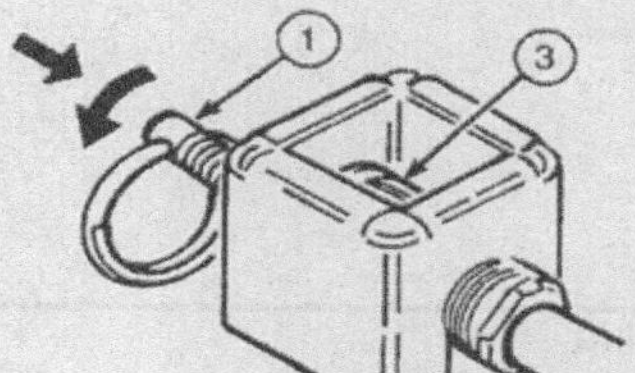

NOTE

Check for tightness after firing approximately 50 blank rounds.

b. REMOVAL

CAUTION

Do not use tools to tighten the blank firing attachment. USE HANDS ONLY.

1 Unscrew slide (1) to remove from compensator (2).

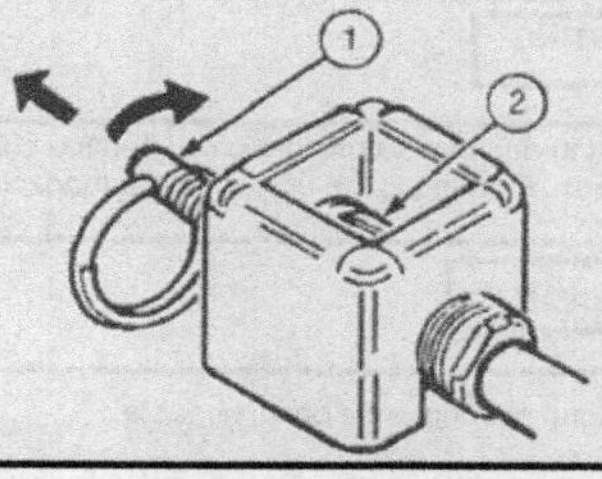

2. Unhook blank firing attachment (3) from behind the first groove of the compensator (2)

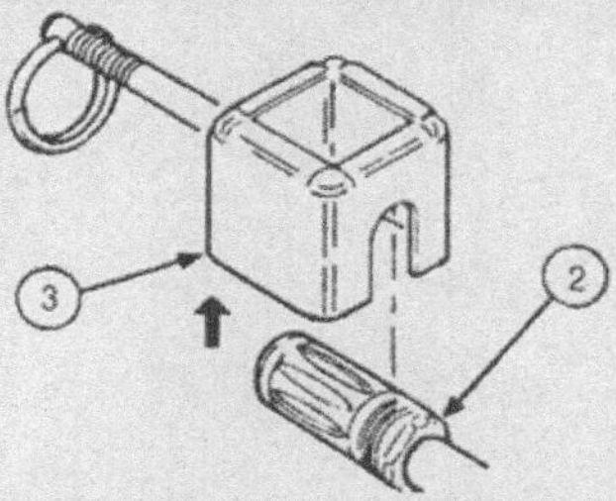

3 Screw slide (1) all the way in on blank firing attachment (3).

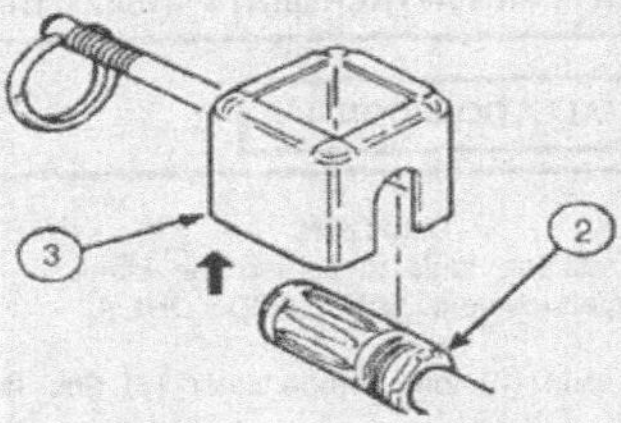

c. CLEANING

Clean blank firing attachment with CLP, wipe dry, and coat with CLP.

d. INSPECTION

Inspect blank firing attachment for cracks or distortion. Be sure the parts in the slide are clear and clean. If blank firing attachment is cracked or distorted, it is unserviceable.

e. REPAINTING

Repaint blank firing attachment using enamel coating compound (Red for M16A2 rifle or Yellow for M4/M4A1 carbine). Painting is the only repair authorized.

f. REPLACEMENT

Replace blank firing attachment if unserviceable.

APPENDIX A
REFERENCES

A-1. TECHNICAL BULLETINSINSTRUCTIONS/MANUALS/ORDERS.

TB 9-1000-247-34 .... Standards for Overseas Shipment or Domestic Issue of Small Arms, Aircraft Armament, Toward Howitzers, Mortars, Recoilless Rifles, Rocker Launchers and Associated Fire Control Equipment

TB 43-0002-73 .... Maintenance Expenditures Limits for FSC Group 10; FSC Classes 1005, 1010, 1015, 1030, 1055, 1090, and 1095

TI 05538A-15/2A .... Early Warning Indications of Malfunctions

TI 05538A-35/12A .... Checking Muzzle and Breech Erosion in Barrel

TI 8005-24/20 .... Prefire Inspection, Small Arms Weapons, Ordnance Material

TM 9-1005-237-23&P .... Organizational and Direct Support Maintenance Manual (Including Repair Parts and Special Tools List) for Bayonet-Knife M6 and M7, with Bayonet-Knife Scabbard M10 and M9 Multipurpose Bayonet System

TM 9-1005-319-10 .... Operator's Manual for 5.56mm, M16A2 Rifle and M4/M4A1 Carbine

TM 9-1010-221-23&P .... Organizational, Direct Support and General Support Maintenance Manual (Including Repair Parts and Special Tools List) for Launcher, Grenade: 40mm M203 W/E (NSN 1010-00-179-6447)

TM 9-6920-363-12&P .... M261 Conversion Kit

TM 9-6920-746-12&P .... M2 Practice Bolt

TM 11-5855-213-23&P .... Night Vision Sight, Individual Served Weapon, AN/PVS-4

TM 750-244-7 .... Procedures for Destruction of Equipment in Federal Supply Classifications 1000, 1005, 1015, 1020, 1025, 1030, 1055, 1090, and 1095 to Prevent Enemy Use

TM 4700-15/1 .... Equipment Record Procedures

TO 00-35D-54 .... Materiel Deficiency Reporting and Investigating System

TO 11 A-13-10-7 .... Storage and Maintenance Procedures for Small Arms Ammunition

TO 11W-1-10 .... Historical Data Recording of Inspection, Maintenance, and Firing Data for Ground Weapons

TO 33K-1-100 .... (TMDE) Interval Calibration and Repair Reference Guide and Work Unit Code Manual

A-2. FIELD MANUALS.

FM 21-11 ....First Aid for Soldiers

FM 23-9 ....M16A1 and M16A2 Rifle and Rifle Marksmanship

A-3 PAMPHLETS.

DA PAM 25-30..Consolidated Index of Army Publications and Blank Forms

DA PAM 738-750....................................The Army Maintenance Management System (TAMMS)

A-4 RELATED PUBLICATIONS.

AFM 36-2227, Vol 1.................................Combat Arms Training Management and Range Operation

AMC-P 310-9..Equipment Publications Listing

AR 725-50 ....Requisitioning, Receipt, and Issue System

AR 750-1 ....Army Materiel Policies

CTA 8-100 ....Army Medical Department Expendable/Durable Items

CTA 50-970 ....Expendable/Durable Items (Except: Medical, Class V, Repair Parts and Herald Items)

DOD 4160.21-M-1 ....Defense Demilitarization Manual

FMFM 1-3A...Field Firing Techniques

SB 746-1...Publications for Packing Army General Supplies

SPI 00-856-6885 ....Special Packaging Instructions for M16 Rifle

A-5 FORMS.

AFTO Form 22 ....Technical Order System Publications Improvement Report

AFTO Form 105 ....Inspection, Maintenance, and Firing Data for Ground Weapons

DA Form 2028 ....Recommended Changes to Publications and Blank Forms

DA Form 2028-2.......................................Recommended Changes to Equipment Technical Manuals

DA Form 2404 ....Equipment Inspection and Maintenance Worksheet

DA Form 2408-9.......................................Equipment Control Record

DD Form 314 ....Preventive Maintenance Schedule and Record

NAVMC 10772..Recommended Changes to Technical Publications

NAVMC 11003..Ordnance Serialized Items Subsidiary Records

SF 364...Report of Discrepancy (ROD)

SF 368...Product Quality Deficiency Report

A-6 SUPPLY CATALOGS/SUPPLY LISTS.

SC 4933-95-CL-A11...............................Shop Set, Small Arms Field Maintenance

SC 5180-95-CL-A07...............................Tool Kit, Small Arms Repair

APPENDIX B

MAINTENANCE ALLOCATION CHART

Section I. INTRODUCTION

B-1. GENERAL.

a. This section provides a general explanation of all maintenance and repair functions authorized at variou maintenance levels

b. The Maintenance Allocation Chart (MAC) In section II designates overall authority and responsibility for the performance of maintenance functions on the identified end Item or component. The application of the maintenance functions to the end Item or component will be consistent with the capacities and capabilities of the designated maintenance levels

c. Section III lists the tools and test equipment (both special tools and common tool sets) required for each maintenance function as referenced from section II.

d. Section IV contains supplemental instructions and explanatory notes for a particular maintenance function.

B-2. MAINTENANCE FUNCTIONS.

Maintenance functions will be limited to and defined as follows: (except for ammunition MAC')

a. Inspect. To determine the serviceability of an Item by comparing Its physical, mechanical, and or electrical characteristics with established standards through examination leg , by sight, sound, or feel).

b. Test. To verify serviceability by measuring the mechanical, pneumatic, hydraulic, or electrical characteristics of an Item and comparing those characteristics with prescribed standards

c. Service. Operations required periodically to keep an Item In proper operating condition, I e to clean (includes decontaminate, when required), to preserve, to drain, to paint, or to replenish fuel, lubricants, chemical fluids. or gases.

d. Adjust. To maintain or regulate, within prescribed limits, by bringing Into proper or exact position, or by setting the operating characteristics to specified parameters

---

'Exception is authorized for ammunition MAC to permit the redesignation/redefinition of maintenance function headings to more adequately identify ammunition maintenance functions The heading designations and definitions will be Included In the appropriate technical manual for each category of ammunition

B-2. MAINTENANCE FUNCTIONS (CONT).

e. Align. To adjust specified variable elements of an item to bring about optimum or desired performance

NOTE

Air Force Precision Measurement Equipment Laboratory (PMEL) person will calibrate small arms Inspection gages In accordance with TO 33K-1-100

f. Calibrate. To determine and cause corrections to be made or to be adjusted on Instruments or test, measuring, and diagnostic equipments used In precision measurement Consists of comparisons of two Instruments, one of which Is a certified standard of known accuracy, to detect and adjust any discrepancy in the accuracy of the Instrument being compared

g. Remove/Install. To remove and Install the same item when required to perform service or other maintenance functions. Install may be the act of emplacing, seating, or fixing Into position a spare, repair part, or module (component or assembly) In a manner to allow the proper functioning of an equipment or system.

h. Replace. To remove an unserviceable Item and Install a serviceable counterpart In Its place "Replace" is authorized by the MAC and is shown as the 3d position code of the SMR code.

i. Repair. The application of maintenance services2, including fault location troubleshooting3, removal 'installation, and disassembly'assembly4 procedures, and maintenance actions" to Identify troubles and restore serviceability to an Item by correcting specific damage, fault, malfunction, or failure In a part, subassembly, module (component or assembly), end Item, or system.

j. Overhaul. That maintenance effort (service/action) prescribed to restore an item to a completely serviceable operational condition as required by maintenance standards In appropriate technical publications (i.e , DMWR). Overhaul is normally the highest degree of maintenance performed by the Army Overhaul does not normally return an Item to like new condition

k. Rebuild. Consists of those services/actions necessary for the restoration of unserviceable equipment to a like new condition In accordance with original manufacturing standards Rebuild is the highest degree of materiel maintenance applied to Army equipment The rebuild operation Includes the act of returning to zero those age measurements (hours miles, etc) considered In classifying Army equipment/components.

---

[2]Servlces--Inspect, test, service, adjust, align, calibrate, and or replace
[3]Fault locate troubleshoot The process of investigating and detecting the cause of equipment malfunctioning; the act of isolating a fault within a system or unit under test (UUT).
[4]Disassemble assemble encompasses the step-by-step taking apart (or breakdown) of a spare/functional group coded Item to the level of Its least competency identified as maintenance significant (.e , assigned an SMR code) for the level of maintenance under consideration '
[5]Actionswelding, grinding, riveting, straightening, facing, remachinery, and or resurfacing

B-3. EXPLANATION OF COLUMNS IN THE MAC, SECTION II.

a. Column 1, Group Number. Column 1 lists functional group code numbers, the purpose of which is to identify maintenance significant components, assemblies, subassemblies, and modules with the next higher a assembly. End item group number shall be "OO."

b. Column 2, Component Assembly. Column 2 contains the names of components. Assemblies, subassemblies, and modules forwhich maintenance is authorized.

c. Column 3, Maintenance Function. Column 3 lists the functions to be performed on the it em listed in Column 21. (for detailes explanation of these functions, see paragraph B-2.)

d. Column 4, Maintenance Level. Column 4 specifies, by the list ing of a work time figure in the appropriate subcolumn(s), the level of maintenance. If the number of complexity of the t asks within the listed maintenance function vary at different maintenance levels, appropriate work time figures will be shown for each level. The work time figure represents the average time required to restore an item (assembly, subassembly, component, module, end it4m, or system) or a serviceable condition under typicalfield operating conditions. This time includes preparation time (including any necessary disassembly assembly time), troubleshooting fault location time, and quality assurance quality control time in addition to the time required to perform the specific tasks identified for the maintenance functions authorized i the maintenance allocation chart. The symbol designations for the varies maintenance levels are as follows:

C..Operator or Crew
O..Unit Maintenance
F..Direct Support Maintenance
H..General Support Maintenance
L ....Specialized Repair Activity (SRA)[6]
D..Depot Maintenance

e. Column 5, Tools and Equipment. Column 5 specifies, by code, those common tool sets (not individual tools) and special tools., TMDE, and support equipment required to perform the designated function.

f. Column 6, Remarks. This column shall, when applicable, contain a lett4er code, in alphabetic order, which shall be keyed to the remarks contained in Section IV.

---

[6]This maintenance level is not included in Section II, column (4) of the Maintenance Allocation Chart. To identify functions to this level of maintenance, enter a work time figure in the "H" column of section II, column (4), and use an associated reference code in the Remarks column (6). Key the c ode to Section IV . Remarks, and explain the SRA complete repair application there. The explanatory remarks)s) shall reference the specific Repair Parts and Special Tools List (RPSTL) TM which contains additional SRA criteria and the authorized spare/repair parts.

B-4. EXPLANATION OF COLUMNS IN TOOL AND TEST EQUIPMENT REQUIREMENTS, SECTION III.

a. **Column 1, Reference Code**. The tool and test equipment reference code correlates with a code used In the MAC, Section II, Column 5.

b. **Column 2, Maintenance Level**. The lowest level of maintenance authorized to use the tool or test equipment.

c. **Column 3, Nomenclature**. Name or identification of the tool or test equipment.

d. **Column 4, National Stock Number**. The National stock number of the tool or test equipment.

e. **Column 5, Tool Number**. The manufacturer's part number.

B-5. EXPLANATION OF COLUMNS IN REMARKS, SECTION IV.

a. **Column 1, Reference Code**. The code recorded In column 6, Section II.

b. **Column 2, Remarks**. This column lists information pertinent to the maintenance function being performed as indicated In the MAC, Section II.

Section II. MAINTENANCE ALLOCATION CHART FOR M16A2 RIFLE AND M4/M4A1 CARBINES

(1) GROUP NUMBER	(2) COMPONENT ASSEMBLY	(3) MAINTENANCE FUNCTION	(4) MAINTENANCE LEVEL UNIT C	O	DS F	DS H	DEPOT D	(5) TOOLS AND EQPT	(6) REMARKS CODE
00	M16A2 5.56MM RIFLE, AND M4/ M4A1 5.56MM CARBINES	Inspect	0.1	0.3	0.7			2	
		Test			0.2			2,3	
		Service	0.2	0.3	0.3				
		Replace		0.1					
		Overhaul							
00.1	CARRYING HANDLE ASSEMBLY M4/M4A1	Inspect	0.1	0.1	0.1				
		Service	0.1						
		Remove/ Install	0.1	0.1	0.1				
		Replace			0.3			1,3	
		Repair		0.1	0.2			1,3	
00.101	Rear Sight Assembly M4/ M4A1	Inspect			0.1				
		Replace			0.3			1,3	
		Repair			0.2			1,3	
01	BOLT CARRIER ASSEMBLY	Inspect	0.1	0.1	0.1				
		Service	0.1	0.1					
		Remove/ Install	0.1						
		Replace			0.1			1,2	
		Repair	01	0.1				3	
0101	Bolt Assembly	Inspect	0.1	0.1	0.1				
		Test			0.1			2	
		Service	0.1						
		Remove/ Install	0.1						
		Replace			0.1			2	
		Repair		0.1	0.1			3	

*Worktimes are included in DMWR 9-1005-249

MAINTENANCE ALLOCATION CHART (CONT)

(1) GROUP NUMBER	(2) COMPONENT ASSEMBLY	(3) MAINTENANCE FUNCTION	(4) MAINTENANCE LEVEL UNIT C	UNIT O	DS F	DS H	DEPOT D	(5) TOOLS AND EQPT	(6) REMARKS CODE
0102	Key and Bolt Carrier Assembly	Inspect	0.1		0.1				
		Service	0.1						
		Remove/ Install	0.1						
		Replace			0.1			1	
		Repair		0.1	0.1			1,3	
02	HANDLE ASSEMBLY	Inspect	0.1						
		Service	0.1						
		Remove/ Install	0.1						
		Replace			0.1		**		
		Repair			0.1			3	
03	UPPER RECEIVER AND BARREL ASSEMBLY	Inspect	0.1	0.2	0.1				
		Test			0.2			2	
		Service	0.2		0.1			1	
		Remove/ Install	0.1						
		Replace			0.5			1,2	
		Repair		0.2	0.5			1,2,3	A,B
0301	M16A2 Barrel Assembly and M4/ M4A1 Replacement Barrel	Inspect		0.1	0.1				
		Replace			0.3			1,2	
		Repair	0.1					3	
0302	Upper Receiver Assembly	Inspect			0.1				
		Replace			0 5			1.2	
		Repair			0.3			1.3	
030201	Forward Assist Assembly	Inspect			0.1				
		Replace			0.1				
		Repair			0 1			1.3	
030202	Rear Sight Assembly M16A2	Inspect			0 1				
		Replace			0.3			1,3	
		Repair			0 2			1,3	

MAINTENANCE ALLOCATION CHART (CONT)

(1) GROUP NUMBER	(2) COMPONENT ASSEMBLY	(3) MAINTENANCE FUNCTION	(4) MAINTENANCE LEVEL UNIT C	O	DS F	DS H	DEPOT D	(5) TOOLS AND EQPT	(6) REMARKS CODE
04	LOWER RECEIVER AND BUTTSTOCK ASSEMBLY	Inspect	0.1	0.1	0.1				
		Test			0.1			2	
		Service	0.2	0.2					
		Repair		0.2	0.3			1,2,3	C,D,E
0401and 0401A	Buttstock Assembly	Inspect		0.1					
		Remove/ Install		0.1				3	
		Replace		0.1				3	
		Repair		0.1				3	
0402	Hammer Assembly	Inspect			0.1				
		Remove/ Install			0.1				
		Replace			0.1				
		Repair			0.1				
0403	Trigger Assembly	Inspect			0.1				
		Remove/ Install			0.1			1,3	
		Replace			0.1			1,3	
		Repair			0.1			1,3	
040301	Trigger Subassembly M16A2 and M4	Inspect			0.1				
		Replace			0.1			1,3	
		Repair			0.1			1,3	
0404 and 0404A	Lower Receiver and Receiver Extension Assembly	Inspect			0.1				
		Test			0.1			2	
		Repair			0.2			1,3	

Section III. TOOL AND TEST EQUIPMENT REQUIREMENTS FOR M16A2 RIFLE AND M4/M4A1 CARBINES

TOOL OR TEST EQUIPMENT REF CODE	(2) MAINTENANCE CATEGORY	(3) NOMENCLATURE	(4) NATIONAL/NATO STOCK NUMBER	(5) TOOL NUMBER
1 (ARMY ONLY)	F	Shop Set, Small	4933-00-754-0664	SC 4933-95-CL-A11
2 (ARMY ONLY)	F	Tool and gage Set, DS/GS Maintenance for 5.56mm Rifle, M16 Series	4933-00-056-7106	8426685
3 (ARMY ONLY)	O	Tool Kit, Small Arms Repairman	5180-00-357-7770	SC 5180-95-CL-A07
4 (A.F ONLY)	F	Torque Wrench, ft-lb	5120-00-640-6365	A-A-411
5 (A.F ONLY)	F in.-lb	Torque Wrench,	5120-00-230-6380	T-E-12A
6 A.F ONLY)	F	Trigger Weights	4933-00-647-3696	7274758

Section IV. REMARKS

Reference Code	Remarks
A	Tool, Front Sight Post Removal and Installation
B	Depressor, Front Sight Detent
C	Tool, Pivot Pin Removal
D	Tool, Pivot Pin Installation
E	Pin, Slave

APPENDIX C

REPAIR PARTS AND SPECIAL TOOLS LIST

Section I. INTRODUCTION

C-1. SCOPE.

This RPSTL lists and authorizes spares and repair parts; special tools; special test, measurement, and diagnostic equipment (TMDE); and other special support equipment required for performance of unit and direct support maintenance of the M16A2 rifle and M4/M4A1 carbines. It authorizes the requisitioning, issue, and disposition of spares, repair parts, and special tools as indicated by the source, maintenance, and recoverability (SMR) codes.

C-2. GENERAL.

In addition to Section I, Introduction, this Repair Parts and Special Tools List is divided into the following sections:

a. Section II Repair Parts List. A list of spares and repair parts authorized by this RPSTL for use in the performance of maintenance. The list also includes parts which must be removed for replacement of the authorized parts. Parts lists are composed of functional groups in ascending alphanumeric sequence, with the parts in each group listed in ascending figure and item number sequence. Bulk materials are listed by item name in FIG BULK at the end of the section. Repair parts kits or sets are listed separately in their own functional group within Section II. Repair parts for repairable special tools are also listed in the section.

b. Section III Special Tools List. A list of special tools, special TMDE, and other special support equipment authorized by this RPSTL (as indicated by Basis of Issue (BOI) information in DESCRIPTION AND USABLE ON CODE (UOC) column) for the performance of maintenance.

c. Section IV Cross-Reference Indexes. A list, in National item identification number (NIIN) sequence, of all National stock numbered items appearing in the listings, followed by a list in alphanumeric sequence of all part numbers appearing in the listing. National stock numbers and part numbers are cross-referenced to each illustration figure and item number in alphanumeric sequence and cross-references NSN, CAGEC, and part number.

C-3. EXPLANATION OF COLUMNS (SECTIONS II AND III).

a. ITEM NO. (Column (1)). Indicates the number used to identify items called out in the illustration.

b. SMR CODE (Column (2)). The Source, Maintenance, and Recoverability (SMR) code is a 5-position code containing supply/requisitioning information, maintenance level authorization criteria, and disposition instruction, as shown In the following breakout:

C-3. EXPLANATION OF COLUMNS (SECTION II AND III) (CONT).

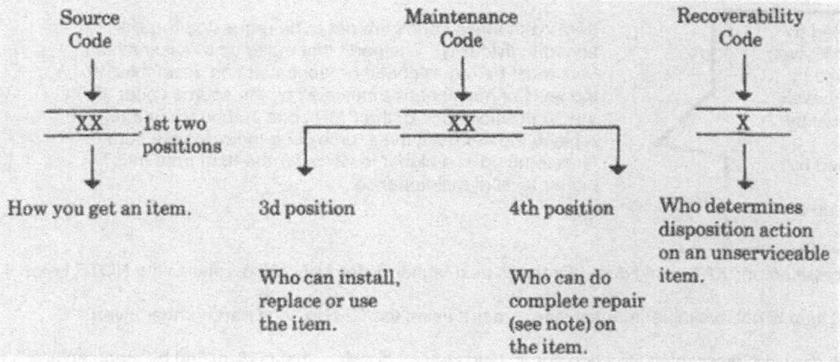

*Complete Repair: Maintenance capacity, capability, and authority to perform all corrective maintenance tasks of the "Repair" function in a use/user environment In order to restore serviceability to a failed item.

(1) Source Code. The source code tells you how to get an item needed for maintenance, repair, or overhaul of an end item/equipment. Explanations of source codes follow:

Code	Explanation
PA PB PC* * PD PE PF PG	Stocked items; use the applicable NSN to request/requisition Items with these source codes They are authorized to the level indicated by the code entered In the 3d position of the SMR code. * *NOTE: Items coded PC are subject to deterioration.
KD KF KB	Items with these codes are not to be requested/requisitoned Individually They are part of a kit which is author->zed to the maintenance level indicated In the 3d positon of the SMR code The complete kit must be requisitioned and applied.
MO-(Made at Unit/ AVUM Level) MF- (Made at DS/ AVIM Level) MH-(Made at GS Level) ML- (Made at Specialized Repair Act (SRA)) MD-(Made at Depot)	Items with these codes are not to be requested/requisitioned Individually They must be made from bulk material which Is identified by the part number In the DESCRIPTION AND USABLE ON CODE (UOC) column and listed In the Bulk Material group of the repair parts list in this RPSTL If the Item Is authorized to you by the 3d position code of the SMR code, but the source code Indicates It is made at a higher level, order the Item from the higher level of maintenance.

Code	Explanation
AO- (Assembled by unit/AVUM Level) AF- (Assembled by DS/AVIM Level) AH- (Assembled by GS Level) AL- (Assembled by SRA) AD- (Assembled by Depot)	Items with these codes are not to be requested/requisitioned Individually. The parts that make up the assembled Item must be requisitioned or fabricated and assembled at the level of maintenance indicated by the source code. If the 3d position code of the SMR code authorizes you to replace the Item, but the source code Indicates the Item Is assembled at a higher level, order the Item from the higher level of maintenance.

XA-Do not requisition an "XA"-coded item. Order Its next higher assembly. (Also, refer to the NOTE below.)

XB--If an "XB" item is not available from salvage, order It using the CAGEC and part number given.

XC-installation drawing, diagram, instruction sheet, field service drawing, that is identified by manufacturer's part number.

XD-Item is not stocked. Order an "XD"-coded Item through normal supply channels using the CAGEC and part number given, If no NSN Is available.

NOTE

Cannibalization or controlled exchange, when authorized, may be used as a source of supply for items with the above source codes, except for those source coded "XA" or those aircraft support Items restricted by requirements of AR 750-1.

(2) Maintenance Code. Maintenance codes tell you the level(s) of maintenance authorized to USE and REPAIR support items. The maintenance codes are entered In the third and fourth positions of the SMR code as follows:

(a) The maintenance code entered in the third position tells you the lowest maintenance level authorized to remove, replace, and use an Item. The maintenance code entered In the third position will indicate authorization to one of the following levels of maintenance

Code	Application/Explanation
C	-Crew or operator maintenance done within unit or aviation unit maintenance.
O	-Unit or aviation unit level can remove, replace, and use the Item.
F	-Direct support or aviation intermediate level can remove, replace, and use the Item.
H	-General support level can remove, replace, and use the Item.
L	-Specialized repair activity can remove, replace, and use the Item.
D	-Depot level can remove, replace, and use the Item

C-3. EXPLANATION OF COLUMNS (SECTIONS II AND III) (CONT).

(b) The maintenance code entered In the fourth position tells whether or not the Item is to be repaired and identifies the lowest maintenance level with the capability to do complete repair (I e, perform all authorized repair functions). (NOTE: Some limited repair may be done on the Item at a lower level of maintenance, If authorized by the Maintenance Allocation Chart (MAC) and SMR codes I This position will contain one of the following maintenance codes

Code	Application/Explanation
O	-Unit or aviation unit is the lowest level that can do complete repair of the item
F	-Direct support or aviation Intermediate is the lowest level that can do complete repair of the Item
H	-General support is the lowest level that can do complete repair of the item .
L	-Specialized repair activity is the lowest level that can do complete repair of the item
D	-Depot is the lowest level that can do complete repair of the Item
Z	-Nonreparable. No repair Is authorized
B	-No repair is authorized. (No parts or special tools are authorized for the maintenance of a "B" coded item) However, the Item may be reconditioned by adjusting, lubricating, etc , at the user level

(3) Recoverability Code. Recoverability codes are assigned to Items to Indicate the disposition action on unserviceable items The recoverability code is entered In the fifth position of the SMR code as follows.

Recoverability Codes	Application/Explanation
Z	Nonreparable item unserviceable, condemn and dispose of the item at the level of maintenance shown In 3d position) of SMR code
O	Reparable Item. When uneconomically reparable condemn and dispose of the item at unit or aviation unit level
F	Reparable Item. When uneconomically reparable, condemn and dispose of the Item at the direct support or aviation level
H	Reparable Item. When uneconomically reparable, condemn and dispose of the Item at the .general support level

Recoverability Codes	Application/Explanation
D	-Reparable item. When beyond lower level repair capability, return to depot. Condemnation and disposal of item not authorized below depot level.
L	-Reparable item. Condemnation and disposal not authorized below Specialized Repair Activity (SRA).
A	-Item requires special handling or condemnation procedures because of specific reasons (e.g., precious metal content, high dollar value, critical material, or hazardous material). Refer to appropriate manuals/directives for specific instructions.

c. NSN (Column (3)). Indicates the National stock number (NSN) assigned to the item and which will be used for requisitioning.

d. CAGEC (Column (4)). The Contractor and Government Entity Code (CAGEC) is a 5-digit numeric code which is used to identify the manufacturer, distributor, or Government agency, etc., that supplies the Item.

e. PART NUMBER (Column (5)). Indicates the primary number used by the manufacturer (individual, company, firm, corporation, or Government activity), which controls the design and characteristics of the item by means of its engineering drawings, specifications standards, and inspection requirements to identify an item or range of items.

NOTE

When you use an NSN to requisition an item, the item you receive may have a different part number from the part ordered.

f. DESCRIPTION AND USABLE ON CODE (UOC) (Column (6)). This column includes the following Information:

(1) The Federal item name and, when required, a minimum description to identify the Item.

(2) The physical security classification of the item is indicated by the parenthetical entry which is a physical security classification abbreviation, e.g., Phy Sec C1 (C) Confidential, Phy Sec C1 (S) Secret, Phy Sec C1 (T) Top Secret.

(3) Items that are included in kits and sets are listed below the name of the kit or set.

(4) Spare/repair parts that make up an assembled item are listed immediately following the assembled item line entry.

(5) Part numbers for bulk materials are referenced in this column In the line item entry for the Item to be manufacture red/fabricated.

(6) When the item is not used with all serial numbers of the same model, the effective serial numbers are shown on the last line(s) of the description (before UOC).

(7) The usable on code, when applicable (see paragraph 5, Special Information).

C-3. EXPLANATION OF COLUMNS (SECTION II AND III) (CONT).

(8) In the Special Tools List section, the basis of issue (BOI) appears as the last line(s) in the entry for each special tool, special TMDE, and other special support equipment. When density of equipment supported exceeds density spread indicated in the basis of Issue, the total authorization is increased proportionately.

(9) The statement "END OF FIGURE" appears just below the last item description in Column 5 for a given figure In both Section II and Section III.

g. Oty (Column (7)) . The Qty (quantity per figure column) indicates the quantity of the item used in the breakout shown on the illustration figure, which is prepared for a functional group, subfunctional group, or an assembly. A "V" appearing In this column in lieu of a quantity indicates that the quantity is variable and the quantity may vary from application to application.

C-4. EXPLANATION OF COLUMNS (SECTION IV).

a. NATIONAL STOCK NUMBER (NSN) INDEX.

(1) STOCK NUMBER column. This column lists the NSN by National Item Identification Number (NIIN) sequence. The NIIN consists of the last nine digits of the NSN.

NSN
(i.e., 5305-01-674-1467). When using this column to locate an item, ignore the first 4 digits NIIN of the NSN. However, the complete NSN should be used when ordering items by stock number.

(2) FIG. column. This column lists the number of the figure where the item Is Identified/located. The figures are in numerical order in Section II and Section III.

(3) ITEM column. The item number identifies the item associated with the figure listed in the adjacent FIG. column. This item is also identified by the NSN listed on the same line.

b. PART NUMBER INDEX. Part numbers in this index are listed by part number in ascending alphanumeric sequence (i.e., vertical arrangement of letter and number combination which places the first letter or digit of each group In order A through Z, followed by the numbers 0 through 9 and each following letter or digit in like order).

(1) CAGEC column. The Contractor and Government Entity Code (CAGEC) is a 5-digit numeric code used to identify the manufacturer, distributor, or Government agency, etc., that supplies the item.

(2) PART NUMBER column. Indicates the primary number used by the manufacturer (individual, firm, corporation, or Government activity), which controls the design and characteristics of the item by means of Its engineering drawings, specifications standards, and inspection requirements to identify an Item or range of Items.

(3) STOCK NUMBER column. This column lists the NSN for the associated part number and manufacturer identified In the PART NUMBER and CAGEC columns to the left.

(4) FIG. column. This column lists the number of the figure where the item is identified/located in Sections II and III.

(5) ITEM column. The item number is that number assigned to the item as it appears in the figure referenced in the adjacent figure number column.

C-5. SPECIAL INFORMATION.

a. ASSEMBLY INSTRUCTION. Detailed assembly instructions for items source coded to be assembled from component spare/repair parts are found in Chapter 3. Items that make up the assembly are listed immediately following the assembly item entry or reference is made to an applicable figure.

b. ASSOCIATED PUBLICATIONS. The publications listed below pertain to the M16A2 rifle, M4/M4A1 carbine, and its components:

Publication	Short Title
TM 9-1005-319-10	M16A2 Rifle and M4/M4A1 Carbine

c. USABLE ON CODE. The usable on code appears in the lower left corner of the DESCRIPTION column heading. Usable on codes are shown as "UOC..." in DESCRIPTION column (justified left) on the first line applicable item description/nomenclature. Uncoded items are applicable to all models. Identification of the usable on codes used in the RPSTL are:

Code	Used On
AR8	M16A2 Rifle
AS1	M4 Carbine
AY6	M4A1 Carbine

C-6. HOW TO LOCATE REPAIR PARTS.

a. When National Stock Number or Part Number is Not Known.

(1) First. Using the table of contents, determine the assembly group or subassembly group to which the item belongs. This is necessary since figures are prepared for assembly groups and subassembly groups, and listings are divided into the same groups.

C-6. HOW TO LOCATE REPAIR PARTS (CONT).

(2) Second. Find the figure covering the assembly group or subassembly group to which the item belongs.

(3) Third. Identify the item on the figure and use the Figure and Item Number Index to find the NSN.

b. When National Stock Number or Part Number is Known.

(1) First. Using the National Stock Number or the Part Number Index, find the pertinent National Stock Number or Part Number. The NSN index is in National Item Identification Number (NIIN) sequence. (See C4.a.(1).) The pat numbers in the Part Number Index are listed in ascending alpha numeric sequence. (See C4.b.) Both indexes cross-reference you to the illustration/figure and item number of the item you are looking for.

(2) Second. Turn to the figure and item number, verify that the item is the one you're looking for, then locate the item number in the repair parts list for the figure.

C-7. ABBREVIATIONS.

Not applicable.

(1) ITEM NO QTY	(2) SMR CODE	(3) NSN	(4) CAGEC	(5) PART NUMBER	(6) DESCRIPTION AND USABLE	(7) ON CODE (UOC
					GROUP 00 (M16A2) RIFLE 5.56MM 9349000,(M4) CARBINE 5.56MM 9390000, AND (M4A1) CARBINE 5.56MM 12972700 FIG. C-1. (M16A2) RIFLE, (M4) AND CM4A1) CARBINES	
1	PAOOO	1005000179546	19204	8448517	HANDLE ASSEMBLY,CHA CHARGING	1
2	AFFFF		19204	8448501	BOLT CARRIER ASSEMBLY (M16A2) UOC:AR8	1
2A	AFFFF	19200	12011849	BOLT	AND BOLT CARRI ASSEMBLY (M4 AND M4A1) UOC:AS1,AY6	1
3	AFFFF	1005011879716	19200	9349050	UPPER RECEIVER AND BARREL ASSEMBLY (M16A2) UOC:AR8	1
3A	AFFFF	19200	12972680	UPPER	RECEIVER AND BARREL ASSEMBLY (M4 AND M4A1) UOC:AS1,AY6	1
4	PACZZ	1005009215004	19200	8448670	MAGAZINE,CARTRIDGE (30 ROUND)	1
5	PACZZ	1005012164510	19204	12624561	SLING,SMALL ARMS (M16A2) UOC:AR8	1
5A	PACZZ	1005013689852	19200	12011996	SLING,SMALL ARMS (M4 AND M4A1) UOC:AS1,AY6	1
6	XAFFA	19200	9349100	LOWER	RECEIVER AND BUTTSTOCK ASSEMBLY (M16A2) UOC:AR8	1
6A	XAFFA		19200	9390011	LOWER RECEIVER AND BUTTSTOCK ASSEMBLY (4) UOC:AS1	1
6B	XAFFA		19200	12972690	LOWER RECEIVER AND BUTTSTOCK ASSEMBLY (M4A1) UOC:AY6	1
7	AFFFF		19200	12951011	CARRYING HANDLE ASS (M4 AND M4A1) UOC:AS1,AY6	1

END OF FIGURE

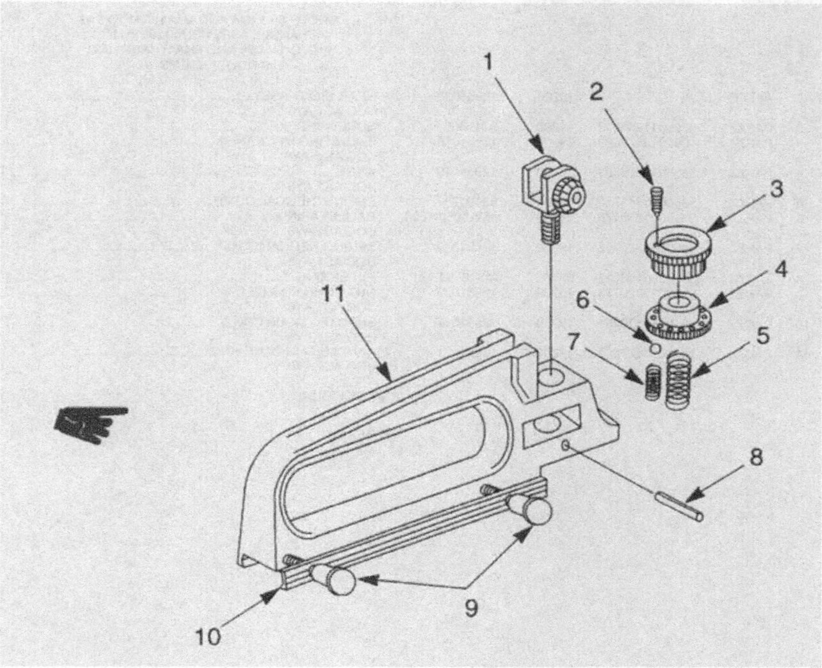

Figure C-1A. (M4 and M4A1) Carrying Handle Assembly 12951011.

(1) ITEM NO	(2) SMR CODE	(3) NSN	(4) CAGEC	(5) PART NUMBER	(6) DESCRIPTION AND USABLE ON CODE (UOC)	(7) QTY
					GROUP 00.1 (M4 AND M4A1) CARRYING HANDLE ASSEMBLY 12951011 FIG. C-1A. (M4 AND M4A1) CARRYING HANDLE ASSEMBLY	
1	AFFFF		19200	12951026	REAR SIGHT ASSY UOC:AS1,AY6	1
2	PAFZZ	5305011343622	19200	9349065	SCREW,INDEX	1
3	PAFZZ	1005013827089	19200	12951018	ELEVATING MECHANISM UOC:AS1 ,AY6	1
4	PAFZZ	5355013826801	19200	12951019	KNOB UOC:AS1,AY6	1
5	PAFZZ	5360011343710	19200	9349070	SPRING ,HELICAL,COMP	1
6	PAFZZ	3110001839175	96906	MS19060-4808	BALL,BEARING UOC:AS1,AY6	1
7	PAFZZ	5360013826802	19200	12951020	SPRING,HELICAL,COMP UOC:AS1,AY6	1
8	PAFZZ	5315008403812	96906	MS16562-121	PIN,SPRING	1
9	PAOZZ	5310013826793	19200	12951023	NUT,PLAIN,KNURLED UOC:AS1,AY6	2
10	PAOZZ	5340013823201	19200	12951017	BRACKET,MOUNTING UOC:AS1,AY6	1
11	PAFZZ	1005013827083	19200	12951021	HANDLE,GUN CARRYING UOC:AS1,AY6	1

END OF FIGURE

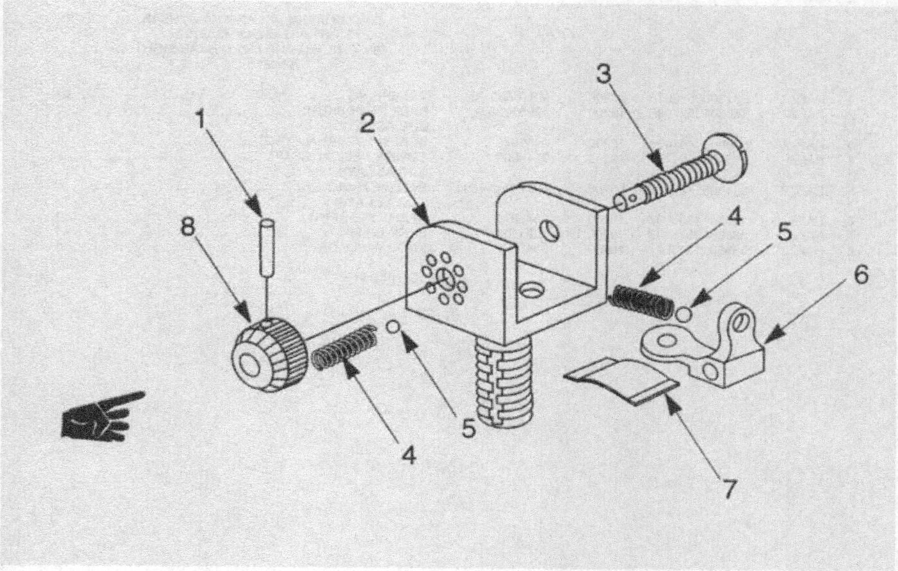

Figure C-1B. (M4 and M4A1) Rear Sight Assembly 12951026.

(1) ITEM NO	(2) SMR CODE	(3) NSN	(4) CAGEC	(5) PART NUMBER	(6) DESCRIPTION AND USABLE ON CODE (UOC)	(7) QTY
					GROUP 00.101 (M4 AND M4A1) REAR SIGHT ASSEMBLY 12951026 FIG. C-18. (M4 AND M4A1) REAR SIGHT ASSEMBLY	
1	PAFZZ	5315000586678	96906	MS16562-103	PIN,SPRING	1
2	PAFZZ	1005013827086	19200	12951028	BASE,REAR SIGHT UOC:AS1,AY6	1
3	PAFZZ	5305011441490	19200	9349076	SCREW,EXTERNALLY RE	1
4	PAFZZ	5360011481751	19200	9349069	SPRING,HELICAL,COMP UOC:AS1,AY6	2
5	PAFZZ	3110001839175	96906	MS19060-4808	BALL,BEARING UOC:AS1,AY6	2
6	PAFZZ	1005011353697	19200	9349075	APERTURE,SIGHT	1
7	PAFZZ	5360013816183	19200	12011987	SPRING,FLAT	1
8	PAFZZ	5355011343627	19200	9349077	KNOB,WINDAGE	1

END OF FIGURE

SECTION II

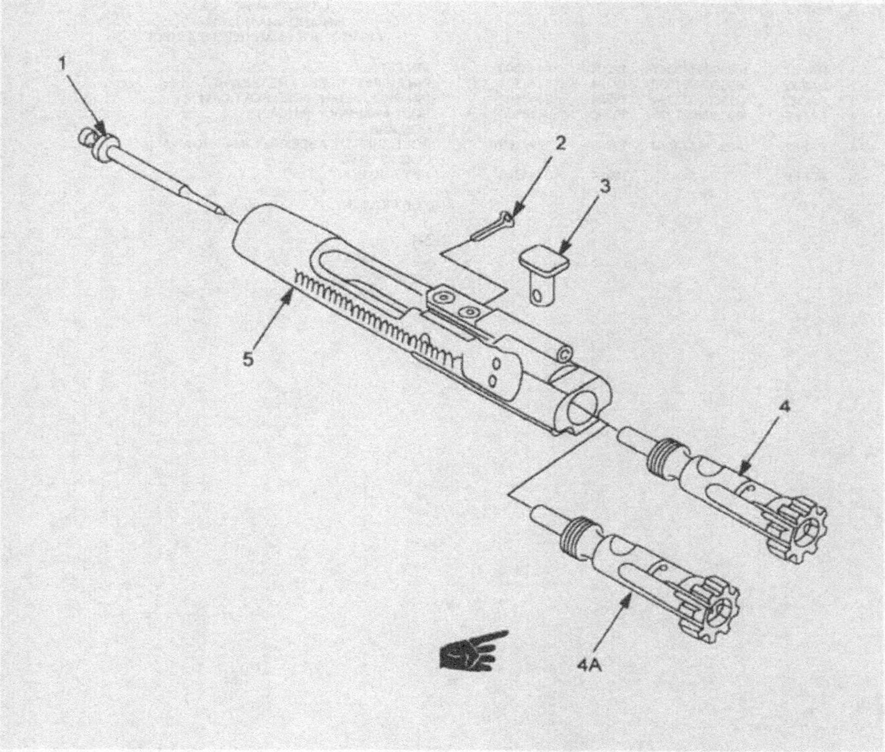

Figure C-2. Bolt Carrier Assembly (M16A2) 8448501 and (M4 and M4A1) 12011849.

(1) ITEM NO	(2) SMR CODE	(3) NSN	(4) CAGEC	(5) PART NUMBER	(6) DESCRIPTION AND USABLE ON CODE (UOC)	(7) QTY
					GROUP 01 BOLT CARRIER ASSEMBLY (M16A2) 8448501 AND (M4 AND M4A1) 12011849 FIG. C-2. BOLT CARRIER ASSEMBLY	
1	PAFZZ	1005000179547	19204	8448503	PIN,FIRING	1
2	PAOZZ	1005009991509	19204	8448504	PIN,FIRING PIN RETA RETAINING	1
3	PAOZZ	5315009927294	19204	8448502	PIN,GROOVED,HEADED BOLT CAM	1
4	PAFFF	1005009927285	19200	8448509	BOLT ASSEMBLY (M16A2) UOC:AR8	1
4A	PAFFF	1005014223770	19200	12972691	BOLT,BREECH ASSEMBLY (M4 AND M4A1) UOC:AS1,AY6	1
5	AFFFF		19204	8448505	KEY AND BOLT	1

END OF FIGURE

SECTION II

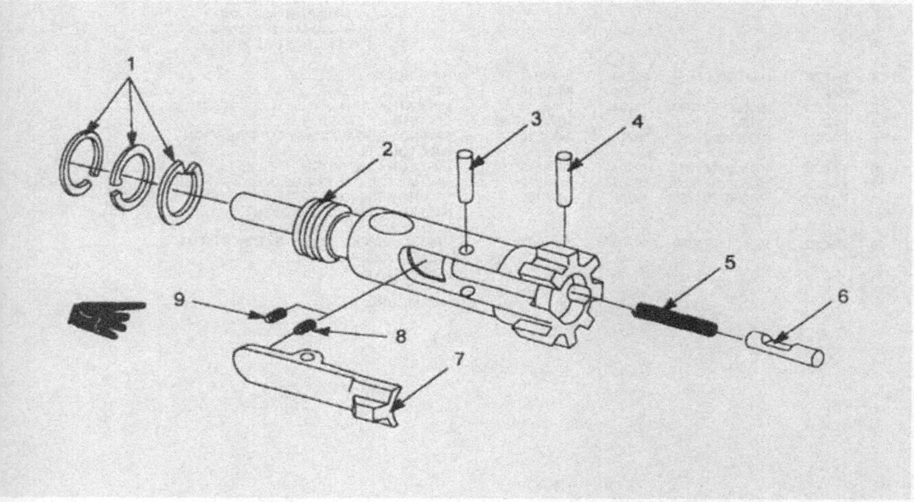

Figure C-3. Bolt Assembly (M16A2) 8448509 and (M4 and M4A1) 12972691.

(1) ITEM NO	(2) SMR CODE	(3) NSN	(4) CAGEC	(5) PART NUMBER	(6) DESCRIPTION AND USABLE ON CODE (UOC)	(7) QTY
					GROUP 0101 BOLT ASSEMBLY (M16A2) 8448509 AND (M4 AND M4A1) 12972691 FIG. C-3. BOLT ASSEMBLY	
1	PAFZZ	1005009927287	19204	8448511	RING,BOLT	
2	XAFZZ		19204	8448510	BOLT	
3	PAOZZ	1005009927290	19204	8448513	PIN,EXTRACTOR	
4	PAOZZ	5315005975086	96906	MS16562-9B	PIN,SPRING EJECTOR	
5	PAOZZ	5360009927292	19204	8448516	SPRING,HELICAL,COMP COMPRESSION, EJECTOR	
6	PAOZZ	1005009927291	19204	8448515	EJECTOR,CARTRIDGE	
7	PAOZZ	1005009927288	19204	8448512	EXTRACTOR,CARTRIDGE	
8	PAOZZ	1005007603768	19200	8448755	SPRING ASSEMBLY,EXT EXTRACTOR (M16A2) UOC:AR8	
9	PAOZZ	1005014245899	19200	12972692	SPRING ASSEMBLY,EXT EXTRACTOR (M4 AND M4A1)(BLACK) UOC:AS1 ,AY6	

END OF FIGURE

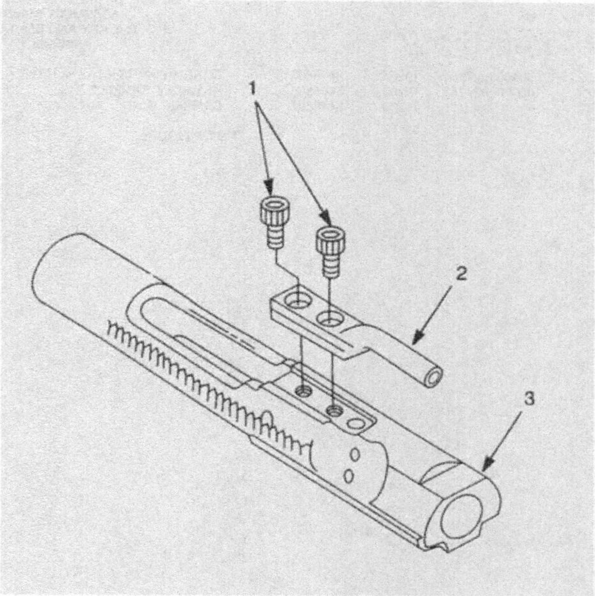

Figure C-4. Key and Bolt Carrier Assembly 8448505.

(1) ITEM NO	(2) SMR CODE	(3) NSN	(4) CAGEC	(5) PART NUMBER	(6) DESCRIPTION AND USABLE ON CODE (UOC)	(7) QTY
					GROUP 0102 KEY AND BOLT CARRIER ASSEMBLY 8448505 FIG. C-4. KEY AND BOLT CARRIER ASSEMBLY	
1	PAFZZ	5305009927284	19204	8448508	SCREW,CARRIER AND K KEY	2
2	PAFZZ	1005009927283	19200	8448506	KEY,BOLT CARRIER	1
3	PAFZZ		19200	8448507	CARRIER,BOLT	1

END OF FIGURE

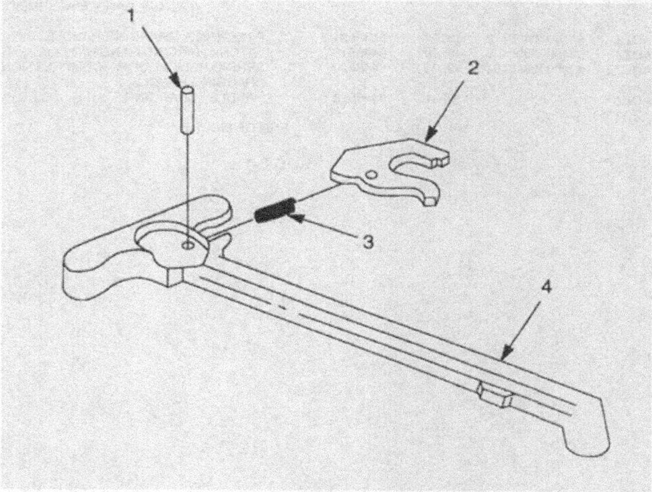

Figure C-5. Handle Assembly 8448517.

(1) ITEM NO	(2) SMR CODE	(3) NSN	(4) CAGEC	(5) PART NUMBER	(6) DESCRIPTION AND USABLE ON CODE (UOC)	(7) QTY
					GROUP 02 HANDLE ASSEMBLY 8448517 FIG. C-5. HANDLE ASSEMBLY	
1	PAOZZ	5315010489372	19204	8448521-2	PIN,SPRING CHARGING HANDLE....................................	1
2	PAOZZ	5340009990405	19200	8448519	LATCH,CHARGING HAND HANDLE....................................	1
3	PAOZZ	5360009990404	19204	8448520	SPRING,HELICAL,COMP COMPRESSION, ... CHARGING HANDLE...	1
4	XAOZZ		19204	8448518	HANDLE ,CHARGING..	1

END OF FIGURE

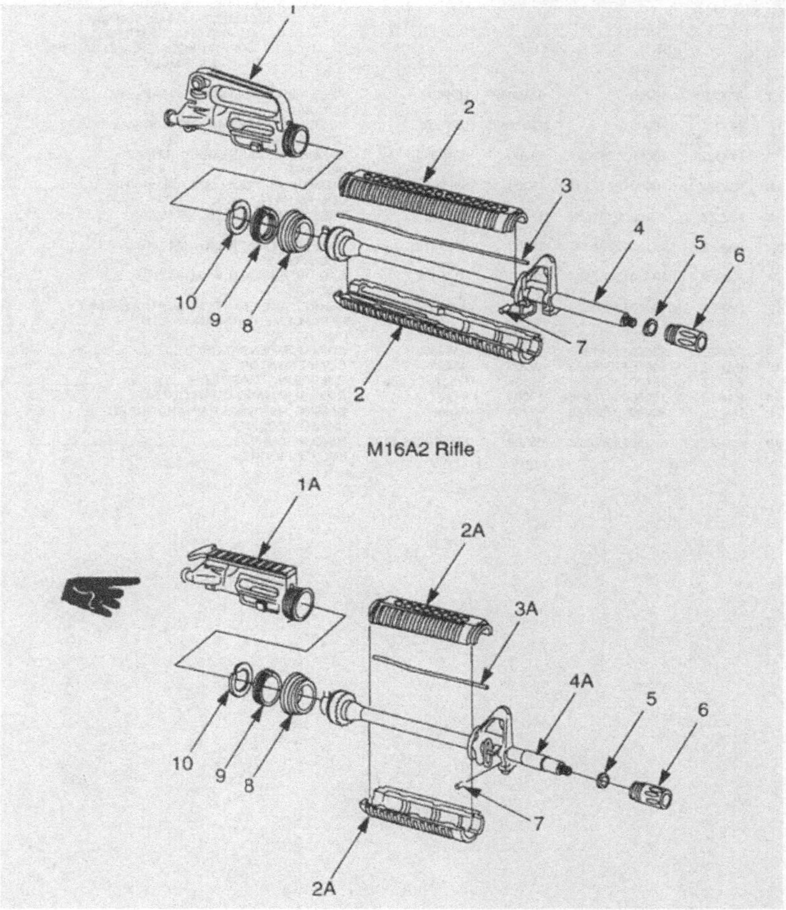

M4/M4A1 Carbine

Figure C-6. Upper Receiver and Barrel Assembly (M16A2) 9349050, and (M4 and M4A1) 12972680.

(1) ITEM NO	(2) SMR CODE	(3) NSN	(4) CAGEC	(5) PART NUMBER	(6) DESCRIPTION AND USABLE ON CODE (UOC)	(7) QTY
					GROUP 03 UPPER RECEIVER AND BARREL ASSEMBLY (M16A2) 9349050, (M4 AND M4A1] 12972680 FIG. C-6. UPPER RECEIVER AND BARREL ASSEMBLY	
1	AFFFF	19200	9349062	UPPER	RECEIVER ASSEMBLY (M16A2)... UOC:AR8	
1A	AFFFF	19200	12972675	UPPER	RECEIVER ASSY (M4 AND M4A1)....................................... UOC:AS1,AY6	
2	PAOZZ	1005011343629	19200	9349059	HANDGUARD ASSEMBLY (M16A2)....................................... UOC:AR8	
2A	PAOZZ	1005012342297	19200	9390003	GUARD,HAND,GUN (M4 AND M4A1)...................................... UOC:AS1,AY6	
3	PAFZZ	4710009781038	19200	8448567	TUBE,BENT,METALLIC (M16A2)... UOC:AR8	
3A	PAFZZ	4710012338637	19200	9390016	TUBE,BENT,METALLIC (M4 AND M4A1) UOC:AS1,AY6	
4	PAFFF	1005011467684	19200	9349124	BARREL ASSEMBLY (M16A2)... UOC:AR8	
4A	PAFFF	1005012338529	19200	9390007	BARREL AND FRONT SI SIGHT ASSEMBLY, ... REPLACEMENT (M4 AND M4A1).. UOC:AS1,AY6	
5	PAFZZ	5365011441496	19200	9349052	SHIM COMPENSATOR... 	
6	PAFZZ	1005011343633	19200	9349051	COMPENSATOR..	
7	PAFZZ	5315000586044	96906	MS16562-106	PIN,SPRING ,GAS TUBE	
8	PAFZZ	1005000878998	19204	8448712	RING,SLIP,HAND GUAR GUARD	
9	PAFZZ	5360009781036	19204	8448555	SPRING,SLIP RING,HA HANDGUARD................................. UPPER RECEIVER	
10	PAFZZ	5325009990863	96906	M5S16626-3137	RING,RETAINING .... END OF FIGURE	

SECTION II

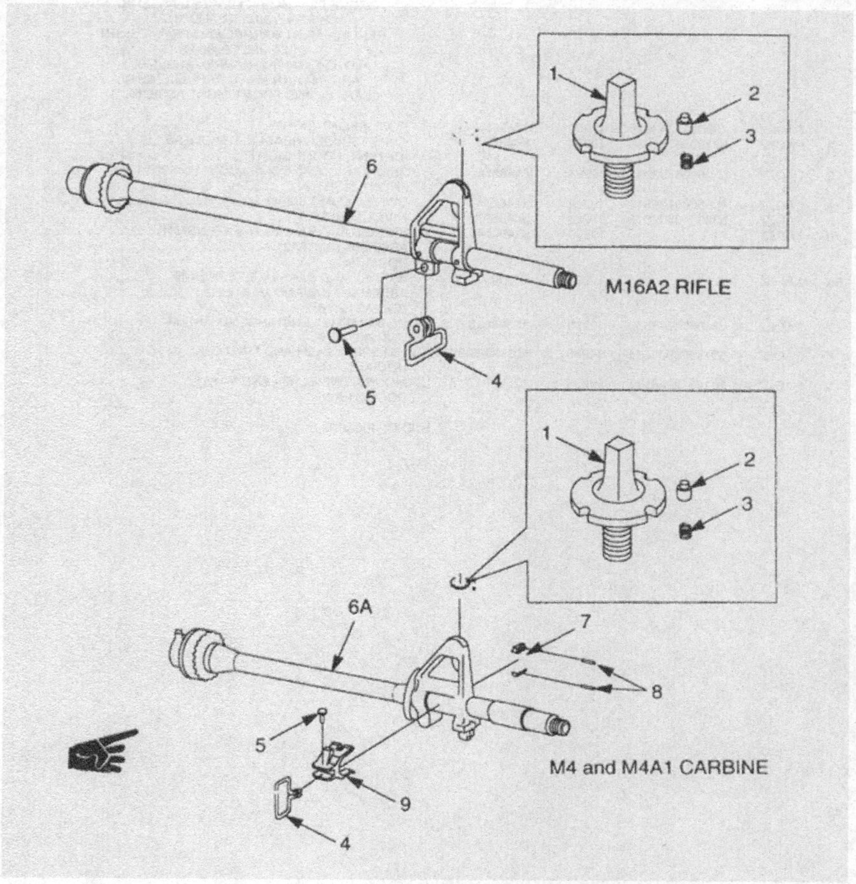

Figure C-7. (M16A2) Barrel Assembly 9349124 and (M4 and M4A1) Replacement Barrel and Front Sight Assembly 9390007.

(1) ITEM NO	(2) SMR CODE	(3) NSN	(4) CAGEC	(5) PART NUMBER	(6) DESCRIPTION AND USABLE ON CODE (UOC)	(7) QTY
					GROUP 0301 (M16A2) BARREL ASSEMBLY 9349124, AND (M4 AND M4A1) REPLACEMENT BARREL AND FRONT SIGHT ASSEMBLY 9390007 FIG. C-7. (M16A2) BARREL ASSEMBLY, AND (M4 AND M4A1) REPLACEMENT BARREL AND FRONT SIGHT ASSEMBLY	
1	PAOZZ	1005011343625	19200	9349056	POST,FRONT SIGHT	1
2	PAOZZ	5315009793930	19204	8448573	PIN,SHOULDER,HEADLE HEADLESS (DETENT, FRONT SIGHT)	1
3	PAOZZ	5360009793931	19204	8448574	SPRING,HELICAL,COMP ,COMPRESSION, FRONT SIGHT	1
4	PAOZZ	1005000179543	19204	8448571	SWIVEL,SLING,SMALL	1
5	PAOZZ	5320010637635	19204	8448697	RIVET,TUBULAR	1
6	XAFZZ		19200	9349054	BARREL AND BARREL E EXTENSION ASSEMBLY (M16A2) UOC:AR8	1
6A	XAFZZ		19200	9390009	BARREL AND BARREL EXTENSION ASSEMBLY (M4 AND M4A1) UOC:AS1,AY6	1
7	PAOZZ	5340012646530	19200	12598618	CLAMP,RIM CLENCHING (M4 AND M4A1) UOC:AS1,AY6	1
8	PAOZZ	5315006900544	96906	MS39086-93	PIN,SPRING (M4 AND M4A1) UOC:AS1,AY6	2
9	PAOZZ	1010012646517	19200	12598617	MOUNT,SWIVEL (M4 AND M4A1) UOC:AS1,AY6	1

END OF FIGURE

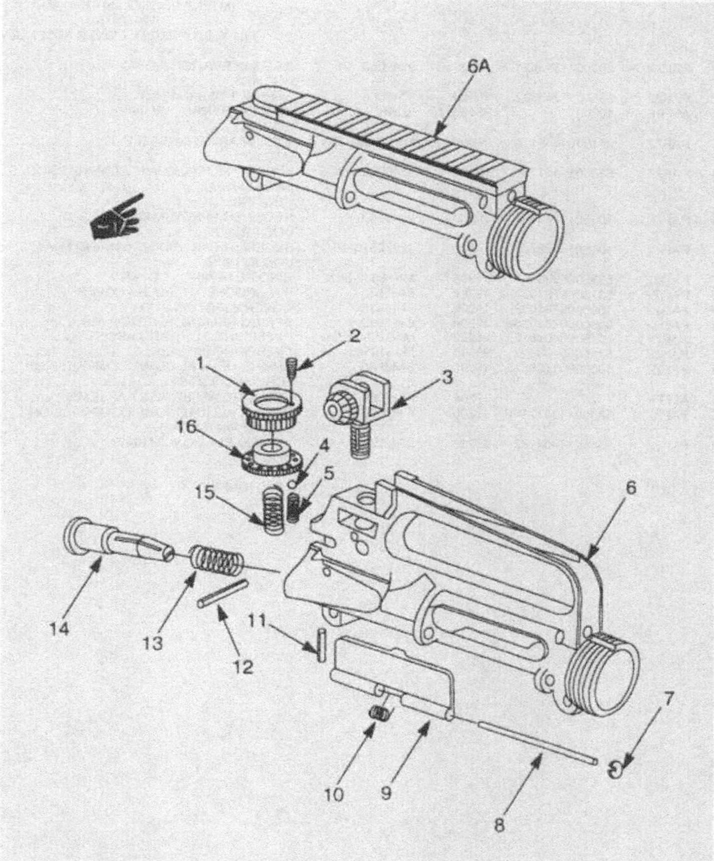

Figure C-8. Upper Receiver Assembly (M16A2) 9349062, and (M4 and M4A1) 12972675.

(1) ITEM NO	(2) SMR CODE	(3) NSN	(4) CAGEC	(5) PART NUMBER	(6) DESCRIPTION AND USABLE ON CODE (UOC)	(7) QTY
					GROUP 0302 UPPER RECEIVER ASSEMBLY (M16A2) 9349062, (M4 AND M4A1) 12972675 FIG. C-8. UPPER RECEIVER ASSEMBLY	
1	PAFZZ	1005011343621	19200	9349066	INDEX,ELEVATION (M16A2) UOC:AR8	1
2	PAFZZ	5305011343622	19200	9349065	SCREW,INDEX (M16A2)	1
3	AFFFF	19200	9349072	REAR	SIGHT ASSEMBLY (M16A2) UOC:AR8	1
4	PAFZZ	3110001839175	96906	MS19060-4808	BALL,BEARING (M16A2) UOC:AR8	1
5	PAFZZ	5360011481751	19200	9349069	SPRING,HELICAL,COMP COMPRESSION, ... INDEX (M16A2) UOC:AR8	1
6	PAFZZ	1005011343701	19200	9349063	RECEIVER,UPPER (M16A2) UOC:AR8	1
6A	PAFZZ	1005013826795	19200	12972670	RECEIVER,CARTRIDGE (M4 AND M4A1) UOC:AS1,AY6	1
7	PAOZZ	5325009990864	96906	MS16632-3012	RING,RETAINING ,COVER	1
8	PAOZZ	5315009781023	19204	8448533	PIN,GROOVED,HEADLES COVER	1
9	PAOZZ	1005009781022	19204	8448525	COVER,EJECTION PORT	1
10	PAOZZ	5360009781025	19204	8448532	SPRING,HELICAL,TORS ,COVER	1
11	PAFZZ	5315008403812	96906	M5S16562-121	PIN,SPRING FORWARD ASSIST	1
12	PAFZZ	5315008403812	96906	M5S16562-121	PIN,SPRING (M16A2)	1
13	PAFZZ	5360000179541	19200	8448540	SPRING,HELICAL,COMP COMPRESSION .... FORWARD ASSIST	1
14	AFFFF		19204	9349086	FORWARD ASSIST ASSY ASSEMBLY	1
15	PAFZZ	5360011343710	19200	9349070	SPRING,HELICAL,COMP COMPRESSION, ... ELEVATION (M16A2)	1
16	PAFZZ	5355011354972	19200	9349067	KNOB,ELEVATION (M16A2) UOC:AR8	1

END OF FIGURE

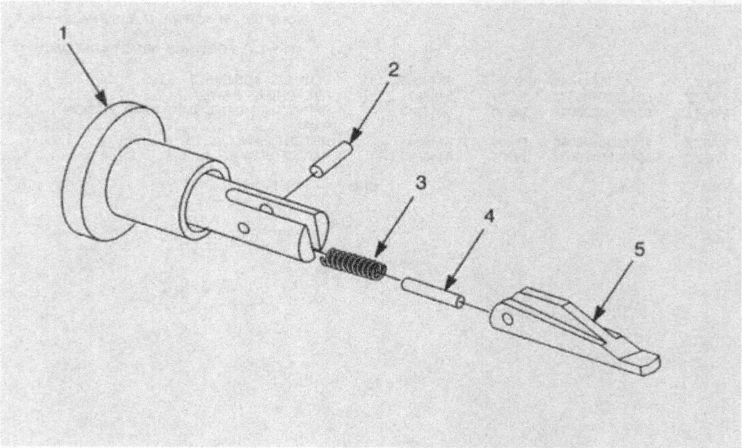

Figure C-9. Forward Assist Assembly 9349086.

(1) ITEM NO	(2) SMR CODE	(3) NSN	(4) CAGEC	(5) PART NUMBER	(6) DESCRIPTION AND USABLE ON CODE (UOC)	(7) QTY
					GROUP 030201 FORWARD ASSIST ASSEMBLY 9349086	
					FIG. C-9. FORWARD ASSIST ASSEMBLY	
1	PAFZZ	1005011441468	19200	9349085	PLUNGER ASSEMBLY	
2	PAFZZ	5315010489372	19204	8448521-2	PIN,SPRING ,PAWL	
3	PAFZZ	5360005238084	19200	8448542	SPRING,HELICAL,COMP COMPRESSION .... PAWL	
4	PAFZZ	1005000179540	19204	8448544	DETENT,PAWL	
5	PAFZZ	3040000179539	19204	8448543	PAWL,FORWARD ASSIST	

END OF FIGURE

SECTION II

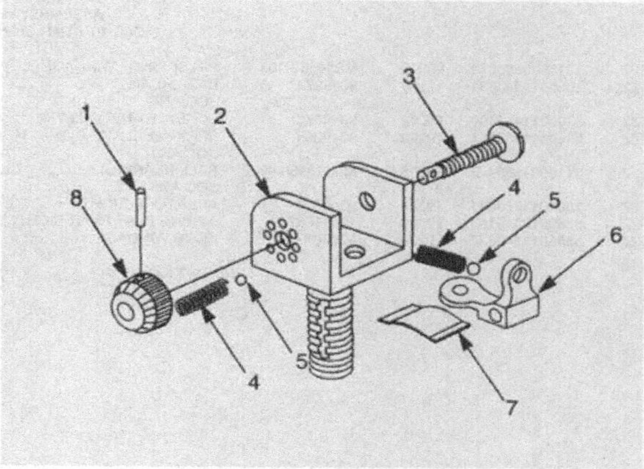

Figure C-10. (M16A2) Rear Sight Assembly 9349072.

(1) ITEM NO	(2) SMR CODE	(3) NSN	(4) CAGEC	(5) PART NUMBER	(6) DESCRIPTION AND USABLE ON CODE (UOC)	(7) QTY
					GROUP 030202 (M16A2) REAR SIGHT ASSEMBLY 9349072 FIG. C-10. REAR SIGHT ASSEMBLY	
1	PAFZZ	5315000586678	96906	MS16562-103	PIN,SPRING ,WINDAGE	1
2	PAFZZ	1005011343631	19200	9349074	BASE,REAR SIGHT UOC:ARB	1
3	PAFZZ	5305011441490	19200	9349076	SCREW,EXTERNALLY RE	1
4	PAFZZ	5360011481751	19200	9349069	SPRING,HELICAL,COMP ,REAR SIGHT UOC:AR8	2
5	PAFZZ	3110001839175	96906	M5S19060-4808	BALL,BEARING UOC:AR8	2
6	PAFZZ	1005011353697	19200	9349075	APERTURE,SIGHT	1
7	PAFZZ	5360013816183	19200	12011987	SPRING,FLAT REAR SIGHT	1
8	PAFZZ	5355011343627	19200	9349077	KNOB,WINDAGE	1

END OF FIGURE

SECTION II

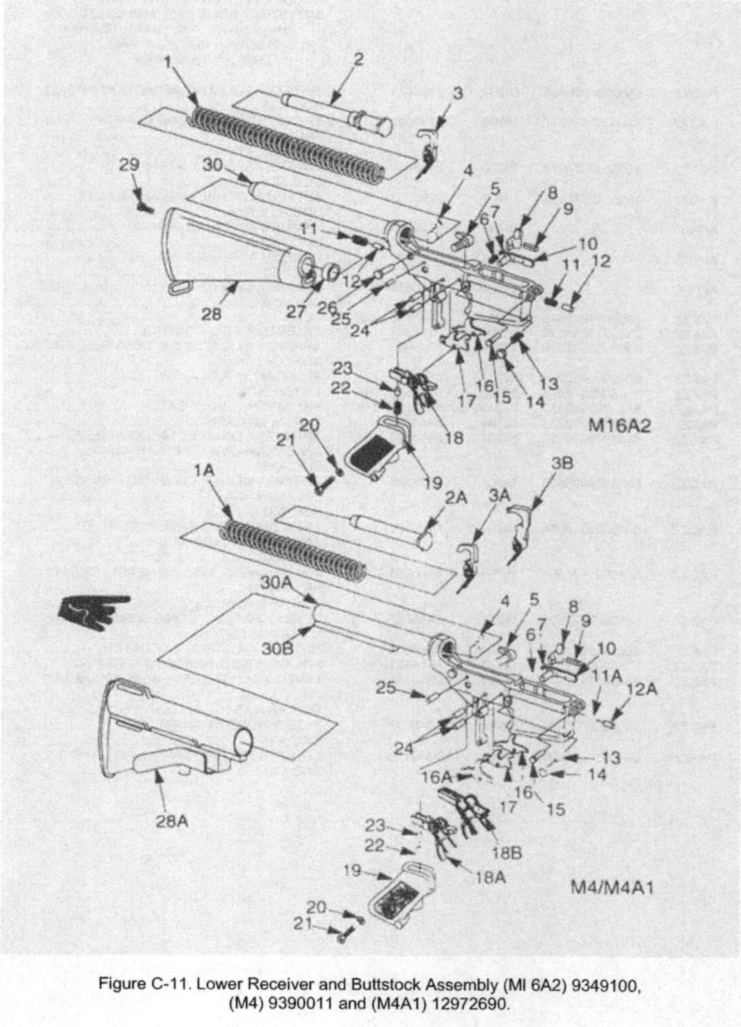

Figure C-11. Lower Receiver and Buttstock Assembly (MI 6A2) 9349100, (M4) 9390011 and (M4A1) 12972690.

(1) ITEM NO	(2) SMR CODE	(3) NSN	(4) CAGEC	(5) PART NUMBER	(6) DESCRIPTION AND USABLE ON CODE (UOC)	(7) QTY
					GROUP 04 LOWER RECEIVER AND BUTTSTOCK ASSEMBLY (M16A2) 9349100, (M4) 9390011, AND (M4A1) 12972690 FIG C-11 LOWER RECEIVER AND BUTTSTOCK ASSEMBLY	
1	PAOZZ	5360009926665	19204	8448629	SPRING, HELICAL, COMP ACTION (M16A2) UOC:AR8	1
1A	PAOZZ	5360012338617	19200	9390022	SPRING, HELICAL, COMP COMPRESSION (M4 AND M4A1) UOC:AS1, AY6	1
2	PAOZZ	1005009373078	19200	8448615	BUFFER ASSEMBLY (MI6A2) UOC:AR8	1
2A	PAOZZ	1005012313138	19200	9390023	BUFFER ASSEMBLY (M4 AND M4AI) UOC:AS1, AY6	1
3	AFFFF		19200	9349106	HAMMER ASSEMBLY (M16A2) UOC:AR8	1
3A	AFFFD		19200	9390032	HAMMER ASSEMBLY (M4) UOC:AS1	1
3B	AFFFF		19204	8448610	HAMMER ASSY (M4A1) UOC:AY6	1
4	PAFZZ	1005009926649	19200	8448595	SEAR	1
5	PAFZZ	5340012258339	19200	9381367	SELECTOR, FIRE CONTROL	1
6	PAFZZ	5360000562246	19204	8448633	SPRING, HELICAL, COMP COMPRESSION, BOLT CATCH	1
7	PAFZZ	1005000562247	19204	8448634	PLUNGER, BOLT CATCH	1
8	PAFZZ	1005000179548	19200	8448628	CATCH, BOLT	1
9	PAFZZ	5315008123312	96906119	MS16562-119	PIN, SPRING , BOLT CATCH	1
10	PAFZZ	1005000562201	19204	8448638	CATCH, MAGAZINE	1
11	PAOZZ	5360009926655	19204	8448586	SPRING, HELICAL, COMP COMPRESSION, TAKE DOWN/PIVOT PIN (M16A2) UOC:AR8	2
11A	PAOZZ	5360009926655	19204	8448586	SPRING, HELICAL, COMP COMPRESSION (M4 AND M4Al) UOC:AS1	1
12	PAOZZ	5315009926654	19204	8448585	PIN, STRAIGHT, HEADLE DETENT, TAKEDOWN PIN (M16A2) UOC:ARS	2
12A	PAOZZ	5315009926654	19204	8448585	PIN, STRAIGHT, HEADLE HEADLESS (M4 AND M4A1) UOC:AS1, AY6	1
13	PAFZZ	5360009927301	19204	8448637	SPRING, HELICAL, COMP COMPRESSION, MAGAZINE CATCH	1
14	PAFZZ	1005009927302	19204	8448636	BUTTON, MAGAZINE CAT CATCH	1
15	PAOZZ	5315000179537	19204	8448621	PIN, GROOVED, HEADED (PIVOT PIN)	1
16	PAFZZ	5340011457910	19200	9349114	LEVER, LOCK-RELEASE SEMI (M16A2 AND M4) UOC:AR8, AS1	1
16A	PAFZZ	1005009990406	19200	8448635	DISCONNECTOR (M4A1) UOC:AY6	1
17	PAFZZ	5340011441499	19200	9349113	LEVER, LOCK-RELEASE , BURST (MI6A2 AND M4)	1

(1) TEM NO	(2) SMR CODE	(3) NSN	(4) CAGEC	(5) PART NUMBER	(6) DESCRIPTION AND USABLE ON CODE (UOC)	(7) QTY
					UOC:AR8, AS1	
18	AFFFF		19200	9349115	TRIGGER ASSEMBLY (M16A2)	1
					UOC:AR8	
18A	AFFFF		19200	12972697	TRIGGER ASSEMBLY (M4)	1
					UOC:AS1	
18B	AFFFF		19200	12972698	TRIGGER ASSEMBLY (M4A1)	1
					UOC:AY6	
19	PAOZZ	1005011484805	19200	9349127	GRIP, RIFLE PLASTIC, BLACK	1
20	PAOZZ	5310005273634	96906	MS35335-61	WASHER, LOCK , RIFLE GRIP	1
21	PAOZZ	5305012681191	88044	AN501D416-18	SCREW, MACHINE , RIFLE GRIP	1
22	PAOZZ	5360009927292	19204	8448516	SPRING, HELICAL, COMP SAFETY	1
23	PAOZZ	1005009926667	19204	8448631	DETENT, SAFETY	1
24	PAFZZ	5315009927309	19204	8448609	PIN, GROOVED, HEADLES TRIGGER AND HAMMER	2
25	PAFZZ	5315009926650	19204	8448599	PIN, GROOVED, HEADLES , AUTOMATIC SEAR	1
26	PAFZZ	5315009926653	19204	8448584	PIN, GROOVED, HEADED (M16A2)	1
27	PAOZZ	5365012672169	19200	12597640	SPACER, STEPPED (M16A2)	1
					UOC:AR8	
28	PAOOO	1005011354973	19200	9349119	BUTTSTOCK ASSEMBLY, (M16A2)	1
					UOC:AR8	
28A	AOOOO		19200	9390012	BUTTSTOCK ASSY (M4 AND M4A1)	1
					UOC:AS1, AY6	
29	PAOZZ	5305011478585	19200	9349128	SCREW, MACHINE , BUTTCAP (M16A2)	1
					UOC:AR8	
30	XAFFA		19200	9349101	LOWER RECEIVER AND RECEIVER EXTENSION ASSEMBLY (M16A2)	1
					UOC:AR8	
30A	XAFFA		99999	NPN	LOWER RECEIVER ASSY ASSEMBLY (M4)	1
					UOC:AS1	
30B	XAFFA		99999	NPN	LOWER RECEIVER ASSY (M4Al)	1
					UOC:AY6	
					END OF FIGURE	

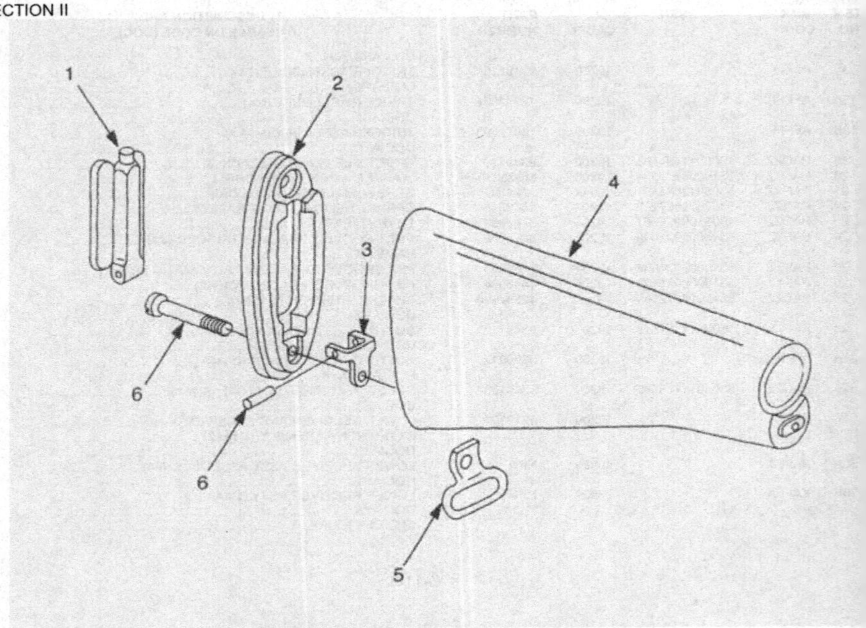

Figure C-12. (M16A2) Buttstock Assembly 9349119.

(1) ITEM NO	(2) SMR CODE	(3) NSN	(4) CAGEC	(5) PART NUMBER	(6) DESCRIPTION AND USABLE ON CODE (UOC)	(7) QTY
					GROUP 0401 (M16A2) BUTTSTOCK ASSEMBLY 9349119	
					FIG C-12 (M16A2) BUTTSTOCK ASSEMBLY	
1	PAOZZ	1005012288504	19200	9381380	DOOR ASSEMBLY, THUMB UOC:AR8	1
2	PAOZZ	1005011467685	19200	9349130	PLATE, BUTT, SHOULDER GUN STOCK UOC:AR8	1
3	PAOZZ	5340004633892	19200	8448653	HINGE, ACCESS DOOR BUTT PLATE UOC:AR8	1
4	XAOZZ		19200	9349121	BUTTSTOCK UOC:AR8	1
5	PAOZZ	1005004030964	19204	8448652	SWIVEL, SLING, SMALL UOC:AR8	1
6	PAOZZ	5315004633894	19204	8448655	PIN, STRAIGHT, HEADLE HEADLESS, ACCESS DOOR UOC:AR8	1
7	PAOZZ	5305011441494	19200	9349120	SCREW, MACHINE BUTT PLATE UOC:AR8	1

END OF FIGURE

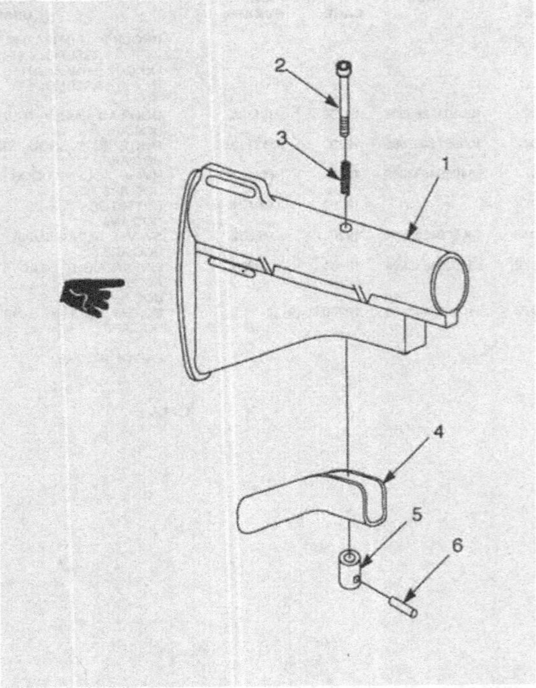

Figure C12A. (M4 and M4A1) Buttstock Assembly 9390012.

(1) ITEM NO	(2) SMR CODE	(3) NSN	(4) CAGEC	(5) PART NUMBER	(6) DESCRIPTION AND USABLE ON CODE (UOC)	(7) QTY
					GROUP 0401A (M4 AND M4A1) 9390012 BUTTSTOCK ASSEMBLY FIG C12A (M4 AND M4A1) BUTTSTOCK ASSEMBLY	
1	PAOZZ	1005012338636	19200	9390013	STOCK, GUN, SHOULDER UOC:AS1, AY6	1
2	PAOZZ	5315012338608	19200	9390025	PIN, SHOULDER, HEADLE UOC:AS1, AY6	1
3	PAOZZ	5360012338616	19200	9390027	SPRING, HELICAL, COMP COMPRESSION UOC:AS1, AY6	1
4	PAOZZ	5340012338638	19200	9390014	LEVER, LOCK-RELEASE UOC:AS1, AY6	1
5	PAOZZ	5310012338626	19200	9390026	NUT, SELF-LOCKING, RO UOC:AS1, AY6	1
6	PAOZZ	5315008439487	96906	MS16562-202	PIN, SPRING UOC:AS1, AY6	1

END OF FIGURE

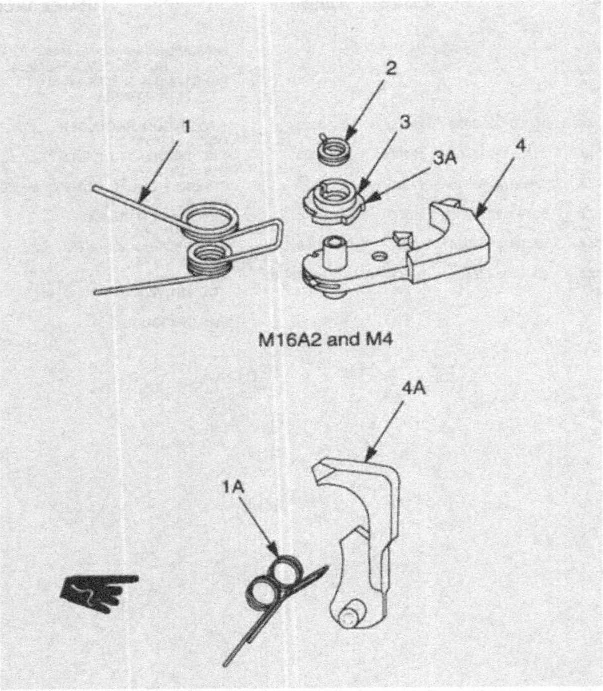

Figure C-13. Hammer Assembly(M16A2) 9349106, (M4) 9390032, and (M4A1) 8448610.

(1) TEM NO	(2) SMR CODE	(3) NSN	(4) CAGEC	(5) PART NUMBER	(6) DESCRIPTION AND USABLE ON CODE (UOC)	(7) QTY
					GROUP 0402 HAMMER ASSEMBLY (M16A2) 9349106, (M4) 9390032, AND (M4A1) 8448610 FIG C-13 HAMMER ASSEMBLY	
1	PAFZZ	5360011441492	19200	9349107	SPRING, HELICAL, TORS HAMMER (MI6A2 AND M4) UOC:AR8, AS1	1
1A	PAFZZ	5360009926648	19204	8448611	SPRING, HELICAL, TORS (M4A1) UOC:AY6	1
2	PAFZZ	5360011365471	19200	9349109	SPRING, HELICAL, TORS BURST CAM (M16A2 AND M4) UOC:AR8, AS1	1
3	PAFZZ	1005011480172	19200	9349108	CAM, BURST (M16A2) (BLACK) UOC:AR8	1
3A	PAFZZ	3040012477969	19200	9390031	CAM, CONTROL (M4) (NICKEL/SHINY) UOC:AS1	1
4	PAFZZ	1005011343630	19200	9349110	HAMMER AND HAMMER P PIN RETAINER ASSEMBLY (MI6A2 AND M4) UOC:AR8, AS1	1
4A	PAFZZ	1005000179551	19200	8448612	HAMMER, FIRING, SMALL (M4A1) UOC:AY6	1

END OF FIGURE

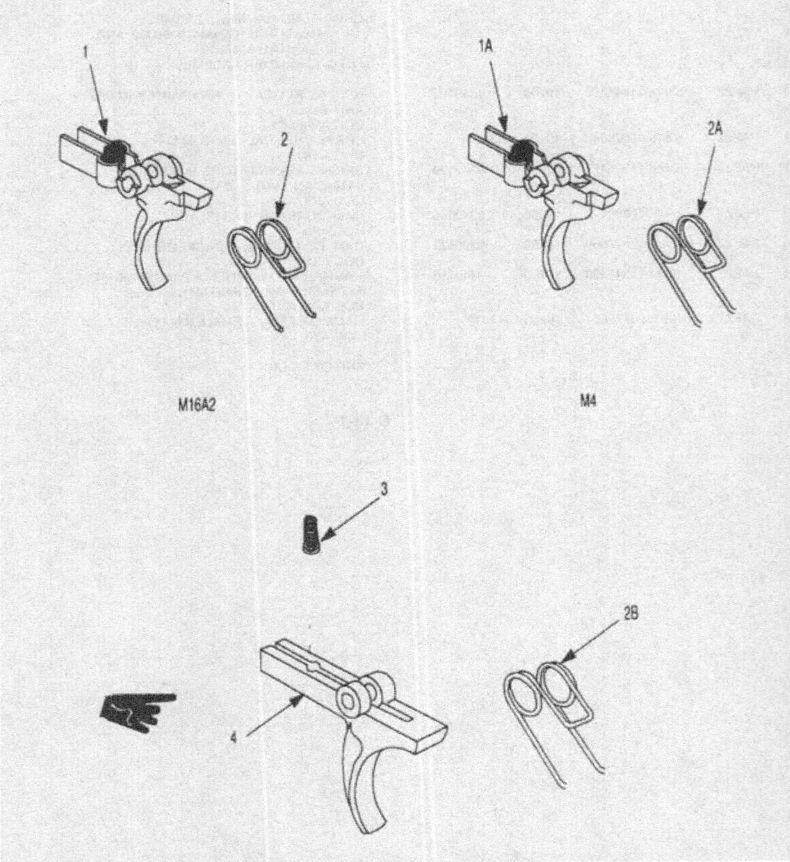

Figure C-14. Trigger Assembly(M16A2) 9349115, (M4) 12972697, and (M4A1) 12972676

(1) ITEM NO	(2) SMR CODE	(3) NSN	(4) CAGEC	(5) PART NUMBER	(6) DESCRIPTION AND USABLE ON CODE (UOC)	(7) QTY
					GROUP 0403 TRIGGER ASSEMBLY (M16A2) 9349115, (M4) 12972697, AND (M4A1) 12972676 FIG C-14 TRIGGER ASSEMBLY	
1	PAFFF	1005012192402	19200	9392518	TRIGGER SUB ASSEMBL (M16A2)...................................... UOC:AR8	1
1A	PAFFF	1005013954257	19200	12972696	TRIGGER SUBASSEMBLY (M4).. UOC:AS1	1
2	PAFZZ	5360009927308	19204	8448593	SPRING, HELICAL, TORS TORSION, ... TRIGGER (MI6A2).. UOC:AR8	1
2A	PAFZZ	5360009927308	19204	8448593	SPRING, HELICAL, TORS (M4)... UOC:ASI	1
2B	PAFZZ	5360009927308	19204	8448593	SPRING, HELICAL, TORS (M4A1)...................................... UOC:AY6	1
3	PAFZZ	5360013960256	19200	12972695	SPRING, HELICAL, COMP DISCONNECT .... (M4A1)... UOC:AY6	1
4	PAFZZ	1005009927307	19204	8448592	TRIGGER (M4A1)... UOC:AY6	1

END OF FIGURE

SECTION II

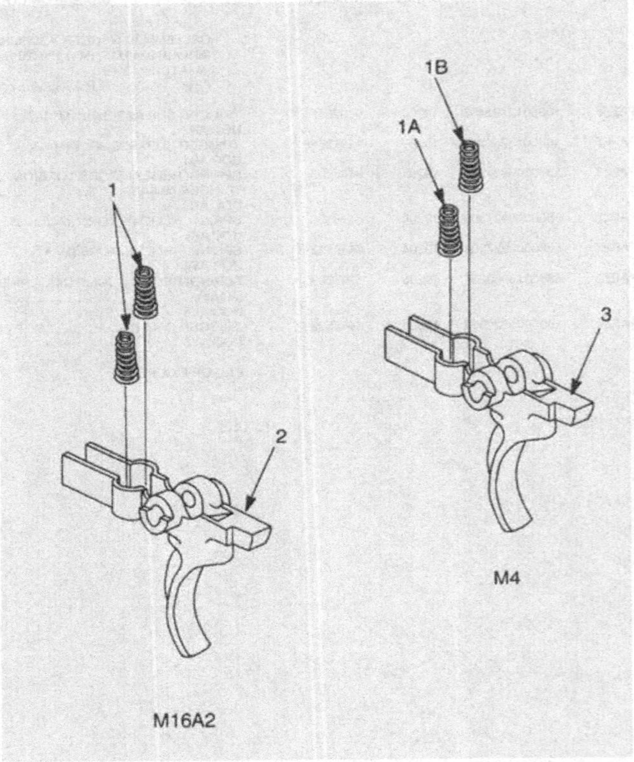

Figure C-15. Trigger Subassembly(M16A2) 9392518, (M4) 12972696.

(1) ITEM NO	(2) SMR CODE	(3) NSN	(4) CAGEC	(5) PART NUMBER	(6) DESCRIPTION AND USABLE ON CODE (UOC)	(7) QTY
					GROUP 040301 TRIGGER SUBASSEMBLY (M6A2) 9392518 AND (M4) 12972696 FIG C-15 TRIGGER SUBASSEMBLY	
1	PAFZZ	5360011350353	19200	9349116	SPRING, HELICAL, COMP DISCONNECT (M16A2) UOC:AR8	2
1A	PAFZZ	5360011350353	19200	9349116	SPRING, HELICAL, COMP DISCONNECT (M4) (NICKEL/SHINY) UOC:AS1	1
1B	PAFZZ	5360013960256	19200	12972695	SPRING, HELICAL, COMP DISCONNECT (M4) (BLACK) UOC:AS1	1
2	XAFZZ		19200	9390736	TRIGGER (M16A2) UOC:AR8	1
3	XAFZZ		19200	9390736	TRIGGER (M4) UOC:AS1	1
					END OF FIGURE	

SECTION II

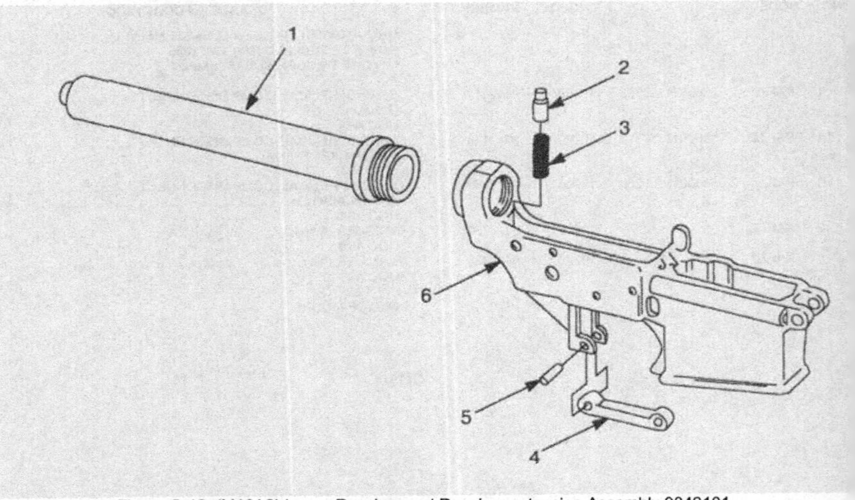

Figure C-16. (M16A2) Lower Receiver and Receiver extension Assembly 9349101.

(1) ITEM NO	(2) SMR CODE	(3) NSN	(4) CAGEC	(5) PART NUMBER	(6) DESCRIPTION AND USABLE ON CODE (UOC)	(7) QTY
					GROUP 0404 (M16A2) LOWER RECEIVER AND RECEIVER EXTENSION ASSEMBLY 9349101	
					FIG C-16 (M16A2) LOWER RECEIVER AND RECEIVER EXTENSION ASSEMBLY	
1	PAFZZ	5340009927297	19200	8448581	EXTENSION, LOWER REC RECEIVER (LOWER RECEIVER EXTENSION) (M1SA2) UOC:AR8	1
2	PAFZZ	5315009926651	19204	8448582	PIN, SHOULDER, HEADLE BUFFER RETAINER (M16A2) UOC:AR8	1
3	PAFZZ	5360009926652	19200	8448583	SPRING, HELICAL, COMP COMPRESSION, BUFFER RETAINER (NI6A2) UOC:AR8	1
4	PAFZZ	1005009927299	19204	8448587	GUARD, TRIGGER (M16A2)	1
5	PAFZZ	5315000586081	96906	MS16562-129	PIN, SPRING TRIGGER GUARD (M16A2)	1
6	XAFDA		19200	9349102	RECEIVER, (M16A2) UOC:AR8	1

END OF FIGURE

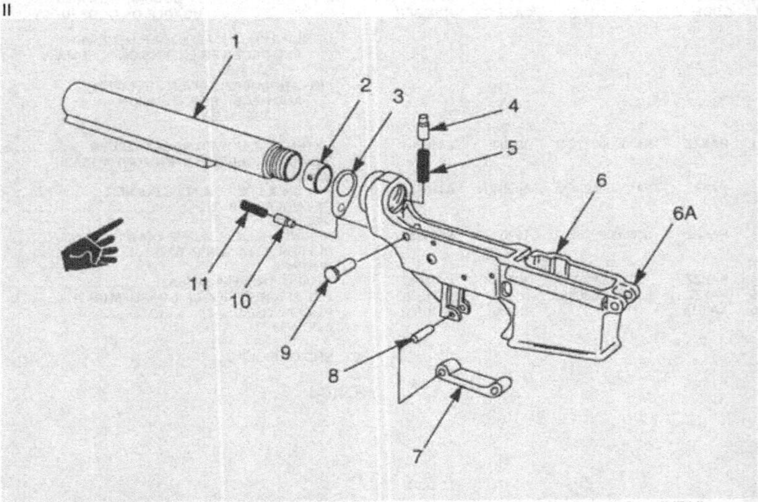

Figure C16A. (M4 and M4A1) Lower Receiver and Receiver extension Assembly, (NPN)

(1) ITEM NO	(2) SMR CODE	(3) NSN	(4) CAGEC	(5) PART NUMBER	(6) DESCRIPTION AND USABLE ON CODE (UOC)	(7) QTY
					GROUP 0404A (M4 AND M4A1) LOWER RECEIVER AND RECEIVER EXTENSION ASSEMBLY (NPN) FIG C16A (M4 AND M4A1) LOWER RECEIVER AND RECEIVER EXTENSION ASSEMBLY	
1	PAFZZ	1005012338531	19200	9390019	EXTENSION, LOWER RECEIVER (M4 AND M4A1).................... UOC:AS1, AY6	1
2	PAFZZ	5310012338625	19200	9390020	NUT, PLAIN, ROUND (M4 AND M4A1).................... UOC:AS1, AY6	1
3	PAFZZ	105012338530	19200	9390021	PLATE, RECEIVER END (M4 AND M4A1).................... UOC:AS1, AY6	1
4	PAFZZ	5315009926651	19204	8448582	PIN, SHOULDER, HEADLE (M4 AND M4A1).................... UOC:AS1, AY6	1
5	PAFZZ	5360009926652	19200	8448583	SPRING, HELICAL, COMP (M4 AND M4A1).................... UOC:AS1, AY6	1
6	XAFDA		19200	9390015	RECEIVER, (M4).................... UOC:AS1	1
6A	XAFDA		19200	12972652	LOWER RECEIVER (M4A1).................... UOC:AY6	1
7	PAFZZ	1005009927299	19204	8448587	GUARD, TRIGGER (M4 AND M4A1)....................	1
8	PAFZZ	5315000586081	96906	MS16562-129	PIN, SPRING (TRIGGER GUARD PIN) (M4 AND M4A1)....................	1
9	PAFZZ	5315009926653	19204	8448584	PIN, GROOVED, HEADED (TAKE DOWN) (M4 AND M4A1)....................	1
10	PAFZZ	5315009926654	19204	8448585	PIN, STRAIGHT, HEADLE HEADLESS (M4 AND M4A1).................... UOC:AS1, AY6	1
11	PAFZZ	5360009926655	19204	8448586	SPRING, HELICAL, COMP COMPRESSION (M4 AND M4A1).................... UOC:AS1, AY6	1
					END OF FIGURE	

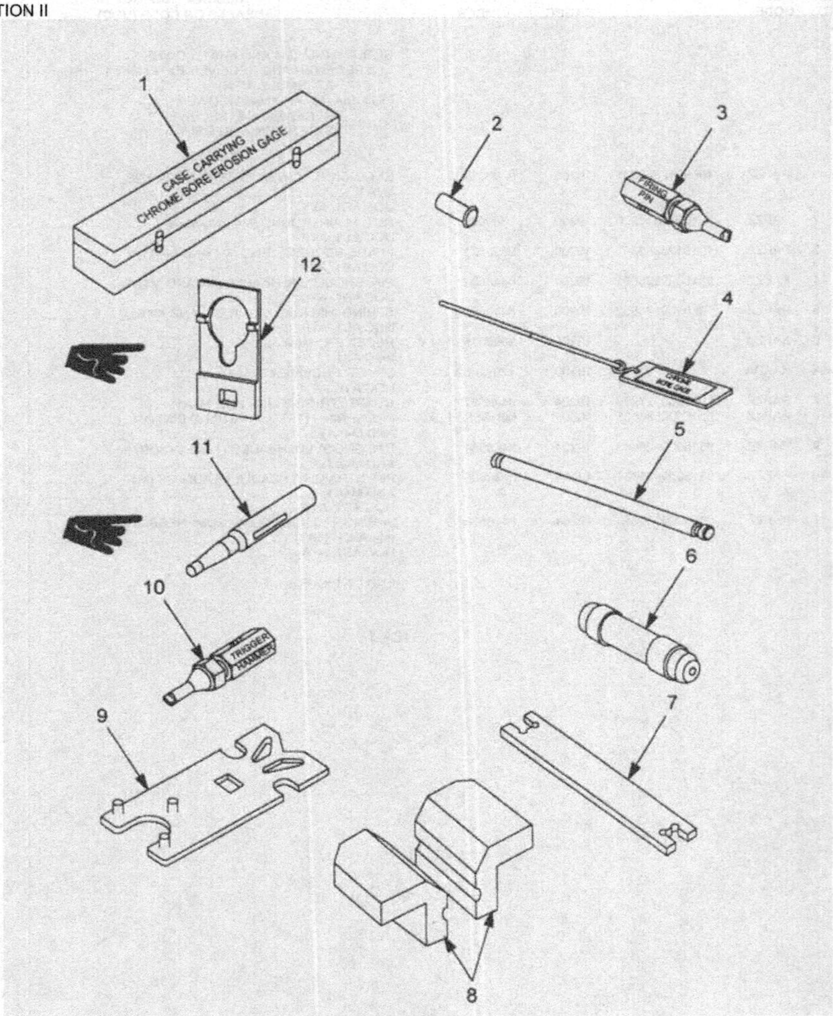

Figure C-17. Special Tools

(1) ITEM NO	(2) SMR CODE	(3) NSN	(4) CAGEC	(5) PART NUMBER	(6) DESCRIPTION AND USABLE ON CODE (UOC)	(7) QTY
					GROUP 9500 SPECIAL TOOLS FIG C-17 SPECIAL TOOLS	
	PEFZZ	1005000567106	19204	8426685	MAINTENANCE KIT, GUN DS/GS SUPPORT MAINTENANCE FOR 5.S6MM RIFLE, M16 RIFLE SERIES BOI: 2 PER SUPPORTING DSU/GSU	
1	PAFZZ	4933010355607	19204	12006359	CASE, BORE GAGE PART OF KIT P/N 8426685..	
2	PAFZZ	4933008007508	19204	8448201	..REFLECTOR TOOL, CHAMBER, PART OF KIT P/N 8426685	
3	PAFZZ	5220010755004	19200	12620101	..GAGE, PLUG, PLAIN PART OF KIT P/N 8426685..	
4	PAFZZ	5220010148183	19204	8448496	..GAGE, BARREL, EROSION BARREL EROSION PART OF KIT P/N 8426685(CHROME BARREL)..	
5	PAFZZ	5220002219391	19204	8448202	..GAGE, STRAIGHTNESS PART OF KIT P/ N 8426685..	
6	PAFZZ	5220000707814	19204	7799734	..GAGE, HEADSPACE PART OF KIT P/N 8426685..	
7	PAFZZ	6695000707815	19204	7799735	..GAGE, FIRING PIN PROTRUSION PART OF KIT P/N 8426685	
8	PAFZZ	4933000709151	19204	11010032	..FIXTURE, BARREL REMOVAL PART OF KIT P/N 8426685	
9	PAFZZ	5120000709152	19204	11010033	..WRENCH, COMBINATION PART OF KIT P/ N 8426685	
10	PAFZZ	5220010439473	19204	12006472	GAGE, PLUG, TAPER CYLINDER PART OF KIT P/N 8426685	
11	PAOZZ	5315013100370	19200	12926769	KEY, MACHINE KEY, 2 PER ORGANIZATIONAL OR DIRECT SUPPORT SHOP	
12	PAFZZ	5120013246631	19200	9390035	WRENCH, SPANNER 2 PER DS/GS SUPPORT SHOP A/R FOR M4/M4A1 CARBINE UOC:AS1, AY6	
					END OF FIGURE	

CROSS-REFERENCE INDEXES

NATIONAL STOCK NUMBER INDEX

STOCK NUMBER	FIG	ITEM
5315-00-017-9537	C-11	15
3040-00-017-9539	C-9	5
1005-00-017-9540	C-9	4
5360-00-017-9541	C-8	13
1005-00-017-9543	C-7	4
1005-00-017-9546	C-1	1
1005-00-017-9547	C-2	1
1005-00-017-9548	C-11	8
1005-00-017-9551	C-13	4A
1005-00-056-2201	C-11	10
5360-00-056-2246	C-11	6
1005-00-056-2247	C-11	7
1005-00-056-7106	C-17	
5315-00-058-6044	C-6	7
5315-00-058-6081	C-16	5
	C16A	8
5315-00-058-6678	C-1B	1
	C-10	1
5220-00-070-7814	C-17	6
6695-00-070-7815	C-17	7
4933-00-070-9151	C-17	8
5120-00-070-9152	C-17	9
1005-00-087-8998	C-6	8
3110-00-183-9175	C-8	4
	C-1A	6
	C-1B	5
	C-10	5
5220-00-221-9391	C-17	5
1005-00-403-0964	C-12	5
5340-00-463-3892	C-12	3
5315-00-463-3894	C-12	6
5360-00-523-8084	C-9	3
5310-00-527-3634	C-11	20
5315-00-597-5086	C-3	4
5315-00-690-0544	C-7	8
1005-00-738-6213	C-4	3
1005-00-760-3768	C-3	8
4933-00-800-7508	C-17	2
5315-00-812-3312	C-11	9
5315-00-840-3812	C-8	11
	C-8	12
	C-1A	8
5315-00-843-9487	C12A	6
1005-00-921-5004	C-1	4
1005-00-937-3078	C-11	2
1005-00-978-1022	C-8	9
5315-00-978-1023	C-8	8
5360-00-978-1025	C-8	10
5360-00-978-1036	C-6	9
4710-00-978-1038	C-6	3

STOCK NUMBER	FIG	ITEM
5315-00-979-3930	C-7	2
5360-00-979-3931	C-7	3
5360-00-992-6648	C-13	1A
1005-00-992-6649	C-11	4
5315-00-992-6650	C-11	25
5315-00-992-6651	C-16	2
	C16A	4
5360-00-992-6652	C-16	3
	C16A	5
5315-00-992-6653	C-11	26
	C16A	9
5315-00-992-6654	C-11	12
	C-11	12A
	C16A	10
5360-00-992-6655	C-11	11
	C-11	11A
	C16A	11
5360-00-992-6665	C-11	1
1005-00-992-6667	C-11	23
1005-00-992-7283	C-4	2
5305-00-992-7284	C-4	1
1005-00-992-7285	C-2	4
1005-00-992-7287	C-3	1
1005-00-992-7288	C-3	7
1005-00-992-7290	C-3	3
1005-00-992-7291	C-3	6
5360-00-992-7292	C-3	5
	C-11	22
5315-00-992-7294	C-2	3
5340-00-992-7297	C-16	1
1005-00-992-7299	C-16	4
	C16A	7
5360-00-992-7301	C-11	13
1005-00-992-7302	C-11	14
1005-00-992-7307	C-14	4
5360-00-992-7308	C-14	2
	C-14	2A
	C-14	2B
5315-00-992-7309	C-11	24
5360-00-999-0404	C-5	3
5340-00-999-0405	C-5	2
1005-00-999-0406	C-11	16A
5325-00-999-0863	C-6	10
5325-00-999-0864	C-8	7
1005-00-999-1509	'C-2	2
5220-01-014-8183	C-17	4
4933-01-035-5607	C-17	1
5220-01-043-9473	C-17	10
5315-01-048-9372	C-5	1
	C-9	2

CROSS-REFERENCE INDEXES

NATIONAL STOCK NUMBER INDEX

STOCK NUMBER	FIG	ITEM
5320-01-063-7635	C-7	5
5220-01-075-5004	C-17	3
1005-01-134-3621	C-8	1
5305-01-134-3622	C-8	2
	C-1A	2
1005-01-134-3625	C-7	1
5355-01-134-3627	C-1B	8
	C-10	8
1005-01-134-3629	C-6	2
1005-01-134-3630	C-13	4
1005-01-134-3631	C-10	2
1005-01-134-3633	C-6	6
1005-01-134-3701	C-8	6
5360-01-134-3710	C-8	15
	C-1A	5
5360-01-135-0353	C-15	1
	C-15	1A
1005-01-135-3697	C-1B	6
	C-10	6
5355-01-135-4972	C-8	16
1005-01-135-4973	C-11	28
5360-01-136-5471	C-13	2
1005-01-144-1468	C-9	1
5305-01-144-1490	C-1B	3
	C-10	3
5360-01-144-1492	C-13	1
5305-01-144-1494	C-12	7
5365-01-144-1496	C-6	5
5340-01-144-1499	C-11	17
5340-01-145-7910	C-11	16
1005-01-146-7684	C-6	4
1005-01-146-7685	C-12	2
5305-01-147-8585	C-11	29
1005-01-148-0172	C-13	3
5360-01-148-1751	C-8	5
	C-1B	4
	C-10	4
1005-01-148-4805	C-11	19
1005-01-216-4510	C-1	5
1005-01-219-2402	C-14	1
5340-01-225-8339	C-11	5
1005-01-228-8504	C-12	1
1005-01-231-3138	C-11	2A
1005-01-233-8529	C-6	4A
1005-01-233-8530	C16A	3
1005-01-233-8531	C16A	1
5315-01-233-8608	C12A	2
5360-01-233-8616	C12A	3
5360-01-233-8617	C-11	1A
5310-01-233-8625	C16A	2
5310-01-233-8626	C12A	5
1005-01-233-8636	C12A	1
4710-01-233-8637	C-6	3A
5340-01-233-8638	C12A	4
1005-01-234-2297	C-6	2A
3040-01-247-7969	C-13	3A
1010-01-264-6517	C-7	9
5340-01-264-6530	C-7	7
5365-01-267-2169	C-11	27
5305-01-268-1191	C-11	21
5315-01-310-0370	C-17	11
5120-01-324-6631	C-17	12
1005-01-368-9852	C-1	5A
5360-01-381-6183	C-1B	7
	C-10	7
5340-01-382-3201	C-1A	10
5310-01-382-6793	C-1A	9
1005-01-382-6795	C-8	6A
5355-01-382-6801	C-1A	4
5360-01-382-6802	C-1A	7
1005-01-382-7083	C-1A	11
1005-01-382-7086	C-1B	2
1005-01-382-7089	C-1A	3
1005-01-395-4257	C-14	1A
5360-01-396-0256	C-14	3
	C-15	1B
1005-01-422-3770	C-2	4A
1005-01-424-5899	C-3	9

CROSS-REFERENCE INDEXES

NATIONAL STOCK NUMBER INDEX

STOCK NUMBER	FIG	ITEM	STOCK NUMBER	FIG	ITEM
88044	AN5010416-18		5305-01-268-1191	C-11	21
96906	MS16562-103		5315-00-058-6678	C-1B	1
				C-10	1
96906	MS16562-106		5315-00-058-6044	C-6	7
96906	MS16562-119		5315-00-812-3312	C-11	9
96906	MS16562-121		5315-00-840-3812	C-8	11
				C-8	12
				C-1A	8
96906	MS16562-129		5315-00-058-6081	C-16	5
				C16A	8
96906	MS16562-202		5315-00-843-9487	C12A	6
96906	MS16562-98		5315-00-597-5086	C-3	4
96906	MS16626-3137		5325-00-999-0863	C-6	10
96906	MS16632-3012		5325-00-999-0864	C-8	7
96906	MS19060-4808		3110-00-183-9175	C-8	4
				C-1A	6
				C-1B	5
				C-10	5
96906	MS35335-61		5310-00-527-3634	C-11	20
96906	MS39086-93		5315-00-690-0544	C-7	8
99999	NPN			C-11	30A
				C-11	30B
19204	11010032		4933-00-070-9151	C-17	8
19204	11010033		5120-00-070-9152	C-17	9
19204	12006359		4933-01-035-5607	C-17	1
19204	12006472		5220-01-043-9473	C-17	10
19200	12011849			C-1	2A
19200	12011987		5360-01-381-6183	C-1B	7
				C-10	7
19200	12011996		1005-01-368-9852	C-1	5A
19200	12597640		5365-01-267-2169	C-11	27
19200	12598617		1010-01-264-6517	C-7	9
19200	12598618		5340-01-264-6530	C-7	7
19200	12620101		5220-01-075-5004	C-17	3
19204	12624561		1005-01-216-4510	C-1	5
19200	12926769		5315-01-310-0370	C-17	11
19200	12951011			C-1	7
19200	12951017		5340-01-382-3201	C-1A	10
19200	12951018		1005-01-382-7089	C-1A	3
19200	12951019		5355-01-382-6801	C-1A	4
19200	12951020		5360-01-382-6802	C-1A	7
19200	12951021		1005-01-382-7083	C-1A	11
19200	12951023		5310-01-382-6793	C-1A	9
19200	12951026			C-1A	1
19200	12951028		1005-01-382-7086	C-1B	2
19200	12972652			C16A	6A
19200	12972670		1005-01-382-6795	C-8	6A
19200	12972675			C-6	1A
19200	12972680			C-1	3A
19200	12972690			C-1	6B
19200	12972691		1005-01-422-3770	C-2	4A

CROSS-REFERENCE INDEXES

PART NUMBER INDEX

CAGEC	PART NUMBER	STOCK NUMBER	FIG	ITEM
19200	12972692	1005-01-424-5899	C-3	9
19200	12972695	5360-01-396-0256	C-14	3
			C-15	1B
19200	12972696	1005-01-395-4257	C-14	1A
19200	12972697		C-11	18A
19200	12972698		C-11	18B
19204	7799734	5220-00-070-7814	C-17	6
19204	7799735	6695-00-070-7815	C-17	7
19204	8426685	1005-00-056-7106	C-17	
19204	8448201	4933-00-800-7508	C-17	2
19204	8448202	5220-00-221-9391	C-17	5
19204	8448496	5220-01-014-8183	C-17	4
19204	8448501		C-1	2
19204	8448502	5315-00-992-7294	C-2	3
19204	8448503	1005-00-017-9547	C-2	1
19204	8448504	1005-00-999-1509	C-2	2
19204	8448505		C-2	5
19200	8448506	1005-00-992-7283	C-4	2
19200	8448507	1005-00-738-6213	C-4	3
19204	8448508	5305-00-992-7284	C-4	1
19200	8448509	1005-00-992-7285	C-2	4
19204	8448510		C-3	2
19204	8448511	1005-00-992-7287	C-3	1
19204	8448512	1005-00-992-7288	C-3	7
19204	8448513	1005-00-992-7290	C-3	3
19204	8448515	1005-00-992-7291	C-3	6
19204	8448516	5360-00-992-7292	C-3	5
			C-11	22
19204	8448517	1005-00-017-9546	C-1	1
19204	8448518		C-5	4
19200	8448519	5340-00-999-0405	C-5	2
19204	8448520	5360-00-999-0404	C-5	3
19204	8448521-2	5315-01-048-9372	C-5	1
			C-9	2
19204	8448525	1005-00-978-1022	C-8	9
19204	8448532	5360-00-978-1025	C-8	10
19204	8448533	5315-00-978-1023	C-8	8
19200	8448540	5360-00-017-9541	C-8	13
19200	8448542	5360-00-523-8084	C-9	3
19204	8448543	3040-00-017-9539	C-9	5
19204	8448544	1005-00-017-9540	C-9	4
19204	8448555	5360-00-978-1036	C-6	9
19200	8448567	4710-00-978-1038	C-6	3
19204	8448571	1005-00-017-9543	C-7	4
19204	8448573	5315-00-979-3930	C-7	2
19204	8448574	5360-00-979-3931	C-7	3
19200	8448581	5340-00-992-7297	C-16	1
19204	8448582	5315-00-992-6651	C-16	2
			C16A	4
19200	8448583	5360-00-992-6652	C-16	3
			C16A	5

CROSS-REFERENCE INDEXES

PART NUMBER INDEX.

CAGEC	PART NUMBER	STOCK NUMBER	FIG	ITEM
19204	8448584	5315-00-992-6653	C-11	26
			C16A	9
19204	8448585	5315-00-992-6654	C-11	12
			C-11	12A
			C16A	10
19204	8448586	5360-00-992-6655	C-11	11
			C-11	11A
			C16A	11
19204	8448587	1005-00-992-7299	C-16	4
			C16A	7
19204	8448592	1005-00-992-7307	C-14	4
19204	8448593	5360-00-992-7308	C-14	2
			C-14	2A
			C-14	2B
19200	8448595	1005-00-992-6649	C-11	4
19204	8448599	5315-00-992-6650	C-11	25
19204	8448609	5315-00-992-7309	C-11	24
19204	8448610		C-11	3B
19204	8448611	5360-00-992-6648	C-13	1A
19200	8448612	1005-00-017-9551	C-13	4A
19200	8448615	1005-00-937-3078	C-11	2
19204	8448621	5315-00-017-9537	C-11	15
19200	8448628	1005-00-017-9548	C-11	8
19204	8448629	5360-00-992-6665	C-11	1
19204	8448631	1005-00-992-6667	C-11	23
19204	8448633	5360-00-056-2246	C-11	6
19204	8448634	1005-00-056-2247	C-11	7
19200	8448635	1005-00-999-0406	C-1I	11A
19204	8448636	1005-00-992-7302	C-11	14
19204	8448637	5360-00-992-7301	C-11	13
19204	8448638	1005-00-056-2201	C-11	10
19204	8448652	1005-00-403-0964	C-12	5
19200	8448653	5340-00-463-3892	C-12	3
19204	8448655	5315-00-463-3894	C-12	6
19200	8448670	1005-00-921-5004	C-1	4
19204	8448697	5320-01-063-7635	C-7	5
19204	8448712	1005-00-087-8998	C-6	8
19200	8448755	1005-00-760-3768	C-3	8
19200	9349050		C-1	3
19200	9349051	1005-01-134-3633	C-6	6
19200	9349052	5365-01-144-1496	C-6	5
19200	9349054		C-7	6
19200	9349056	1005-01-134-3625	C-7	1
19200	9349059	1005-01-134-3629	C-6	2
19200	9349062		C-6	1
19200	9349063	1005-01-134-3701	C-8	6
19200	9349065	5305-01-134-3622	C-8	2
			C-1A	2
19200	9349066	1005-01-134-3621	C-8	1
19200	9349067	5355-01-135-4972	C-8	16
19200	9349069	5360-01-148-1751	C-8	5

CROSS-REFERENCE INDEXES

PART NUMBER INDEX

CAGEC	PART NUMBER	STOCK NUMBER	FIG	ITEM
19200	9349069	5360-01-148-1751	C-1B	4
			C-10	4
19200	9349070	5360-01-134-3710	C-8	15
			C-1A	5
19200	9349072		C-8	3
19200	9349074	1005-01-134-3631	C-10	2
19200	9349075	1005-01-135-3697	C-1B	6
			C-10	6
19200	9349076	5305-01-144-1490	C-1B	3
			C-10	3
19200	9349077	5355-01-134-3627	C-1B	8
			C-10	8
19200	9349085	1005-01-144-1468	C-9	1
19204	9349086		C-8	14
19200	9349100		C-1	6
19200	9349101		C-11	30
19200	9349102		C-16	6
19200	9349106		C-11	3
19200	9349107	5360-01-144-1492	C-13	1
19200	9349108	1005-01-148-0172	C-13	3
19200	9349109	5360-01-136-5471	C-13	2
19200	9349110	1005-01-134-3630	C-13	4
19200	9349113	5340-01-144-1499	C-11	17
19200	9349114	5340-01-145-7910	C-11	16
19200	9349115		C-11	18
19200	9349116	5360-01-135-0353	C-15	1
			C-15	1A
19200	9349119	1005-01-135-4973	C-11	28
19200	9349120	5305-01-144-1494	C-12	7
19200	9349121		C-12	4
19200	9349124	1005-01-146-7684	C-6	4
19200	9349127	1005-01-148-4805	C-11	19
19200	9349128	5305-01-147-8585	C-11	29
19200	9349130	1005-01-146-7685	C-12	2
19200	9381367	5340-01-225-8339	C-11	5
19200	9381380	1005-01-228-8504	C-12	1
19200	9390003	1005-01-234-2297	C-6	2A
19200	9390007	1005-01-233-8529	C-6	4A
19200	9390009		C-7	6A
19200	9390011		C-1	6A
19200	9390012		C-11	28A
19200	9390013	1005-01-233-8636	C12A	1
19200	9390014	5340-01-233-8638	C12A	4
19200	9390015		C16A	6
19200	9390016	4710-01-233-8637	C-6	3A
19200	9390019	1005-01-233-8531	C16A	1
19200	9390020	5310-01-233-8625	C16A	2
19200	9390021	1005-01-233-8530	C16A	3
19200	9390022	5360-01-233-8617	C-11	1A
19200	9390023	1005-01-231-3138	C-11	2A
19200	9390025	5315-01-233-8608	C12A	2

CROSS-REFERENCE INDEXES

PART NUMBER INDEX

CAGEC	PART NUMBER	STOCK NUMBER	FIG	ITEM
19200	9390026	5310-01-233-8626	C12A	5
19200	9390027	5360-01-233-8616	C12A	3
19200	9390031	3040-01-247-7969	C-13	3A
19200	9390032		C-11	3A
19200	9390035	5120-01-324-6631	C-17	12
19200	9390736		C-15	2
			C-15	3
19200	9392518	1005-01-219-2402	C-14	1

APPENDIX D
EXPENDABLE/DURABLE SUPPLIES AND MATERIALS LIST

Section I. INTRODUCTION

-1. **SCOPE**. This appendix lists expendable durable supplies and materials you will need to operate and maintain the 16A2 Rifle This listing is for informational purposes only and Is not authority to requisition the listed Items. These items re authorized to you by CTA 50-970, Expendable, Durable Items (Except Medical, Class V, Repair Parts, and Heraldic ems), or CTA 8-100, Army Medical Department Expendable/Durable Items.

-2. **EXPLANATION OF COLUMNS.**

a. Column (1)-Item Number. This number is assigned to the entry in the listing and is referenced In the narrative structions to identify the material (e.g., "Use cloth, abrasive, crocus, Item 12, app D").

b. Column (2)-Level. This column identifies the lowest level of maintenance that requires the listed Item.

C - Operator/Crew Maintenance
O - Unit Maintenance
F - Direct Support Maintenance

c. Column (3)-National Stock Number. This is the National stock number assigned to the Item, use It to request or equisition the Item.

d. Column (4)-Description. Indicates the Federal Item name and, if required, a description to identify the item. The st line for each Item indicates the Commercial and Government Entity Code (CAGEC) in parentheses followed by the art number.

e. Column (5)-Unit of Measure (U/M). Indicates the measure used in performing the actual maintenance function. his measure is expressed by a two-character alphabetical abbreviation (e g., ea, ln , pr). If the unit of measure differs om the unit of issue, requisition the lowest unit of issue that will satisfy your requirements.

Section II. EXPENDABLE/DURABLE SUPPLIES AND MATERIALS LIST

(1) ITEM NUMBER	(2) LEVEL	(3) NATIONAL STOCK NUMBER	(4) DESCRIPTION PART NUMBER AND CAGE	(5) UNIT OF MEAS
1	F	8040 00-944 7292	ADHESIVE, I<IT, (81348) MMM A 1754	KT
2	0	8020 00 244 0153	BRUSH, ARTIST'S metal ferrule, flat chisel ed(i(e 7 16 w, 1 1/8, exposed bristle (813481 H B 241	EA
3	F	1005-00-716 2702	BRUSH, CLEANING, SMALL ARMS (192051 7162702	EA
4	C	1005-00-903 1296	BRUSH, CLEANING, SMALL ARMS bore (19204) 11686340	EA
5	C	1005-00 999-1435	BRUSH, CLEANING, SMALL ARMS chamber (19204) 8432358	EA
6	C	1005 00-444-6602	BRUSH, CLEANING, SMALL ARMS tooth 1192041 8448462	EA
7	O	7920-00 205-2401	BRUSHCLEANING, TOOLS AND PARTS (813491 MILS43871	EA
8	O	6850-00-965-2332	CARBON REMOVING COMPOUND (813481 P-C-111	GL
9			CLEANER, LUBRICANT AND PRESERVATIVE (27412)	
	O	9150-01-079-6124	CLP- 4 oz (118 30 ml) bottle	EA
	O	9150-01-054 6453	CLP- 1 pt (0 47 1l bottle	EA
	O	9150-01 053-6688	CLP 7 gal (26 50 11 bottle	EA
10	C	9150 01-102 1473	CLEANER, LUBRICANT AND PRESERVATIVE (81349i MIL L 63460 1/2 oz (14 79 ml) bottle	EA
11	C	9920-00-292 9946	CLEANER, TOBACCO PIPE cotton turf, wire core (89855) DILLSPIPE cleaner (6 per pkg)	
			CLEANING COMPOUND, RIFLE BORE small arms bore cleaner solution (RBC) 1813491 MIL C 372	
	C	6850-00-224-6656	2 oz 159.15 mil) bottle	OZ
	O	6850-00-224-6657	8 oz (236 59 ml) can	CN
	O	6850-00-224-6663	1 gal (3.79 l) can	CN

EXPENDABLE/DURABLE SUPPLIES AND MATERIALS LIST (CONT)

(1) ITEM NUMBER	(2) LEVEL	(3) NATIONAL STOCK NUMBER	(4) DESCRIPTION PART NUMBER AND CAGE	(5) UNIT OF MEAS
13	0	5350-00-221-0872	CLOTH, ABRASIVE: (58536) A-A-1206	SH
14			Deleted See Item 23.2	
15	F		DICHLOROMETHANE, TECHNICAL: (81346) ASTM D 4701	
		6810-00-244-0290	5 gal (18.93 1) pail	CN
		6810-00-616-9188	600 lb (272.16 kg) drum	DR
16	0	6850-00-281-1985	DRY CLEANING SOLVENT: (58536) A-A-711 1 gal(3.79 l1) can	GL
17	0	8010-00-297-0560	ENAMEL: olive drab no 3407 (81348) TT-E-527 1 gal (3.791) can	GL
18	0		GLOVES, CHEMICAL AND OIL PROTECTIVE: (81348) ZZ-G-381	
		8415-00-823-7455	Size 9	PR
		8415-00-823-7456	Size10	PR
		8415-00-823-7457	Size 11	PR
19	F	9150-00-754-2595	GREASE, MOLYBDENUM DISULFIDE: (81349) MIL-G-21164	LB
20	C	1005-01-113-0321	HANDLE SECTION, CLEANING ROD, SMALL ARMS: (19204)8436776	EA
21	0	9150-01-260-2534	LUBRICANT, SOLID FILM: (81349) MIL-L-23398 16 OZ (473.18 ml) spray can	OZ
22	C	9150-00-292-9689	LUBRICATING OIL, WEAPONS: (LAW) (81349) MIL-L-14107 1 qt (0.95 1) can	QT

EXPENDABLE/DURABLE SUPPLIES AND MATERIALS LIST (CONT)

(1) ITEM NUMBER	(2) LEVEL	(3) NATIONAL STOCK NUMBER	(4) DESCRIPTION PART NUMBER AND CAGE	(5) UNIT OF MEAS
23			LUBRICATING OIL, WEAPONS: (LSA), semifluid (81349) MIL-L-46000	
	C	9150-00-935-6597	2 oz (59.15 ml) plastic btl	OZ
	C	9150-00-889-3522	4 oz (118.30 ml) plastic btl	OZ
	O	9150-00-687-4241	1 qt (0.95 l1) can	CN
	O	9150-00-753-4686	1 gal(3.79 l1) can	CN
23.1	0	5340-01-230-3181	MOUNTING BRACKET (M4/M4A1 ONLY) (19200)12556036	EA
23.2	0	8010-00-087-0102	PAINT, ENAMEL, SEMIGLOSS: paint for blank firing attachment (M1 5A2) (81348) TT-E-529 1 qt can (RED - M16A2)	EA
23.3	0	8010-01-031-1274	PAINT, ENAMEL, SEMIGLOSS: paint for blank firing attachment (M23) (81348) TT-E-529 1 pt can (YELLOW - M4/M4A1)	EA
24	0	4940-00-795-3595	PAN, WASH: (94453) 1211	EA
25	F	6850-00-826-0981	PENETRANT KIT: (81349) MIL-I-25135	KT
26	F	8135-01-019-1691	POLYETHYLENE SHEET: (84744) PE88-80-2	EA
26.1	0	1005-01-394-7677	PROTECTOR, RAIL (M4/M4A1 ONLY) (19200)12972676	EA
27	C	7920-00-205-1711	RAG, WIPING: (58536) A-A-531 50 lb (22.68 kg) bdl	LB
28	C	1005-00-050-6357	ROD SECTION, CLEANING SMALL ARMS: (19204) 8436775 (3 REQUIRED)	EA
29	F	8030-00-670-8553	SEALING COMPOUND (16059) DEVCONF	KT

EXPENDABLE/DURABLE SUPPLIES AND MATERIALS LIST (CONT)

(1) ITEM NUMBER	(2) LEVEL	(3) NATIONAL STOCK NUMBER	(4) DESCRIPTION PART NUMBER AND CAGE	(5) UNIT OF MEAS
30	C	1005-00-937-2250	SWAB HOLDER SECTION, CLEANING ROD, SMALL ARMS: (19204) 11686327	EA
31	C	1005-00-912-4248	SWAB, SMALL ARMS: (19204)11686408	SH
32	C	6920-01-395-2949	TARGET, 25 METER ZEROING, M16A2: (19200)12012024	BX

APPENDIX E

ILLUSTRATED LIST OF MANUFACTURED ITEMS

E-1. INTRODUCTION.

a. This appendix includes complete instructions for making items authorized to be manufactured or fabricated at unit or direct support maintenance.

b. A part number in alphanumeric order is provided for cross-referencing the part number of the item to be manufactured to the figure which covers fabrication criteria.

c. All bulk matenals needed for manufacture of an item are listed by part number or specification number In a tabular list on the illustration.

E-2. MANUFACTURED ITEMS PART NUMBER INDEX.

INDEX

Item	Figure Number
Front sight detent depressor	E-1
Front sight post removal and installation tool	E-2
Pivot pin removal tool	E-3
Pivot pin installation tool	E-4
Slave pin	E-5
Adapter bar for M12 arms rack	E-6
Modified needle nose pliers	E-7

E-3. MANUFACTURED ITEMS ILLUSTRATIONS - FORMAT AND CONTENT.

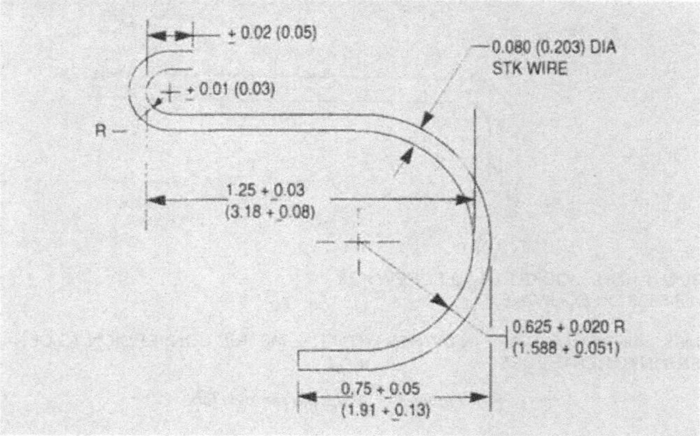

FABRICATE FROM 0.08 IN. MUSIC WIRE OR EQUIVALENT.
FINISH: NO. 5.3.1.2 OR 5.3.2.2 OF MIL-STD 17.

NOTE: ALL DIMENSIONS SHOWN ARE IN INCHES WITH THE METRIC CONVERSION TO CENTIMETERS IN PARENTHESIS.

Figure E-1. Front sight detent depressor.

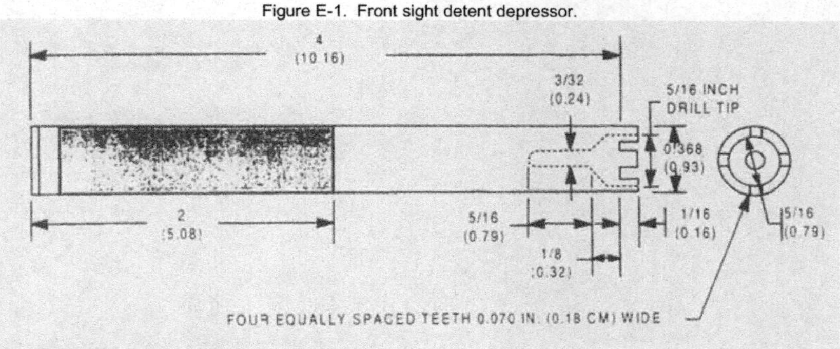

FABRICATE FROM 0.375 INCH ROUND METAL BAR,
ASTM A686. FSCM 81346, GRADE C. CLASS W2-09.
NSN 9510-00-640-4407 OR EQUIVALENT.

NOTES:
1. ALL DIMENSIONS SHOWN ARE IN INCHES WITH THE METRIC CONVERSION TO CENTIMETERS IN PARENTHESIS.
2 TEETH MUST BE HAND FILED TO FIT FRONT SIGHT POST.

Figure E-2. Front sight post removal and Installation tool.

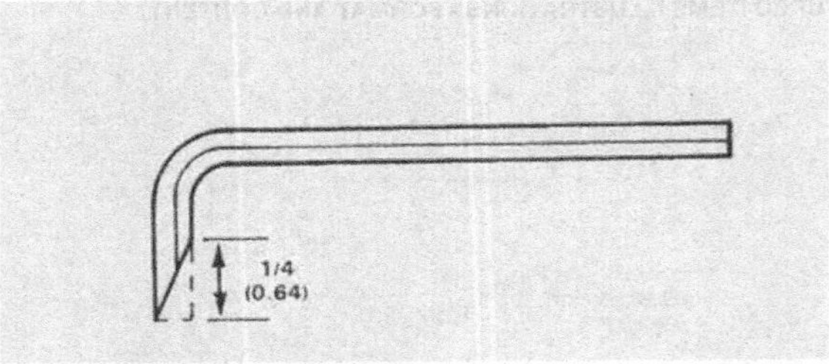

FABRICATE FROM 1'16 IN. SOCKET HEAD SCREW KEY
NSN 5120-00-198-5398 OR EQUIVALENT.

NOTE: ALL DIMENSIONS SHOWN ARE IN INCHES WITH THE METRIC CONVERSION TO CENTIMETERS IN PARENTHESES

Figure E-3. Pivot pin removal tool.

E-3. MANUFACTURED ITEMS ILLUSTRATIONS-FORMAT AND CONTENT (CONT).

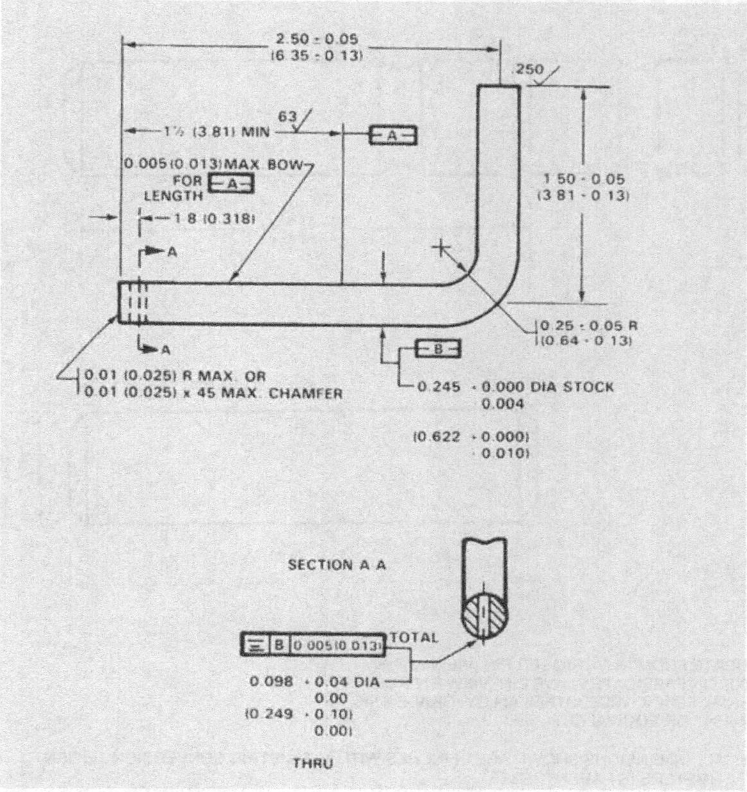

FABRICATE FROM 0 245 IN STEEL AISI 1095 OR EQUIVALENT
HARDEN AND TEMPER TO RC-57-61 FOR LENGTH -A-
FINISH 5 3 1.2 OR 5 3.2.2 OF MIL-STD-171

NOTE: ALL DIMENSIONS SHOWN ARE IN INCHES WITH THE METRIC CONVERSION TO CENTIMETERS IN PARENTHESES

Figure E-4. Pivot pin installation tool.

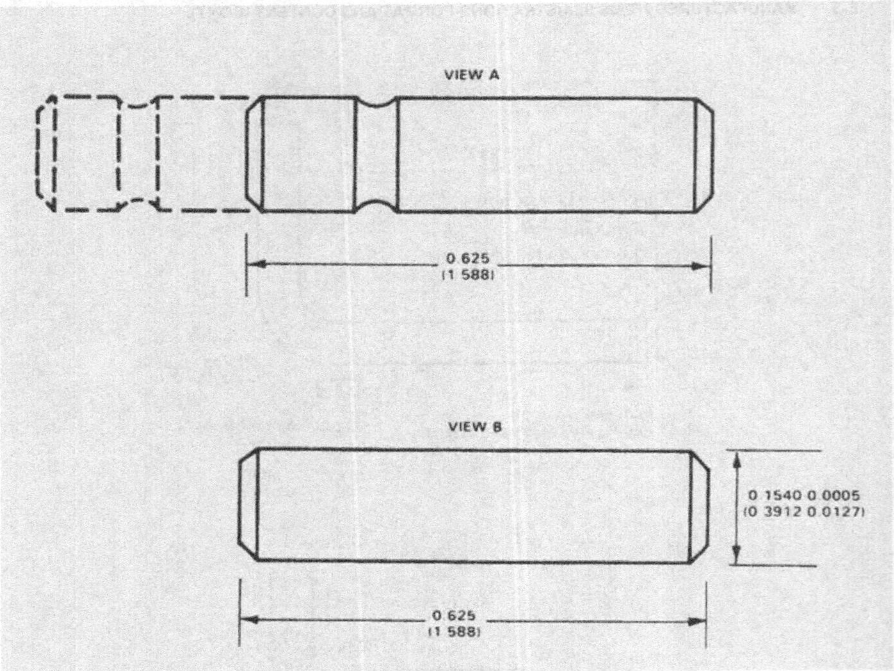

FABRICATE FROM OLD TRIGGER PIN (VIEW AI P,/N 8448609 OR FABRICATE SLAVE PIN (VIEW B) FROM MATERIAL BLOCK, WIRE, STEEL ALLOY, GRADE 4140, ASTM-A547 OR EQUIVALENT

NOTE: ALL DIMENSIONS SHOWN ARE IN INCHES WITH THE METRIC CONVERSION TO CENTIMETERS IN PARENTHESES

Figure E-5. Slave pin.

E-3. MANUFACTURED ITEMS ILLUSTRATIONS -- FORMAT AND CONTENT (CONT).

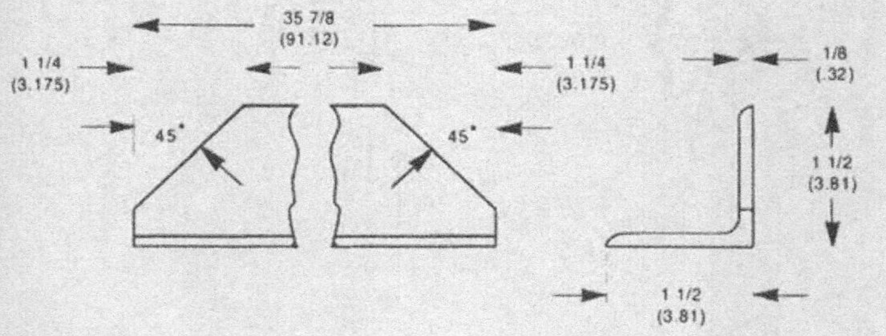

ADAPTER
FABRICATE FROM 1 1/2 INCH BY 1 1/2 INCH
BY 1/8 INCH ANGLE IRON NSN 9520-00-277-4902
OR EQUIVALENT. PAINT WITH OLIVE DRAB
ENAMEL PAINT, NSN 8010-01-350-5249, OR
EQUIVALENT.

NOTE: ALL DIMENSIONS SHOWN ARE IN INCHES WITH THE METRIC CONVERSION TO CENTIMETERS IN PARENTHESES.

Figure E-6. Adapter bar for M12 arms rack.

PIN: 060514-004

***U S Government Printing Office 1995- 648-378**

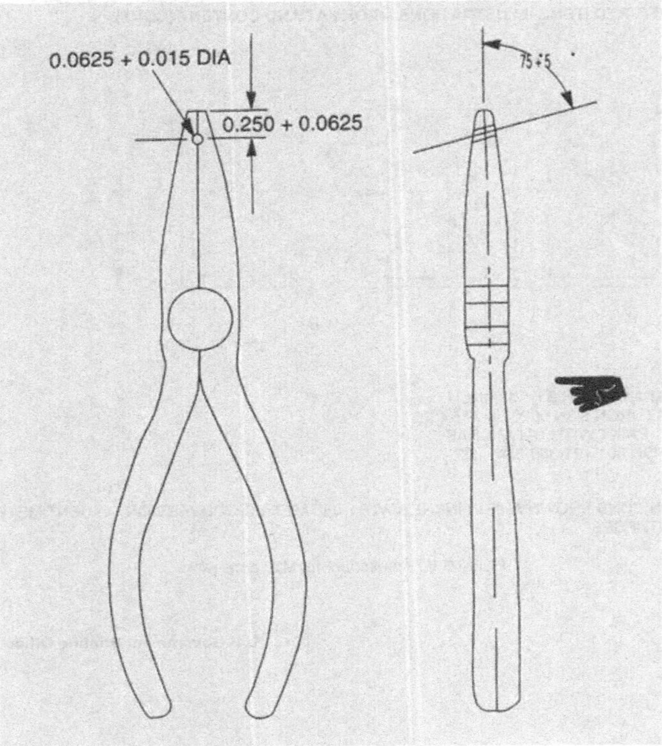

FABRICATE FROM NEEDLE NOSE PLIERS,
NSN 5120-00-268-3579 OR EQUIVALENT.

NOTE: ALL DIMENSIONS ARE IN INCHES.

Figure E-7. Modified needle nose pliers.

***U. S. GOVERNMENT PRINTING OFFICE 1997 - 545-039/4 077**

ALPHABETICAL INDEX

ubject **Page**

A

B

C

ALPHABETICAL INDEX (CONT)

Subject **Page**

ALPHABETICAL INDEX (CONT)

ALPHABETICAL INDEX (CONT)

Subject **Page**

R

Made in United States
Orlando, FL
04 November 2026

90218355R00223